GUIDANCE
OF **UNMANNED**
AERIAL VEHICLES

GUIDANCE
OF UNMANNED
AERIAL VEHICLES

RAFAEL YANUSHEVSKY

CRC Press is an imprint of the
Taylor & Francis Group, an **informa** business

CRC Press
Taylor & Francis Group
6000 Broken Sound Parkway NW, Suite 300
Boca Raton, FL 33487-2742

First issued in paperback 2017

© 2011 by Taylor and Francis Group, LLC
CRC Press is an imprint of Taylor & Francis Group, an Informa business

No claim to original U.S. Government works

ISBN 13: 978-1-4398-5095-4 (hbk)
ISBN 13: 978-1-138-07454-5 (pbk)

Visit the Taylor & Francis Web site at
http://www.taylorandfrancis.com

and the CRC Press Web site at
http://www.crcpress.com

To

DANIEL and CAMILLA

Contents

Preface

Unmanned aerial vehicles (UAVs) represent the fastest growing and the most dynamic growth segment within the aerospace industry. The rapidly increasing fleet of UAVs, along with the widening sphere of their applications, puts new problems before their designers. Although now unmanned aerial vehicles are used mostly in military applications (intelligence, surveillance, and reconnaissance missions; and combat operations—strike missions, suppression and/or destruction of the enemy and its facilities), their future potential civil applications are enormous (e.g., border patrol, forest fire monitoring and firefighting; nonmilitary security work such as surveillance of industrial sites, road/rail infrastructure, mineral exploration, coastal surveillance, pipeline surveillance, spraying of fertilizers, insecticides, aerial photography, land mapping, environmental monitoring, transportation, and gathering scientific data).

Being considered as aerial robots, designers of current UAVs pay close attention to such technologies as imaging, communications, electro-optical sensor systems, sensor fusion, i.e., the technologies that would provide an operator with reliable visual information that allows a pilot of an aircraft to make decisions concerning future flight paths. These topics are widely discussed in many publications. Publications related to the controllers that generate pilot's actions don't contain anything new from the standpoint of control theory.

However, scientific publications focused on trajectory generation and regulation, task allocation, and scheduling, i.e., the problems that help an operator choose an optimal trajectory from one location to another for various possible mission scenarios, contribute to designing the future generation of UAVs in which the operator's involvement will be minimized. Since the flight of autonomously guided UAVs is, in many features, similar to the flight of cruise missiles, their guidance and control systems can be built similar to existing cruise missiles. However, UAVs are employed in high-risk environments. The ability to sense and avoid obstacles, both natural and man-made, and to rebuild its flight path is an important feature that UAVs should possess, and the corresponding algorithms should be imbedded in their guidance and control systems.

In many cases, UAVs are described either as a single air vehicle (with associated surveillance sensors) or a UAV system, which usually consists of three to six air vehicles, a ground control station, and support equipment. One solution to alleviating the problem of controlling the flight of a swarm of UAVs is allowing the UAVs to be operated from the so-called lead UAV, which, in turn, receives commands from the ground mission control unit. To make UAV operators work easier, the UAVs should be endowed with artificial intelligence (AI). It should be noted that missiles, which are unmanned aerial vehicles as well, possess certain AI features, since the guidance laws direct them to hit the target and some of them imitate the behavior of predators pursuing their victims.

Surprisingly, the current policy concerning the future generation of UAVs focuses mostly on the development of more sophisticated sensors, communication and control devices, greater payload capability, endurance, and unmanned lethal capabilities. These areas of research are lavishly funded in contrast to the area of guidance and control of autonomous UAVs. Moreover, future autonomous UAVs are envisaged as controlled by computer programs that imitate a pilot's actions. Of course, future UAVs should possess some AI features that reflect a pilot's reaction in specific situations. But it is improbable to expect, at least in the nearest future, the creation of extremely complex, reliable computational programs to be used with onboard computers that together with related required sensors and other devices would meet payload or other requirements. Moreover, since the unmanned lethal capability problem requires the knowledge of parameters different from what would be used in the mentioned computational programs, it is logical to try to create UAV guidance laws that use the same parameters as missiles, which, we want to underline again, are also unmanned aerial vehicles. Recently, military experts indicated the importance of commonality in unmanned systems control. However, it looks like there is no clear understanding that control commonality starts with guidance laws.

The guidance of UAVs differs from missile guidance, its goal is different and depends on a concrete area of their application. If to perform simple scripted navigation functions, such as waypoint following, UAVs can use missile guidance laws with modifications (e.g., to meet speed requirements), for more complicated problems (e.g., refueling an aircraft or guidance of several UAVs working together) the guidance problem becomes a rendezvous-type problem. The moving objects should be on a fixed distance and their velocities should coincide. To address a wide class of guidance problems for UAVs, a more general guidance problem should be formulated. The interconnection between the missile and UAV guidance problems should not be ignored.

The author developed a class of guidance laws for missiles that implement parallel navigation, the strategy that can be considered as a law of nature; such that predators and later humans have used since antiquity. The approach that allowed him to obtain this class of guidance laws is described in detail in his book *Modern Missile Guidance*. This book generalizes these ideas. As a result, a wide class of guidance laws applicable to missiles and UAVs is considered.

Basic facts about unmanned aerial vehicle guidance are given in Chapter 1. Parallel navigation and the description of the proportional navigation guidance law as a means to control lateral motion are presented in Chapter 2. Proportional navigation is considered here as a control problem. Chapter 3 contains a detailed description of a class of guidance laws obtained based on the Lyapunov approach. It will be shown here that this class of guidance laws improves the effectiveness of the proportional navigation law for maneuvering and nonmaneuvering targets. Moreover, the approach offered can also be considered as another justification of the widely used proportional navigation law. The analytical expressions of the guidance law are given for the generalized planar and three-dimensional engagement models for missiles with and without axial controlled acceleration. The Lyapunov–Bellman approach is used to choose the guidance law parameters. The generalized guidance problem applicable for UAVs is considered.

The analysis of proportional navigation guided systems, which control lateral motion of unmanned aerial vehicles in the time domain based on the method of adjoints, is given in Chapter 4. Chapter 5 contains analysis of the proportional navigation guided systems in the frequency domain. The obtained analytical expressions for the miss distance can be used for guidance and control systems design. They enable the analysis of the influence of the guidance system parameters on its performance. The generalized guidance system model that also includes the target model is considered. The relationship between the frequency response and the miss step response is discussed. The procedure for determining the optimal frequency for which the amplitude of the miss distance has a maximum is presented.

The modification of the proportional navigation guidance law using the results of classical control theory is considered in Chapter 6. The approach is based on utilizing feedforward/feedback control signals to make the real vehicle acceleration close to the commanded acceleration generated by the guidance law. The effectiveness of these guidance laws against highly maneuvering targets is demonstrated. The analysis of the proportional navigation (PN) guidance systems performance under various types of noises is considered in Chapter 7. Analytical expressions for analysis of the proportional navigation guided systems are obtained.

Guidance of UAVs, whose practical application in various areas continues growing, is considered in Chapter 8. The guidance laws applied to a wide class of problems with UAVs are developed. The computational algorithms realizing these laws are tested in three applications: for the surveillance problem, the refueling problem, and the motion control of a swarm of UAVs. Chapter 9 deals with simulation models that can be used effectively for analysis of the guidance laws performance and for the comparative analysis of various guidance laws.

In Chapter 10 an attempt is made to discuss the problem of the integrated design of guidance and control laws. This problem is considered because of an increasing interest in integrated design of flight vehicle systems. Chapter 11 demonstrates how to apply the guidance laws discussed earlier to boost-phase intercept systems. This class of systems should be equipped with the future generation of interceptors. Special attention is paid to airborne interceptors launched from UAVs. Their specific features and various approaches to determine the best guidance laws are discussed.

The last chapter focuses especially on engineers who usually meet any new theoretical approaches with distrust. Lectures for scientists and engineers in the guidance and control area persuaded the author that a detailed practical example should accompany any theoretical course.

The following sentences from *Notebooks* (1508–1518) of Leonardo da Vinci did not lose their meaning in our days: "Those who are enamoured of practice without science are like a pilot who goes into a ship without rudder or compass and never has any certainty where he is going. Practice should always be based upon a sound knowledge of theory."

The offered guidance laws are in full compliance with the laws invented by the well-known science fiction writer Isaac Asimov in his novel *I, Robot*:

1. A robot may not injure a human being or, through inaction, allow a human being to come to harm.
2. A robot must obey orders given to it by human beings except where such orders would conflict with the First Law.
3. A robot must protect its own existence as long as such protection does not conflict with the First or Second Law.

For civil applications, the above statement is obvious. As to their military applications, the words "a human being" in the First Law should be interpreted as those who use the guidance law to protect themselves. It's unlikely that Isaac Asimov would object to such an interpretation.

The attractiveness of the guidance laws considered in the book is in their simplicity. They are as simple as the proportional navigation guidance law, which is widely used in practice mostly because of its simplicity.

It is known that the defense and aerospace industries are experiencing significant difficulty attracting and retaining scientists and engineers and that about 13,000 of the Department of Defense scientists will be eligible to retire in the next decade without sufficient numbers of graduating students to replace them. The material of this book can serve as a basis for several graduate courses in aerospace departments. It can be used by researchers and engineers in their everyday practice. The author hopes that this book will supply aerospace scientists and engineers with new ideas that, when crystallized, will bring significant improvement to unmanned aerial systems performance. Readers can find additional useful information and help on the Web site www.randtc.com.

About the Author

Rafael Yanushevsky was born in Kiev, Ukraine. He received a BS in mathematics and a MS (with honors) in electromechanical engineering from Kiev University and the Kiev Polytechnic Institute, respectively, and his PhD in optimization of multivariable systems in 1968 from the Institute of Control Sciences of the USSR Academy of Sciences, Moscow, Russia.

From 1968 he worked at the Institute of Control Sciences. His research interests were in optimal control theory and its applications (especially, in aerospace), optimal control of differential-difference systems, signal processing, game theory, and operations research. He has published over 40 papers in these areas and two books, *Theory of Linear Optimal Multivariable Control Systems* and *Control Systems with Time-Lag*. He was an editor for 14 books from the Nauka Publishing House. In 1971, the Presidium of the USSR Academy of Sciences gave him the rank of senior scientist in automatic control.

After immigration to the United States in December 1987, he started teaching at the University of Maryland, first in the Department of Electrical Engineering, then in the Department of Mechanical Engineering, and at the University of the District of Columbia in the Department of Mathematics. Since 1999, Dr. Yanushevsky has been involved in projects related to the aerospace industry. He participated in the development of an engagement model as a part of a Battlespace Engineering Assessment Tool, weapon control system software, new guidance laws, and wrote sections of the *Modeling and Simulation Handbook* related to the weapon control system and the fire control system of SM-3 missiles. In 2002, he received a Letter of Appreciation from the Department of the Navy—the Navy Area Theater Ballistic Missile Program.

Dr. Yanushevsky's research interests include guidance and control, signal processing and control, tracking of maneuvering targets, and integrated guidance and control systems design. He has published over 80 papers, was the Chair of the Lyapunov Session of the Second and Fourth World Congress on Nonlinear Analyst, and a member of the Organizing Committee of the Fourth Congress. He is included in *Who's Who in America*, *Who's Who in Science and Engineering*, and *Who's Who in American Education*.

1 Basics of Guidance

1.1 INTRODUCTION

The natural process of improvement of all aspects of our life also includes advances in development of sophisticated systems, the means to defend ourselves from enemies, those who consider wars as a way to improve their living conditions, and the means to help people make their work easier. Webster's dictionary defines robot as "an automatic device that performs functions ordinary ascribed to human beings." With this definition, unmanned aerial vehicles, also called drones, can be considered robots. Moreover, since human beings' functions assume the presence of intellect, missiles can be attributed to stretch the category *robotics* as well. In lieu of the thrown stone, the cast spear, the flying bullet, the dropping bomb, and the launched rocket, the defensive or destructive functions are better performed by missiles, because they possess a certain artificial intelligence that enables them to change their trajectory depending upon the behavior of the maneuvering target.

It is assumed that scientific terminology has precise meaning since scientific terms are offered by experts in certain fields. However, with the development of scientific disciplines and current narrow specialization in science, frequently the terms, which were applied to a definite area, become applicable to a different area that, in turn, operates with its own terms. Such terminological overlapping creates difficulties for a clear and rigorous presentation. For example, unmanned aerial vehicles (UAVs) have been referred to in many ways: RPV (remotely piloted vehicle), drone, robot plane, and pilotless aircraft are a few such names. The UAVs are also called unmanned aircraft systems and then it is not clear why the abbreviation UAV is used; it would be better and more precise to use unmanned aircraft vehicles. In turn, in many publications the UAV acronym stands for unmanned aerial vehicle(s), unmanned air vehicle, uncrewed aerial vehicle, unmanned autonomous vehicle, unmanned airborne vehicle, and unmanned aircraft vehicle. Despite the mentioned terminological mess, missiles—unmanned aerial vehicles—were excluded from this class.

In the future, we'll use the following terminology to describe a wide class of unmanned aerial vehicles.

An *unmanned aerial vehicle* is defined as a space-traversing vehicle that flies without a human crew on board and that can be remotely controlled or can fly autonomously.

The so-called UAV is defined by the Department of Defense as a powered aerial vehicle that does not carry a human operator, uses aerodynamic forces to provide vehicle lift, can fly autonomously or be piloted remotely, can be expendable or recoverable, and can carry a lethal or nonlethal payload. The five general categories of the UAVs depending upon their configurations are: fixed-wing, vertical takeoff and landing (VTOL), short takeoff and landing (STOL), rotary-wing or rotocraft, and helicopters.

A *missile* is defined as a space-traversing unmanned vehicle that contains the means for controlling its flight path and carries a lethal device. A guided missile is considered to operate only above the surface of the Earth, so guided torpedoes do not meet the above definition. The missiles are classified by the physical areas of launching and the physical areas containing the target. The four general categories of the guided missiles are: surface-to-surface, surface-to-air, air-to-surface, and air-to-air.

The *cruise missile* is defined as a dispensable, pilotless, self-guided aerial vehicle that flies like an airplane and carries a lethal device. Similar to conventional missiles, there exist the following versions of cruise missiles classified by the physical areas of launching: land-based or ground-launched cruise missiles, sea-based or sea-launched cruise missiles, and air launched cruise missiles. Cruise missiles are differentiated from UAVs in that the weapon is integrated into the vehicle, and the vehicle is intended to be sacrificed in the mission. Being, in essence, flying bombs, cruise missiles are guided missiles that use a lifting wing and most often a jet propulsion system to allow sustained flight and carry a warhead many hundreds of miles with excellent accuracy. To distinguish UAVs from cruise missiles, a UAV is defined as a reusable unmanned guided vehicle. A cruise missile is a weapon that is not reused, even though it is also unmanned and in some cases remotely guided. Modern cruise missiles normally travel at supersonic or at high subsonic speeds, are self-navigating, and fly in a nonballistic, very low altitude trajectory in order to avoid radar detection.

The first attempt to use a pilotless plane was guided toward a target and then crashed into the target in a power dive—as an airborne counterpart of the naval torpedo—took place in the United States during World War I. In 1916–1917, a prototype called the Hewitt-Sperry Automatic Airplane made a number of short test flights proving that the idea was sound. But it was not used in combat by the United States during that war. Twenty pilotless aircraft, called Bugs, were produced and a successful test flight was made in October of 1918. After World War I ended, all projects were discontinued except for some experiments with Bugs. This project was dropped in 1925. The first pilotless aircraft led to the development of the radio-controlled, pilotless target aircraft in Britain and the United States in the 1930s. Although during this decade there was little missile research, progress in aviation and developments in electronics produced results, which were later applied to missiles. In 1936, the Navy began another pilotless aircraft program intended to provide realistic targets for antiaircraft gunnery practice but which directly influenced missile development. In 1937, the radio-controlled drone Curtiss N2C-2 was tested. During the war the Air Force acquired hundreds of Culver PQ-8 and improved Culver PQ-14 target drones, which were radio-controlled versions of the tidy little Culver Cadet two-seat civil sport plane and used for antiaircraft training.

The United States also used radio-controlled aircraft, including modified B-17 and B-24 bombers, in combat on a small scale during World War II as aerial torpedoes, though without noticeable success. In January 1941, work begun on the conversion of a TG-2 (torpedo plane) and a BG-1 (dive bomber) into missiles. Many missile research and development programs were initiated during World War II. The most advanced were the German surface-to-surface missiles, the V-1 (German FZG-76) and the V-2 (German A-4). The V stood for *Vergeltungswaffe* (vengeance weapon). The advent and rapid development of jet aircraft following World War II changed forever the character of air-to-air combat. It was believed that the high speed and maneuverability of jet aircraft signaled the end of the dogfight and a requirement to engage targets at beyond visual ranges. The solution to this problem was the air-air or surface-to-air missiles. Postwar research in the upper atmosphere gained a new tool with the advent of high-altitude rockets. The improvement of missile guidance and its accuracy became the most important problem of research and development. As to the postwar unmanned aerial vehicle programs in the United States, the various target drone series were developed; but it was not until the Vietnam War that UAVs such as the Firebee drones were used in a surveillance role.

Over the last few years, it has been Israel that has contributed to the development of the UAV sector. There exists a wide variety of UAVs with different sizes, configurations, and characteristics. Some of them are controlled from a remote location and others fly autonomously based on preprogrammed flight plans using more complex guidance and control systems. The United States currently possesses five major UAVs: the Air Force's Predator and Global Hawk, the Navy and Marine Corps's Pioneer, and the Army's Hunter and Shadow. The Hunter and the Pioneer, which are used extensively by the U.S. military, are direct derivatives of Israeli systems. The Pioneer was used successfully in the Gulf War. Following the Gulf War, the importance of unmanned systems became obvious. The Predator, first an advanced technology demonstration project, demonstrated its worth in the skies over the Balkans. Some of the current versions of the Predator are loaded with Hellfire missiles for attack purposes. The Global Hawk is a jet-powered UAV that was used effectively in Afghanistan. The UAVs that are in use and under development are both long-range and high-endurance vehicles. Now numerous UAVs are targeted for multimission roles from persistent surveillance to search-and-destroy.

A guided unmanned aerial vehicle system is defined as a combination of an unmanned aerial vehicle and its launching, guidance, test, and handling equipment.

An unmanned aerial vehicle guidance system is defined as a group of components that measures the position of the guided vehicle with respect to the target(s) and changes its flight path in accordance with a guidance law to achieve the flight mission goal. Usually, the missile guidance system includes sensing, computing, and control components, and the flight goal is to destroy a target. The UAV guidance system also includes an operator of a mission control unit, and targets may be dummy and generated by the operator to control the vehicle's flight.

A guidance law is defined as an algorithm that determines the required commanded unmanned aerial vehicle acceleration.

Guided missile systems have similar tactical duties as the conventional weapons (guns, rockets, and bombs). However, in conventional weapon systems information concerning the target is gathered by observation. After it is evaluated, the weapon is aimed and the projectile is fired. From the time the bullet or rocket is aimed or the bomb is dropped, the trajectory is strictly dependent upon gravity, wind, and the ballistics of the projectile. The time from launching of the projectile till the hit at the target is called the *time of flight.* In contrast to the bullet, rocket, or bomb, the missile in flight is constantly reaimed based on the target information obtained by sensors. The target is tracked to gain intelligence as to its current position, as well as its future behavior. Advanced guidance systems operate with data estimating also the target acceleration and the predicted intercept point.

Guided UAV systems have similar tactical duties as the conventional aircraft. Being pilotless they are usually operated, partially or completely, from the ground mission control station. Its operator determines the future positions of the UAV and controls the flight path. Stationing the mission control element of the UAV system in another aircraft instead of on the ground—a new trend in UAV design—will reduce the reliance on satellites for beyond line-of-sight communication and simplify the UAV guidance and control system.

In order to guide and control an unmanned aerial vehicle several functions must be achieved:

1. The launch function monitors the launch events sequence and establishes the initial vehicle position and velocity after launch.
2. The targeting function establishes the basic geometry between the vehicle and target(s) and operates in the coordinate system relative

to which the targeting and guidance must be performed. For UAVs and cruise missiles, the predetermined vehicle positions, the so-called waypoints, serve as dummy intermediate targets that allow the vehicle to correct its flight path.

3. The guidance function generates guidance commands directing a vehicle toward the target(s). In contrast to traditional missiles, for which a guidance law realized by a computational algorithm directs the missile to hit the target, the flight path of current UAVs and cruise missiles are remotely controlled or preprogrammed so that their guidance function simply generates the desired flight trajectory.

4. The flight control function converts guidance commands into vehicle response; this function is performed by autopilots. The control actuator of a missile generally consists of thrusters that control the direction of the propulsion subsystem's thrust vector and/or mechanical devices that move external surfaces of the missile in order to alter the aerodynamic forces acting on it. The control actuator of a UAV consists of mechanical devices that move control surfaces in the case of the fixed-wing UAV or change the pitch angle of the rotorcraft UAV's blades. Cruise missiles usually use a jet propulsion system. Turbojets are used for subsonic tactical cruise missiles, turbofans for subsonic strategic cruise missiles, and ramjets or mixed turbojet/rocket designs for supersonic tactical cruise missiles. UAVs use conventional aircraft propulsion (e.g., piston or turbojet engines); and for micro UAVs, with less than a 10 cm (6 inch) wingspan, flight experiments are under way with new forms of propulsion using microwave beamed power, rechargeable batteries, or solar cells.

Discussion of the principles of unmanned aerial vehicle guidance involves many fields and subfields of science, which is impossible to cover in one book. Our main attention will focus on the guidance function—the guidance laws that serve as control actions guiding an unmanned aerial vehicle. All other functions mentioned can be considered as auxiliary ones because, on the one hand, they create conditions for the guidance function operation and, on the other hand, make an unmanned aerial vehicle motion possible in accordance with the guidance function commands.

Guidance is a dynamic process of directing an object toward a given point that may be stationary or moving. Usually, in the case of the stationary point, the guidance process is called *navigation*. Until the twentieth century, this term referred mainly to guiding ships across the seas. The word "navigate" comes from the Latin *navis*, meaning "ship," and *agree*, meaning "to move or direct." Today, however, the word also encompasses the guidance of travel on land, in the air, and in inner or outer space. It means finding the way from one place to another (i.e., guiding toward the stationary point). In the future, we will not distinguish the cases of stationary and moving points and consider the general case of a moving point, which will be called a target.

1.2 GUIDANCE PROCESS

The goal of guidance is to reach a target. When getting to a target, an object position coincides with a target position. Additional requirements to an object velocity and possibly acceleration specify various types of guidance. *Rendezvous* is a guidance when an object velocity equals a target velocity. *Conditional rendezvous* is a guidance to a certain position situated at a given distance from a target when an object velocity equals a target velocity.

Applied to missiles, as guided objects, the goal of guidance is to *intercept* a target. It means that at a certain moment of time a missile position should coincide with a target position and a target velocity should be sufficient to destroy a target. The goal of guidance, expressed mathematically precisely or by using a "humanitarian language," should be supported by an adequate rule that is able to realize this goal.

One of the ancient guidance rules, successfully used by mariners wishing to rendezvous with each other at sea or sea pirates trying to catch a boat, is called the *parallel navigation* (the "constant bearing" and "collision course navigation" are also used to characterize this type of guidance). This antique rule requiring an approach with a constant bearing angle (angle measured horizontally from North to whatever direction is pointed) assumed constant speeds of target and pursuer boats. From a pure geometrical consideration, it is easy to establish what velocity the pursuer should have to reach a target. The parallel navigation is one widely used by animal's navigation strategies, which, in general, depend on the environment and the task they have to solve. For example, predators and organisms pursuing mates commonly adjust their position to maintain constant angle with respect to the target.

The rule, obtained many years ago under the assumption of the constant velocities, is applied now to accelerating moving target and objects. The parallel navigation principle was first used in the Lark missiles in 1950 [2,4]. The so-called proportional navigation (PN) was used to implement the parallel navigation. Since that time the PN is used in almost all of the world's tactical guided missiles.

In the general case, missile flight consists of three phases: the boost, midcourse, and homing stages. Guidance at each of the mentioned stages has its own specifics. The boost stage is a part of the missile flight between initial firing and the time when the missile reached a velocity at which it can be controlled. During the midcourse stage, the missile is guided by an external weapon control system. The homing stage corresponds to the terminal guidance when the missile-contained system controls the missile flight. Currently, the parallel navigation is used mostly at the homing stage. However, it can be applied also at the midcourse stage.

Since the area of UAV applications is significantly wider than that of missiles, the goal of guidance is different. Moreover, during the UAV flight it may be formulated as a sequence of subgoals (to reach a certain space position) and the final goal can correspond to the above definitions of *rendezvous* and *conditional rendezvous.*

In the future we will consider the guidance problem as a control problem and characterize the guidance laws from the position of control theory. Taking into account that a guidance law controls the flight of an object (i.e., presents a controlled input of the moving object), we should characterize the object from the position of control theory. We will formulate the goal of control and introduce the parameters that describe the object's behavior and the parameters that describe the environment including external forces that influence the object's behavior.

Despite the rigorousness and attractiveness of such an approach, it is difficult to present a universal dynamic model of various moving objects pursuing targets. That is why we will first ignore an object's inertia and consider a model of objects ignoring their dynamics. This makes the model, to a certain degree, "universal." However, the guidance law obtained for this model can not be considered as the best for various moving objects because it does not consider their dynamics. It will allow us to establish a kind of universal guidance laws that can be later improved based on information concerning dynamic properties of a concrete moving object.

1.3   MISSILE GUIDANCE

Among external factors influencing an object's behavior, the target information is the most important one. It has been pointed out that the two basic categories of targets are moving and stationary targets. Missile targets are classified into two broad classes: air targets, usually aircraft or other missiles; and surface targets, which include ships and various objects on the ground. To be destroyed the targets must be detected, identified, and tracked by the missile or associated equipment. All guided missiles launched to engage moving targets use units that observe or sense the target. The point of observation may vary. It can be observed from the missile or a station outside the missile. Based on a target observation, its characteristics of behavior can be determined. Stationary targets are usually situated at long range; the information about them is gathered and presented by intelligence so that the missile trajectory is determined in advance and can be only corrected during its flight. When a stationary target is at short range and guided missiles are used to deliver sufficient destructive power, the information about the target can be obtained by the units that observe and sense the target and by intelligence.

As mentioned above, the goal of missile guidance is to hit a target (i.e., to nullify the distance between a missile and the target). However, this obvious goal is usually accompanied by additional conditions. It can be, as an additional criterion, minimization of the time of flight or maximization of a missile terminal velocity. Such criteria dictate the path (optimal trajectory) that the guidance system must direct the missile. In the case of a stationary target, the guidance law, obtained as a solution of an optimal problem, enables us to generate and analyze the optimal trajectory that will require only insignificant corrections during the missile flight. The solution of an optimal problem for a moving target requires the information about its future behavior. In the general case such information is not available so that optimal guidance problems for moving targets have a limited application.

Early missile systems used a variety of guidance laws including beam riders and pursuit guidance. However, proportional navigation proved to be the most versatile and, with suitable modification or augmentation, still remains in use in most of contemporary guided missile systems. Many current missile guidance laws are generally based on one of several forms of proportional navigation.

The effectiveness of guidance laws depends on parameters of a missile's flight control system that realizes the flight control function and characterizes a missile's dynamics. Aerodynamics is part of the missile's airframe subsystem, the other major parts being propulsion and structure. The autopilot receives guidance commands and processes them to the controls, such as deflections or rates of deflection of control surfaces or jet controls. The control subsystem transfers the autopilot commands to aerodynamic or jet control forces and in moments changes the position of the airframe to attain the commanded maneuver by rotating the body of a missile to a desired angle of attack. The autopilot response should be accomplished quickly with minimum overshoot. Minimal overshoot enables a missile to avoid exceeding structural limitations.

Three types of aerodynamic controls are used: canard (small surface forward on the body), wing (main lifting surface near the body), and tail (small surface far aft on the body) controls [1,3]. In contrast to the canard and wing controls, the tail steering controls initially give acceleration in a direction opposite to the intended one. The airframe reacts to the control commands with speed depending on the airframe inertia and system damping. The influence of the flight control system on guidance accuracy will be examined in details.

1.4 GUIDANCE OF CRUISE MISSILES AND UAVs

Usually, targets of cruise missiles and UAVs are surface targets. Cruise missiles should destroy these targets, whereas UAVs should provide information about them and destroy them only if the UAVs carry a lethal payload—missiles. Various future UAV applications also assume aerial targets (e.g., aircraft refueling by UAVs).

As mentioned earlier, both cruise missiles and UAVs follow a predetermined path from the launch point to the target(s) in accordance with the selected mission. Their trajectories are defined by a series of waypoints, with a particular latitude, longitude, and altitude, to which the vehicles are guided to fly. Guidance commands can be changed only at waypoints. The guidance system controls the vehicles flight along preprogrammed (for cruise missiles) and predetermined flight profiles between waypoints (geographical coordinates). The new generation of cruise missiles has a direct link with an operator, which is necessary used at a preprogrammed point approximately one minute before target impact, where the seeker turns on; the operator views the target scene (after receiving a video image), selects an aim point for the terminal phase, and the missile flies automatically to that point. Current UAVs communicate most of their flight with an operator and provide him with visual information. If the trend for designing new cruise missiles is to provide an operator with visual information about the target during the terminal phase to improve their accuracy, then new UAV designs are focused to make the operator's work easier so that in this case some of his guidance functions should be transferred to UAVs.

Since most cruise missiles and UAVs cover significant distances, to minimize possible accumulated measurement errors (position, velocity, attitude, and altitude) the combination of the global positioning system (GPS) and inertial navigation system (INS) is used. The long-term accuracy of the GPS combined with the short-term accuracy and autonomy of the INS results in a highly effective integrated system. From time to time the vehicles recognize where they are and compare their actual position with where they should be according to their assigned path or trajectory, and the autopilots make appropriate maneuvers to bring the vehicles back to the correct trajectory. Since neither cruise missiles nor UAVs operate against highly maneuvering targets, it is logical to assume that their guidance laws should be simpler than the guidance laws for conventional missiles. According to the guidance law, the autopilot commands are transferred by the control subsystems to aerodynamics or jet control forces and in moments change the position of the vehicle's airframe. Well-designed UAVs should have better dynamic characteristics than missiles because of their smaller size and weight, so that their airframe inertia should be significantly less than that of missiles. Being remotely piloted vehicles, the advanced UAVs, nevertheless, can fly autonomously. However, the more autonomous ability a UAV has, the more complex its guidance and control systems is and, as a result, the higher is its size and weight, the less—its endurance, a combat radius, and/or speed. The desire to reduce the operator's load leads to an increase of a UAV's payload that worsens its dynamic properties, so that it becomes more difficult to reach its best potential performance characteristics.

1.5   REPRESENTATION OF MOTION

We will consider the so-called two-point systems including an unmanned aerial vehicle M and a target T. In the inertial reference frame of coordinates the positions of M and T is given by the vectors $r_M = (R_{M1}, R_{M2}, R_{M3})$ and $r_T = (R_{T1}, R_{T2}, R_{T3})$, respectively, and the vector $r = (R_1, R_2, R_3)$:

$$r = r_T - r_M \tag{1.1}$$

is called the *range*-vector. Its negative derivative that equals the difference between the vehicle $V_M = (V_{M1}, V_{M2}, V_{M3})$ and target $V_T = (V_{T1}, V_{T2}, V_{T3})$ velocities:

$$-\dot{r} = -(\dot{r}_T - \dot{r}_M) = v_M - v_T = v_{cl} \tag{1.2}$$

is called the *closing velocity* vector $v_{cl} = (v_{cl1}, v_{cl2}, v_{cl3})$. In the future we will use also the *range r* and *closing velocity* v_{cl} (scalars) terms when dealing with absolute values of r and $v_{cl} = -\dot{r}$.

It follows from equations (1.1) and (1.2) that

$$R_s = R_{Ts} - R_{Ms}, \quad v_{cls} = V_{Ms} - V_{Ts} = \dot{R}_{Ms} - \dot{R}_{Ts} \quad (s = 1, 2, 3) \tag{1.3}$$

A two-point guidance is said to be *planar* if r_M, v_M and r_T, v_T remain in the same fixed plane. In general, the guidance process takes place in three-dimensional (3-DOF) space. In some cases, it can be presented as a combination of two orthogonal planar guidance processes.

The solution of the intercept problem requires the utilization of several frames of reference (coordinate axes) to specify relative positions and velocities, forces, accelerations, and so on.

An inertial fixed reference plane is a necessary part of every dynamic problem. The inertial coordinate system ignores both the gravitational attraction of the sun, moon, other bodies, and the orbital motion of Earth that exists because of this attraction. In many problems of aerospace dynamics, the orbital motion of Earth can be neglected, and any reference plane fixed to Earth can be used as an inertial frame. However, for hypervelocity and space flights the angular velocity of Earth usually must be taken into account. Two Earth-fixed frames are used: (i) the Earth-centered fixed inertial (ECI) coordinate system with its origin at the center of the Earth and axes directions fixed by a reference point on the equator and Earth's axes; (ii) the Earth-surface fixed (ESF) coordinate system with origin at the arbitrary Earth surface (usually close to the vehicle) with axes directed North, East, and vertically (mostly downward, but sometimes it is convenient to choose the upward direction).

It is more convenient to consider the aerial vehicle and target motions relative to these inertial coordinate systems. However, missile and UAV dynamics are usually analyzed in the missile and UAV body-fixed frame, respectively, and the tracking process requires a different reference frame.

It is known that a moving object has six degrees of freedom: three translations and three rotations. The principal aerial vehicle motions of interest to the guidance problem are:

1. Translation along the longitudinal axis (velocity)
2. Rotation about the longitudinal axis (roll)
3. Rotation about the lateral horizontal axis (pitch)
4. Rotation about the vertical axis (yaw)

The origin of the body-fixed coordinate system is situated at the vehicle center of gravity. The orientation of the axes is usually taken to be coincident with principal axes of inertia. These motions are controlled by autopilot in accordance with the guidance law. The so-called vehicle-carried vertical frame, also called the North-East-Down (NED) frame, usually has the origin situated at the vehicle center of gravity and axes directed North, East, and vertically downward. It is commonly used for tactical missiles. The NED system is not precisely an inertial coordinate system because the missile axes are slowly changing their orientation in space as the missile moves over Earth's surface. However, except for the North Pole, the rotational effects are negligible.

During the midcourse and terminal stages guidance commands are based on measurements obtained in various coordinate systems (in addition to the above-mentioned frames there exist other frames, e.g., atmosphere-fixed and air-trajectory reference frames that are used for specific analysis [1]). There exist transformations from one coordinate system to another.

As indicated above, for analytical investigation the choice of the reference frame is usually a matter of convenience. In atmospheric flights

Earth-fixed and vehicle-fixed coordinate systems are commonly used. The future consideration will be mostly based on analysis in the Earth-surface fixed coordinate system or the NED coordinate system. However, theoretical results should be compared with the simulation results also obtained by using the six-dimensional model that includes and operates with the various coordinate systems.

The position of moving objects is usually determined in polar (spherical) coordinate systems. The position of a target determined by a vehicle's sensors is typically specified by direction cosines (cosines of the angles that the position vector makes with the coordinate axes, respectively) relative to the vehicle body axes that can be transformed to direction cosines $(\Lambda_N, \Lambda_E, \Lambda_D)$, with respect to the NED axes. The target angular position with respect to the NED coordinate system can also be specified by azimuth α and elevation β angles:

$$\alpha = -\sin^{-1}\Lambda_D, \quad \beta = \tan^{-1}(\Lambda_E/\Lambda_N) \tag{1.4}$$

The target coordinates (R_N, R_E, R_D) in the NED Cartesian coordinate system can be obtained based on range r and direction cosines $(\Lambda_N, \Lambda_E, \Lambda_D)$:

$$R_N = r\Lambda_N = r\cos\alpha\cos\beta,$$

$$R_E = r\Lambda_E = r\sin\alpha\cos\beta, \tag{1.5}$$

$$R_D = r\Lambda_D = r\sin\beta$$

here the sign of the elevation angle is defined to be positive in the downward direction.

Polar coordinates (r, α, β) are related to the Cartesian coordinates (R_N, R_E, R_D) by:

$$r = \sqrt{R_N^2 + R_E^2 + R_D^2}, \quad \alpha = \tan^{-1}(R_E/R_N), \quad \beta = -\sin^{-1}(r/R_D) \tag{1.6}$$

In the future, the North, East, and vertical coordinates will be denoted by the low indices 1, 2, and 3, respectively. For ground-based defense systems, the unmanned aerial vehicle and target positions are determined relative to the Earth-surface fixed (ESF) Cartesian coordinate system. The vertical coordinate is a target (unmanned aerial vehicle) altitude. In the case of space-based strategic systems, the ECI coordinate system is most convenient. In general, tracking is performed in Cartesian position coordinates. However, for single sensor systems, such as airborne radar, the option to track in spherical coordinates may be considered.

1.6 LINE-OF-SIGHT

In order to view an object, one must sight along a line at that object. The line-of-sight (LOS) that passes through the objective of the guidance is an important concept of guidance. Its orientation with respect to the reference coordinate system enables one to formulate precisely the guidance rules.

For the three-dimensional case and the Earth-based coordinate system, the line-of-sight can be represented as:

$$\lambda(t) = \lambda_1(t)\boldsymbol{i} + \lambda_2(t)\boldsymbol{j} + \lambda_3(t)\boldsymbol{k} \tag{1.7}$$

where $\boldsymbol{i}$, $\boldsymbol{j}$, and $\boldsymbol{k}$ are unit vectors along to the North, East, and vertical coordinate axis, respectively,

$$\lambda_s(t) = \frac{R_s}{r} \qquad (s = 1, 2, 3) \tag{1.8}$$

The R_s ($s = 1, 2, 3$) are the range-vector coordinates [see equation (1.1) and also equations (1.3)–(1.7)]. Here, for convenience, we assume that $\boldsymbol{k}$ is directed upward.

The LOS-vector can be presented as a sum of two vectors in the horizontal **x-y** (North-East)-plane and the vertical **x-y** resultant-**z** plane (see Figure 1.1). The LOS's position in the vertical plane λ_2 is determined by the elevation angle β. Its position in the horizontal plane is presented by λ_{13} that is determined by the azimuth angle α, so that λ_s coordinates ($s = 1, 2, 3$) are determined by the expressions that can also be obtained directly from equations (1.6) and (1.9), i.e.,

$$\lambda_1 = \cos\alpha\cos\beta, \quad \lambda_2 = \sin\alpha\cos\beta, \quad \lambda_3 = \sin\beta \tag{1.9}$$

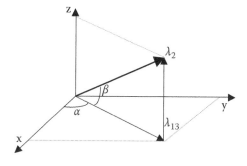

FIGURE 1.1 Three-dimensional presentation of LOS.

The expressions for the LOS rate in the three-dimensional Cartesian coordinate system:

$$\dot{\boldsymbol{\lambda}}(t) = \dot{\lambda}_1(t)\boldsymbol{i} + \dot{\lambda}_2(t)\boldsymbol{j} + \dot{\lambda}_3(t)\boldsymbol{k} \tag{1.10}$$

can be obtained from equations (1.3), (1.7), and (1.8):

$$\dot{\lambda}_s(t) = \frac{\dot{R}_s r - R_s \dot{r}}{r^2} = \frac{V_{Ts} - V_{Ms}}{r} + \frac{R_s v_{cl}}{r^2} \qquad (s = 1, 2, 3) \tag{1.11}$$

where, based on equations (1.3) and (1.8):

$$v_{cl} = -\dot{r} = -\frac{\sum\limits_{s=1}^{3} R_s(V_{Ts} - V_{Ms})}{r} = \frac{\sum\limits_{s=1}^{3} R_s v_{cls}}{r} = \sum\limits_{s=1}^{3} \lambda_s v_{cls} \tag{1.12}$$

When operating with the vertical and horizontal planes, it is convenient to use the vertical R_v and horizontal (ground) R_h ranges and velocities V_h and V_v:

$$R_v = R_3 = r\sin\beta,$$

$$R_h = \sqrt{R_1^2 + R_2^2} = r\cos\beta, \tag{1.13}$$

$$R_1 = R_h\cos\alpha, \quad R_2 = R_h\sin\alpha$$

$$V_v = -v_{cl3} = \dot{R}_v, \quad V_h = -v_{clh} = \dot{R}_h = -\frac{\sum\limits_{s=1}^{2} R_s v_{cls}}{R_h} \tag{1.14}$$

and

$$v_{cl} = -\dot{r} = \frac{R_h v_{clh} + R_v v_{clv}}{r} \tag{1.15}$$

The LOS rate components $\dot{\lambda}_s$ of (1.10) can be presented by the polar coordinates α and β by using equation (1.9):

$$\dot{\lambda}_1(t) = -\dot{\alpha}\sin\alpha\cos\beta - \dot{\beta}\cos\alpha\sin\beta,$$

$$\dot{\lambda}_2(t) = \dot{\alpha}\cos\alpha\cos\beta - \dot{\beta}\sin\alpha\sin\beta,$$

$$\dot{\lambda}_3(t) = \dot{\beta}\cos\beta \tag{1.16}$$

Using the relationship between the vectors $r(t)$ and $\lambda(t)$:

$$r(t) = r(t)\lambda(t) \tag{1.17}$$

$$\dot{r}(t) = (\dot{\lambda}_1(t)r + \dot{r}(t)\lambda_1(t))i$$
$$+ (\dot{\lambda}_2(t)r + \dot{r}(t)\lambda_2(t))j \tag{1.18}$$
$$+ (\dot{\lambda}_3(t)r + \dot{r}(t)\lambda_3(t))k$$

and

$$\ddot{r}(t) = (\ddot{\lambda}_1(t)r(t) + 2\dot{r}(t)\dot{\lambda}_1(t) + \ddot{r}(t)\lambda_1(t))i$$
$$+ (\ddot{\lambda}_2(t)r(t) + 2\dot{r}(t)\dot{\lambda}_2(t) + \ddot{r}(t)\lambda_2(t))j \tag{1.19}$$
$$+ (\ddot{\lambda}_3(t)r(t) + 2\dot{r}(t)\dot{\lambda}_3(t) + \ddot{r}(t)\lambda_3(t))k$$

we can present the equation of motion:

$$\ddot{r}(t) = a_T(t) - a_M(t) \tag{1.20}$$

where $a_M(t) = (a_{M1}, a_{M2}, a_{M3})$ and $a_T(t) = (a_{T1}, a_{T2}, a_{T3})$ are the vectors of the unmanned aerial vehicle and target accelerations created by forces acting on the unmanned aerial vehicle and target, respectively.

The aerial vehicle acceleration is the result of the propulsion forces (thrust), the aerodynamic forces (lift, drag), and gravity forces. In the following chapters this equation will be examined in details.

1.7 LONGITUDINAL AND LATERAL MOTIONS

It is convenient to consider the unmanned aerial vehicle motion consisting of two components: radial (longitudinal) that is directed along the line-of-sight and lateral that is orthogonal to the line-of-sight.

For the three-dimensional case and the Earth-based coordinate system, the target-to-unmanned aerial vehicle range vector $r(t)$ and its derivatives are represented by equations (1.17)–(1.19), so that the dynamic equations of the three-dimensional engagement can be presented in the form (1.20):

$$\ddot{r}(t) = a_T(t) - a_M(t) = a_{Tr}(t) + a_{Tt}(t) - a_{Mr}(t) - a_{Mt}(t) \tag{1.21}$$

where the unmanned aerial vehicle $a_M(t)$ and target $a_T(t)$ accelerations consist of two components: longitudinal and lateral, i.e.,

$$a_M(t) = a_{Mr}(t) + a_{Mt}(t), \quad a_T(t) = a_{Tr}(t) + a_{Tt}(t) \tag{1.22}$$

where $a_{Tr}(t)$, $a_{Mr}(t)$, $a_{Tt}(t)$, and $a_{Mt}(t)$ are the target and vehicle longitudinal (radial) and lateral (tangential) accelerations with the coordinates $a_{Trs}(t)$, $a_{Mrs}(t)$, $a_{Tts}(t)$, and $a_{Mts}(t)$ ($s = 1, 2, 3$), respectively.

As shown later, some guidance laws generate only lateral acceleration (i.e., produce only the lateral motion). Moreover, even if according to a guidance law both longitudinal and lateral motions should be realized, not all existing propulsion systems are able to control the longitudinal motion.

REFERENCES

1. Hemsch, M. *Tactical Missile Aerodynamics, Progress in Astronautics and Aeronautics.* Vol. 141. Washington, DC: American Institute of Astronautics and Aeronautics, Inc., 1992.
2. Shneydor, N. A. *Missile Guidance and Pursuit.* Chichester, UK: Horwood Publishing, 1998.
3. Valavanis, K. P. *Advances in Unmanned Aerial Vehicles.* The Netherlands: Springer, 2007.
4. Zarchan, P. *Tactical and Strategic Missile Guidance, Progress in Astronautics and Aeronautics.* Vol. 176. Washington, DC: American Institute of Astronautics and Aeronautics, Inc., 1997.

2 Control of Lateral Motion

2.1 INTRODUCTION

Decomposition of the unmanned aerial vehicles motion into two parts—longitudinal (radial, axial) and lateral (orthogonal, tangential) motions—enables us to examine these components of motion separately. Watching how predators pursue their victims, one can conclude that they almost never direct themselves at the target. Only in the case of a nonmaneuvering target moving along the line-of-sight, there is no lateral motion and we have pure longitudinal motion. The so-called *pure pursuit* guidance geometric rule requires the pursuer to be directed at the target. This type of guidance has another name: hound-hare pursuit. Its origin is in the note by Dubois-Aymè published in 1811, in which he formulated and solved the intercept problem based on analysis of the traces left by his dog when chasing him on the beach. However, a dog in this case cannot be considered as a predator and this fact can explain low accuracy of the first generation of guided weapons starting from World War II that used the pure pursuit rule. Moreover, when aircraft pilots applied this rule, they actually executed the so-called *lead pursuit* by pointing the fighter's guns at a certain angle ahead of the target (*lead* means in the direction of the future target's position), i.e., the weapon trajectory also contained a lateral component.

Multiple experiments with animal predators showed a substantial lateral component of their motion. Analysis of this component is very important because of acceleration and velocity limits of the existing propulsion systems and inability to control the longitudinal motion of certain types of unmanned aerial vehicles (e.g., missiles without throttleable engines cannot control their axial motion).

2.2 PARALLEL NAVIGATION

According to the parallel navigation (the "constant bearing") rule, the line-of-sight (LOS) direction relative to the inertial coordinate system is kept constant, i.e., during guidance the LOS remains parallel to the initial LOS. Using equations (1.10) and (1.11), this rule can be presented in the form:

$$\dot{\boldsymbol{\lambda}}(t) = \dot{\lambda}_1(t)\boldsymbol{i} + \dot{\lambda}_2(t)\boldsymbol{j} + \dot{\lambda}_3(t)\boldsymbol{k} = 0 \qquad (2.1)$$

or

$$\dot{R}_s r - R_s \dot{r} = 0 \qquad (s = 1, 2, 3) \tag{2.2}$$

The last equations show that for each moment of time guidance, realizing the parallel navigation, keeps constant the ratio of $\dot{R}_s$ and R_s ($s = 1, 2, 3$), i.e.,

$$\frac{\dot{R}_1}{R_1} = \frac{\dot{R}_2}{R_2} = \frac{\dot{R}_3}{R_3} = \frac{\dot{r}}{r} \tag{2.3}$$

It means that the vectors $\dot{r}$ and r are collinear and the vectors $r(t)$, $v_M(t)$, and $v_T(t)$ are instantaneously coplanar (the engagement need not be coplanar). The last statement follows immediately from zero value of the determinant of the 3×3 matrix formed by the above-indicated vectors.

The product of $\dot{r} r = 0.5(r^2)'$ must be negative; otherwise the distance between the pursuer and target will increase rather than decrease. This is equivalent to $\dot{r} < 0$ or $v_{cl} > 0$.

The character of motion in accordance with the parallel navigation rule can be observed vividly on a fixed plane (see Figure 2.1) by assuming that the target is nonmaneuvering, (i.e., is, $a_T(t) = 0$ and the ratio of speeds v_M and v_T is a constant).

Considering the scalar product of $\dot{r}(t)$ and $\lambda(t)$ in (1.2) and taking into account equations (1.10) and (1.18) we obtain:

$$\dot{r} = v_T \cos\theta - v_M \cos\delta \tag{2.4}$$

and the condition $\dot{r} < 0$ is equivalent to:

$$v_M \cos\delta > v_T \cos\theta \tag{2.5}$$

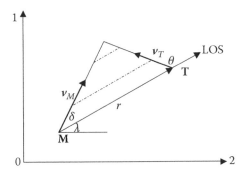

FIGURE 2.1   Geometry of planar engagement.

The collinearity condition (2.3) is equivalent to:

$$v_T \sin \theta - v_M \sin \delta = 0 \qquad (2.6)$$

If the target is moving with a constant speed and the conditions (2.5) and (2.6) are satisfied, a pursuer with a constant speed will intercept the target by moving in a straight line.

Figure 2.1 shows an engagement triangle consisting of the pursuer and target positions, the vectors of their velocities, the LOS, and range vectors. The LOS angle λ is measured with respect to the horizontal reference line 02. The angle δ is called *the lead angle*. The angle $180° - \theta$ is called *the aspect angle*. The size of δ characterizes the internal power of a pursuer (e.g., for animal predators it means how skillful they are within their class, which depends on their age and health; for missiles and other unmanned aerial vehicles it depends on the power and performance of their propulsion and control systems and the vehicle's aerodynamic properties). Smaller lead angles correspond to more powerful pursuers.

If the conditions (2.5) and (2.6) are satisfied, the pursuer with an appropriate constant velocity can intercept the nonaccelerating target. The dash-dotted lines show the position of the pursuer M and target T (the position of the LOS line) according the parallel navigation rule. The triangle in Figure 2.1 is called the *collision triangle*.

2.3   PROPORTIONAL NAVIGATION. PLANAR ENGAGEMENT

Proportional navigation (PN) guidance is the most widely used law in practice. Since it was offered initially for missile guidance, further consideration of this law and related more general laws will be applied to missiles, although these laws can be applied to UAVs directing their motion along the chosen waypoints (see Chapter 8). The basic philosophy behind PN is that missile acceleration should nullify the line-of-sight (LOS) rate between the target and interceptor. Proportional navigation, the guidance law that implements parallel navigation, is based on physical intuition. According to the parallel navigation, the LOS rate must be equal to zero. In reality, it differs from zero, so that the guidance command that is proportional to the rate of the LOS change may decrease the absolute value of the LOS rate and tend it closer to zero value. The PN law states that the commanded acceleration is proportional to the LOS rate; the proportionality constant can be broken down into the product of the effective navigation ratio N times the relative missile-to-target closing velocity, i.e.,

$$a_c(t) = N v_{cl} \dot{\lambda}(t) \qquad (2.7)$$

where $a_c(t)$ is a commanded acceleration acting perpendicular to the instantaneous LOS.

In many tactical endoatmospheric missiles, PN guidance determines the lift that should be created by moving a missile's control surfaces. Exoatmospheric missiles create the acceleration required by the PN law by using thrust vector control, lateral divert engines, or squibs. In tactical radar homing missiles, the LOS rate is measured by the radar seeker. In tactical IR missiles equipped with the imaging IR (infrared) seekers, the LOS rate information is obtained by utilizing pattern image scanning techniques. The measurements of the target IR intensity enables the IR seekers to provide estimates of range and range rate based on intensity of image data. The future generation of the IR seekers will be able to provide these estimates with high accuracy. Now, however, only radar missiles provide reliable estimates of range and range rate (closing velocity).

In UAVs, the LOS rate can be determined based on the waypoint's coordinates and the measured UAV's position and velocity. The corresponding expressions will be considered later. Similar to missiles, the PN guidance determines the lift that should be created by moving a UAV's control surfaces in the case of fixed-wing UAVs or by changing the pitch angle of rotating blades of helicopters.

The PN guidance problem should be formulated as a three-dimensional control problem. However, by assuming that the lateral and longitudinal maneuver planes are decoupled by means of roll control, it is possible to reduce the three-dimensional guidance problem to the equivalent two-dimensional planar problems. That is why we first discuss the planar problem.

Denoting the vertical projection of the range-vector r by y, we can present the LOS angle λ as:

$$\sin \lambda = y/r \tag{2.8}$$

For small λ, equation (2.8) can be presented approximately as:

$$\lambda = y/r \tag{2.9}$$

where $y(t)$ characterizes the displacement between the missile and target at a moment t and is called the miss distance, or simply, *miss*. This expression is widely used in the so-called linearized engagement models.

Analogous to equation (1.11), the approximate value of the LOS rate equals

$$\dot{\lambda} = \frac{\dot{y}r + yv_{cl}}{r^2} \tag{2.10}$$

Assuming the closing velocity to be constant (the missile and target do not maneuver in the future), equations (2.9) and (2.10) can be written as:

$$\lambda(t) = \frac{y(t)}{v_{cl}t_{go}} \tag{2.11}$$

and

$$\dot{\lambda}(t) = \frac{\dot{y}(t)r + y(t)v_{cl}}{r^2} = \frac{\dot{y}(t)t_{go} + y(t)}{v_{cl}t_{go}^2} = \frac{ZEM}{v_{cl}t_{go}^2} \tag{2.12}$$

where $t_{go} = t_F - t$ is the time to go until the end of the flight assuming that it will correspond to intercept (t_F is the time at the end of the flight), and

$$ZEM = \dot{y}(t)t_{go} + y(t) \tag{2.13}$$

is called the *zero-effort miss* (i.e., the miss, the future relative separation between missile and target that would result if the missile does not accelerate and the target does not maneuver after the moment t).

Assuming that by using the acceleration $a_c(t)$ the intercept will take place, ZEM can be considered as the predicted intercept coordinate, and the PN guidance law (2.7) can be rewritten in the following form:

$$a_c(t) = N \frac{\dot{y}(t)t_{go} + y(t)}{t_{go}^2} = N \frac{ZEM}{t_{go}^2} \tag{2.14}$$

By interpreting ZEM as a predicted intercept point that can be calculated based on some knowledge (or assumptions) of the future motion of the target, the PN guidance (2.14) can be considered as a predictive guidance.

The analytical analysis (see [6] and Chapter 4) shows that for the case of ideal dynamics (i.e., no lags between the LOS rate and the commanded acceleration) the LOS rate is a decreasing function of time converging to zero at the pursuit end. When actual dynamics are considered, the PN guidance tends to diverge at the vicinity of interception (i.e., the LOS rate diverges). Despite assertion of some scientists that this divergence may severely affect the miss distance, we will ignore this fact since the simplified models are used only at the initial stage of design.

2.4 PROPORTIONAL NAVIGATION. THREE-DIMENSIONAL ENGAGEMENT

As indicated earlier, the three-dimensional motion can be presented as a combination of the orthogonal planar motions (see Figure 1.1).

Using the azimuth α and elevation β angles instead of λ in equation (2.7) we obtain the following accelerations $a_{ch}(t)$ and $a_{cv}(t)$ in the horizontal and vertical plains, respectively:

$$a_{ch}(t) = Nv_{cl}\dot{\alpha}(t), \quad a_{cv}(t) = Nv_{cl}\dot{\beta}(t) \tag{2.15}$$

The total commanded acceleration $\boldsymbol{a}_c(t) = (a_{c1}(t), a_{c2}(t), a_{c3}(t))$ equals [see equations (1.13) and (2.15)]:

$$a_{c1}(t) = -a_{ch}(t)\sin\alpha - a_{cv}(t)\cos\alpha\sin\beta \tag{2.16}$$

$$a_{c2}(t) = a_{ch}(t)\cos\alpha - a_{cv}(t)\sin\alpha\sin\beta \tag{2.17}$$

and

$$a_{c3}(t) = a_{cv}(t)\cos\beta \tag{2.18}$$

The components of equations (2.16)–(2.18) are written taking into account that $a_{ch}(t)$ and $a_{cv}(t)$ are perpendicular to the LOS projections in the horizontal and vertical planes, respectively (see also Figure 1.1 and Figure 2.1).

The PN acceleration commands that follow directly from equation (1.10) has the following form:

$$\boldsymbol{a}_c(t) = Nv_{cl}\dot{\boldsymbol{\lambda}}(t) \tag{2.19}$$

Taking into account equations (1.16) and (2.15), they can be rewritten as:

$$a_{c1}(t) = -a_{ch}(t)\sin\alpha\cos\beta - a_{cv}(t)\cos\alpha\sin\beta \tag{2.20}$$

$$a_{c2}(t) = a_{ch}(t)\cos\alpha\cos\beta - a_{cv}(t)\sin\alpha\sin\beta \tag{2.21}$$

and

$$a_{c3}(t) = a_{cv}(t)\cos\beta \tag{2.22}$$

A slight difference between equations (2.16)–(2.18) and (2.20)–(2.22) is stipulated by the restriction on the PN guidance components (2.16)–(2.18). In the case of PN guidance (2.15), we operate with two guidance components $a_{ch}(t)$ and $a_{cv}(t)$ acting in two orthogonal planes. The $a_{cs}(t)$ ($s = 1, 2, 3$)

components follow from $a_{ch}(t)$ and $a_{cv}(t)$. The PN guidance (2.19) and the expressions (2.20)–(2.22) are free from these restrictions. For small elevation angles, the above expressions give very close results.

The horizontal and vertical acceleration commands are linked with the NED coordinate system and realized in practice by the roll and pitch autopilots, respectively. The α and β angles are determined by onboard sensors and the acceleration commands are generated by the missile. This process corresponds to the terminal guidance phase. During the midcourse phase the missile relies on off-board sensors. The guidance components are determined by ground (space)-based defense systems in the Earth-fixed coordinate systems.

The PN guidance law in the Earth-fixed coordinate systems is:

$$a_{cs}(t) = Nv_{cl}\dot{\lambda}_s(t) \qquad (s = 1, 2, 3) \tag{2.23}$$

where $\dot{\lambda}_s(t)$ and v_{cl} are determined by equations (1.11) and (1.12), respectively.

Analogous to equation (2.14) we can write:

$$a_{cs}(t) = N\frac{\dot{R}_s(t)t_{go} + R_s(t)}{t_{go}^2} = N\frac{ZEM_s}{t_{go}^2} \qquad (s = 1, 2, 3) \tag{2.24}$$

where the *zero-effort miss* vector $ZEM = (ZEM_1, ZEM_2, ZEM_3)$

$$ZEM_s = \dot{R}_s(t)t_{go} + R_s(t) \qquad (s = 1, 2, 3) \tag{2.25}$$

is perpendicular to the line-of-sight. This property can be established directly from equations (1.8), (1.11), (2.12), (2.14), (2.25), and the equality

$$\sum_{s=1}^{3}\dot{\lambda}_s\lambda_s = 0$$

It means that the proportional navigation guidance law generates the lateral acceleration commands that produce the lateral motion.

2.5 AUGMENTED PROPORTIONAL NAVIGATION

The basic guidance parameter of the PN law is the LOS rate. Knowledge of range and time-to-go are not required, so that proportional navigation can be implemented using only angle sensors on board the missile, which is its great advantage. Although the PN guidance law was not derived rigorously with target acceleration as a major consideration, it is applied against maneuvering targets. It is intuitively clear that additional information concerning the target acceleration can be a source of improving the efficiency of the PN guidance.

The zero-effort miss in equation (2.13) was introduced under assumption that the missile would not accelerate during t_{go}. In the case of a constant target maneuver with the acceleration a_T, the zero-effort miss must be augmented by adding a quadratic term $0.5\,a_T t_{go}^2$, i.e.,

$$ZEM = \dot{y}(t)t_{go} + y(t) + 0.5a_T t_{go}^2 \tag{2.26}$$

By substituting equation (2.26) in equation (2.14) and returning back to the basic variables $\lambda(t)$ and v_{cl} of the PN law [see equations (2.10)–(2.12)], for planar engagements the augmented proportional navigation (APN) law $a_{aug}(t)$ can be presented as:

$$a_{aug}(t) = a_c(t) + 0.5Na_T = Nv_{cl}\dot{\lambda}(t) + 0.5Na_T \tag{2.27}$$

For the three-dimensional case, the coordinates of the missile commanded acceleration $\boldsymbol{a}_{aug}(t) = (a_{aug1}, a_{aug2}, a_{aug3})$ and the target acceleration $\boldsymbol{a}_T(t) = (a_{T1}, a_{T2}, a_{T3})$ are related by the following equation:

$$\boldsymbol{a}_{augs}(t) = a_{cs}(t) + 0.5Na_{Ts} \qquad (s = 1, 2, 3) \tag{2.28}$$

Although the augmented proportional navigation law was derived assuming step-target maneuvers, it was recommended—without rigorous justification—and is used in practice for all types of maneuvering targets.

2.6   PROPORTIONAL NAVIGATION AS A CONTROL PROBLEM

The basic philosophy behind PN guidance that implements parallel navigation is that missile acceleration should nullify the line-of-sight (LOS) rate. However, the realization of this philosophy was done based on physical intuition: when the LOS rate differs from zero, an acceleration command proportional to the deviation from zero is created to eliminate this deviation.

Below we will consider the PN as a control problem that realizes the parallel navigation rule $\dot{\lambda}(t) = 0$. First, the linearized planar model of engagement is considered [see Figure 2.1 and equations (2.10) and (2.11)]. By differentiating equation (2.11) rewritten in the form:

$$\dot{\lambda}(t) = \frac{\dot{y}(t)r(t) - y(t)\dot{r}(t)}{r^2(t)} = \frac{\dot{y}(t)}{r(t)} - \frac{\lambda(t)\dot{r}(t)}{r(t)} \tag{2.29}$$

$$\ddot{\lambda}(t) = \frac{\ddot{y}(t)r(t) - \dot{y}(t)\dot{r}(t)}{r^2(t)} - \frac{(\dot{\lambda}(t)\dot{r}(t) + \lambda(t)\ddot{r}(t))r(t) - \lambda(t)\dot{r}^2(t)}{r^2(t)}$$

$$= \frac{\ddot{y}(t) - \dot{\lambda}(t)\dot{r}(t) - \lambda(t)\ddot{r}(t)}{r(t)} - \frac{\dot{r}(t)}{r(t)}\frac{(\dot{y}(t) - \lambda(t)\dot{r}(t))}{r(t)}$$

$$= \frac{\ddot{y}(t) - \dot{\lambda}(t)\dot{r}(t) - \lambda(t)\ddot{r}(t) - \dot{\lambda}(t)\dot{r}(t)}{r(t)} \qquad (2.30)$$

$$= \frac{\ddot{y}(t) - 2\dot{\lambda}(t)\dot{r}(t) - \lambda(t)\ddot{r}(t)}{r(t)}$$

and introducing the time varying coefficients:

$$a_1(t) = \frac{\ddot{r}(t)}{r(t)} \qquad (2.31)$$

$$a_2(t) = \frac{2\dot{r}(t)}{r(t)} \qquad (2.32)$$

and

$$b(t) = \frac{1}{r(t)} \qquad (2.33)$$

the expression (2.30) can be presented in the form:

$$\ddot{\lambda}(t) = -a_1(t)\lambda(t) - a_2(t)\dot{\lambda}(t) + b(t)\ddot{y}(t) \qquad (2.34)$$

Taking into account that

$$\ddot{y}(t) = -a_M(t) + a_T(t) \qquad (2.35)$$

equation (2.34) can be transformed in

$$\ddot{\lambda}(t) = -a_1(t)\lambda(t) - a_2(t)\dot{\lambda}(t) - b(t)a_M(t) + b(t)a_T(t) \qquad (2.36)$$

Let $x_1 = \lambda(t)$ and $x_2 = \dot{\lambda}(t)$. The missile-target engagement is described by the following system of the first-order differential equations:

$$\dot{x}_1 = x_2$$

$$\dot{x}_2 = -a_1(t)x_1 - a_2(t)x_2 - b(t)u + b(t)f \qquad (2.37)$$

where the control $u = a_M(t)$ and disturbance $f = a_T(t)$.

First, let us consider the case of a nonaccelerating target (i.e., $f = 0$), the assumption that accompanies the main relations used in PN guidance. The asymptotic stability with respect to x_2 (i.e., $\lim x_2 \to 0$), corresponds to the parallel navigation rule, so that the control law that satisfies this condition is the guidance law that implements parallel navigation.

The guidance problem can be formulated as the problem of choosing control u to guarantee the asymptotic stability of the system (2.37) with respect to x_2. (Because in reality we deal with a finite problem, for simplicity and a more rigorous utilization of the term "asymptotic stability" we assume disturbance to be a vanishing function, i.e., contains a factor $e^{-\varepsilon t}$, ε is an infinitely small positive number.)

It is important to mention that the guidance law is determined based on the partial stability of the system dynamics under consideration, only with respect to the LOS derivative [2]. The approach to examine the asymptotic stability is based on the Lyapunov method (see Appendix A). For equation (2.37), it is natural to choose the Lyapunov function Q as a square of the LOS derivative, i.e.,

$$Q = \frac{1}{2} c x_2^2 \tag{2.38}$$

where c is a positive coefficient.

Its derivative along any trajectory of equation (2.37) equals

$$\dot{Q} = c x_2 (-a_1(t) x_1 - a_2(t) x_2 - b(t) u) \tag{2.39}$$

The negative definiteness of the derivative (2.39), i.e., the asymptotic stability of the system (2.35) with respect to x_2, can be presented in the form:

$$\dot{Q} = c x_2 (-a_1(t) x_1 - a_2(t) x_2 - b(t) u) \le -c_1 x_2^2 \tag{2.40}$$

where c_1 is a positive coefficient.

Under the near collision course assumption (collision course assumes nonaccelerating motion), $\ddot{r}(t) = 0$, i.e., $a_1(t) = 0$, and the inequality (2.40) can be written as:

$$(-a_2(t) + c_1/c) x_2^2 - a_1(t) x_1 x_2 - b(t) x_2 u \le 0 \tag{2.41}$$

It follows from the inequalities (2.40) and (2.41) that for $a_1(t) = 0$ and $c_1 \ll c$ the control

$$u = k x_2 = k \dot{\lambda}(t) \tag{2.42}$$

stabilizes the system (2.37) if k satisfies

$$kb(t) + a_2(t) > 0$$

or

$$k > -\frac{a_2(t)}{b(t)} \tag{2.43}$$

Introducing the closing velocity $v_{cl} = -\dot{r}(t)$ and the effective navigation ratio N, the expression (2.43) can be written as $k > 2v_{cl}$ and the control law can be presented as:

$$u = Nv_{cl}\dot{\lambda}(t), \quad N > 2 \tag{2.44}$$

which is the well-known property established for the PN guidance law (2.7).

For the three-dimensional case and the Earth-based coordinate system, the line-of-sight and its derivative are presented by equations (1.10) and (1.11), so that analogous to equation (2.36):

$$\ddot{\lambda}_s(t) = -a_1(t)\lambda_s(t) - a_2(t)\dot{\lambda}_s(t) + b(t)(a_{Ts}(t) - u_s) \quad (s = 1, 2, 3) \tag{2.45}$$

where $\ddot{\lambda}_s$ ($s = 1, 2, 3$) are the LOS second derivative coordinates, $a_{Ts}(t)$ ($s = 1, 2, 3$) are the coordinates of the target acceleration vector, and $u_s(t)$ are the coordinates of the missile acceleration vector that are considered as controls.

The Lyapunov function is chosen as the sum of squares of the LOS derivative components that corresponds to the nature of proportional navigation:

$$Q = \frac{1}{2}\sum_{s=1}^{3} d_s\dot{\lambda}_s^2 \tag{2.46}$$

where d_s are positive coefficients.

Its derivative can be presented in the following form:

$$2\dot{Q} = \sum_{s=1}^{3} d_s\ddot{\lambda}_s\dot{\lambda}_s \tag{2.47}$$

or

$$2\dot{Q} = \sum_{s=1}^{3} d_s(-a_1(t)\lambda_s\dot{\lambda}_s - a_2(t)\dot{\lambda}_s^2 + b_1(t)\dot{\lambda}_s(a_{Ts}(t) - u_s)) \tag{2.48}$$

Analogous to the planar engagement, under the near collision course assumption, the controls $u_s(t)$ that guarantee $\lim \|\dot{\lambda}\| \to 0$, $t \to \infty$, can be presented as:

$$u_s = Nv_{cl}\dot{\lambda}_s, \quad N > 2 \quad (s = 1, 2, 3) \tag{2.49}$$

which coincides with equation (2.23).

2.7 AUGMENTED PROPORTIONAL NAVIGATION AS A CONTROL PROBLEM

For maneuvering targets and the planar engagement, the derivative of the Lyapunov function (2.38) along any trajectory of equation (2.37) equals

$$\dot{Q} = cx_2(-a_1(t)x_1 - a_2(t)x_2 - b(t)u + b(t)f) \tag{2.50}$$

The negative definiteness of the form (2.39) (i.e., the asymptotic stability of the system (2.37), with respect to x_2) can be presented in the form:

$$\dot{Q} = cx_2(-a_1(t)x_1 - a_2(t)x_2 - b(t)u + b(t)f) \leq -c_1x_2^2 \tag{2.51}$$

From the condition of negative definiteness of the derivative of the Lyapunov function (2.51), we can derive the guidance law:

$$u = Nv_{cl}\dot{\lambda} + a_T(t), \quad N > 2 \tag{2.52}$$

where the acceleration term is 0.5 N times less than in the augmented PN law obtained for step maneuvers [see equation (2.27)].

Analogously, from equation (2.48) we can obtain the guidance law for the three-dimensional case:

$$u_s = Nv_{cl}\dot{\lambda}_s + a_{Ts}(t), \quad N > 2 \quad (s = 1, 2, 3) \tag{2.53}$$

where the acceleration term is 0.5 N times less than in the augmented PN law obtained for step maneuvers [see equation (2.28)].

The comparison of equations (2.52) and (2.53) with equations (2.27) and (2.28) shows that the APN gain $N/2$ in the target acceleration terms is larger than obtained above based on the Lyapunov approach.

The law (2.53) is given as a possible law to compare with the existing augmented law. Later more general expressions will be given.

2.8   WHEN IS THE PN LAW OPTIMAL?

The proportional navigation law (2.7) is a result of a simple logical inference. If the LOS rate differs from zero (i.e., a nonzero error exists), an action proportional to this error should be taken to eliminate it. The more rigorous formulation of the problem of nullifying the LOS rate was given earlier in this chapter, where proportional navigation was presented as a control problem. The commanded acceleration was considered as a control and the line-of-sight and its derivative were chosen as the state variables.

In a different way, the proportional navigation guidance law was considered as a control action in [1]. The 1960s were marked with the significant results in the optimal control theory. It was shown that linear controls are optimal in the case of systems characterized by linear differential equations and a quadratic functional as their performance index [3–5]. For the equation of motion (1.20) and (2.35) (for simplicity we consider here the planar case), such performance index should be found, for which the PN guidance law (2.7) is the optimal control. The problems of this kind are called the inverse optimal problems.

Using the near collision course assumptions, i.e., assuming that the missile approaches the target at a constant closing velocity v_{cl} near a collision course, and ignoring missile dynamics, we can write

$$\ddot{y} = -a_M, \quad y = r\lambda << r, \quad r(\tau) = v_{cl}\tau \qquad (2.54)$$

The performance index, or cost functional, is defined as:

$$I = \frac{1}{2}\left(Cy^2(t_F) + \int_0^{t_F} a_M^2(t)dt \right) \qquad (2.55)$$

where C is a constant coefficient, often called the weighting factor, and the initial time of flight is zero. The first term of (2.55) presents the miss distance and the second one characterizes the energy spent during the flight. A high C emphasizes the importance of achieving a small miss distance, where as a small C implies the importance of having sufficient energy at the end of the flight.

The optimal problem consists of finding $a_M(t)$, which minimizes the functional (2.55). The solution of the formulated optimal problem was obtained in [1] (see also Appendix A) as:

$$a_M(t) = \frac{3\tau}{3/C + \tau^3}(y(t) + \dot{y}(t)\tau) \qquad (2.56)$$

Zero miss corresponds to $C \to \infty$, so that the optimal guidance law becomes:

$$a_M(t) = \frac{3}{\tau^2}(y(t) + \dot{y}(t)\tau) \qquad (2.57)$$

Taking into account equation (2.12) that can be rewritten as:

$$\dot{\lambda}(t) = \frac{y(t) + \dot{y}(t)\tau}{v_{cl}\tau^2}$$

instead of equation (2.57) we have:

$$a_M(t) = 3v_{cl}\dot{\lambda}(t) \qquad (2.58)$$

It means that under the above given assumptions the proportional navigation law minimizes the functional (2.55), and the optimal value of the navigation ratio $N = 3$ guarantees zero miss distance. By ignoring missile dynamics and considering a nonmaneuvering target, we excessively simplified the guidance problem so that the above result has a "pure" theoretical rather than practical importance. Difficulties connected with the solution of more realistic guidance problems were mentioned before.

REFERENCES

1. Bryson, A. E. Linear Feedback Solution for Minimal Effort Intercept Rendezvous, and Soft Landing, *AIAA Journal* 3, no. 8 (1965): 1542–48.
2. Rumyantsev, V.V. On Asymptotic Stability and Instability of Motion with Respect to a Part of the Variables, *Journal of Applied Mathematics and Mechanics* 35, no. 1 (1971): 19–30.
3. Yanushevsky, R. *Theory of Optimal Linear Multivariable Control Systems.* Moscow, Russia: Nauka, 1973.
4. Yanushevsky, R., and Boord, W. New Approach to Guidance Law Design, *Journal of Guidance, Control, and Dynamics* 28, no. 1 (2005): 162–66.
5. Zadeh, L., and Desoer, C. *Linear System Theory.* New York, NY: McGraw Hill, 1963.
6. Zarchan, P. *Tactical and Strategic Missile Guidance, Progress in Astronautics and Aeronautics.* Vol. 176. Washington, DC: American Institute of Astronautics and Aeronautics, Inc., 1997.

3 Control of Longitudinal and Lateral Motions

3.1 INTRODUCTION

Proportional navigation (PN) has attracted a considerable amount of interest in the literature related to missile guidance and continues to be a benchmark for new missile guidance laws. The detailed analytical study of this empirical guidance law for nonmaneuvering and maneuvering targets was undertaken in [7,9,16]. Capture regions and conditions for the existence of capture regions were also examined in [3,4].

As mentioned earlier, the basic philosophy behind PN guidance is that missile acceleration should nullify the line-of-sight (LOS) rate. Analysis of PN guidance for the homing stage was usually undertaken for nonmaneuvering targets assuming a constant closing velocity. The so-called augmented PN law and other modifications of the proportional navigation law were obtained based mostly on the relationships established for nonmaneuvering targets. It was discussed in Chapter 2 (see also [9,17]).

Results from the theory of linear multivariable control systems applied to homing guidance (where linear approximation can be justified) enable one to evaluate the performance of the guidance system as well as to generate modified proportional navigation laws [9,17]. Guidance laws utilizing the idea of proportional navigation and based on the results of control theory related to sliding modes and systems with variable structure (see, e.g., [7]) cannot be considered as practical for missile guidance applications. The practical realization of systems with sliding mode is limited because of chatter, and related simplified control laws need rigorous justification and testing. A systematic framework for an almost sliding mode control that eliminates chatter was given in [13]. However, this approach was not used in [7] and other applications of sliding mode controls in the guidance systems. Also, in the presence of a maneuvering target the sliding mode area depends on the target acceleration, and for small LOS derivatives the sliding mode can disappear. A variable structure (different from the ones considered, e.g., in [7]) that requires measurement of target acceleration is needed.

The empirical PN law was obtained also as a solution of an optimization problem (see, e.g., [1,6,15,16]), which justifies this law as an optimal one corresponding to a certain quadratic performance index. The game approach to guidance laws based on the theory of differential games with a quadratic performance index was considered in [2]. The guidance laws developed counteract target maneuvers better than the ordinary PN law.

However, any optimal guidance law assumes that the trajectory of a maneuvering target, as well as time-to-go and/or the intercept point, is known. In practice, such information is unknown and can only be evaluated approximately. The accuracy of prediction significantly influences the accuracy of the intercept.

Taking into account that the PN law is a widely accepted guidance law and has been tested in practice, it is of interest to consider the possibility of its improvement.

The Lyapunov approach, offered in Chapter 2, can also be considered as another justification of the PN law. Moreover, this approach enables us to offer other laws that will improve the effectiveness of the proportional navigation law for maneuvering and nonmaneuvering targets. A new class of the PN guidance laws is obtained as the solution of a stability problem using the Lyapunov method. Analogous to the section of Chapter 2 where the PN guidance was formulated as a control problem, here the Lyapunov function is chosen as a square of the LOS derivative for the planar model and as the sum of the LOS derivative components for the three-dimensional case. The applicability of the laws is determined by the negative definiteness of the derivative of the Lyapunov function. The module of the Lyapunov function derivative is used as the performance index for comparing the PN guidance laws and creating the new ones.

It is important to mention that the guidance laws are determined based on the partial stability of the system dynamics under consideration, only with respect to the LOS derivative [8,10,11].

3.2 GUIDANCE CORRECTION CONTROLS

As discussed in Chapter 2, proportional navigation is the guidance law that implements parallel navigation, which is defined by the rule $\dot{\lambda}(t) = 0$ with an additional requirement $\dot{r}(t) < 0$, where $\lambda(t)$ is the LOS angle with respect to a reference axis and $r(t)$ represents the target-to-missile range.

Traditionally, the PN guidance relates to missiles. Although we will show how to apply missile guidance laws to a wide class of unmanned aerial vehicles, here we use the missile-target terminology.

To describe the missile-target engagement dynamics, first we consider planar engagements and use a Cartesian frame of coordinates (FOC; see Figure 2.1) with the origin O of an inertial reference coordinate system: $y(t)$ is the relative separation between the missile and target perpendicular to the horizontal reference axis; a_M and a_T are the missile and target accelera-tion, respectively. Using a small-angle approximation, the expressions for the second derivative of the LOS angle can be presented in the following form [see equations (2.9), (2.29)–(2.37)]:

$$\ddot{\lambda}(t) = -a_1(t)\lambda(t) - a_2(t)\dot{\lambda}(t) - b(t)a_M(t) + b(t)a_T(t) \qquad (3.1)$$

Let $x_1 = \lambda(t)$ and $x_2 = \dot{\lambda}(t)$. The missile-target engagement is described by the following system of the first-order differential equations:

$$\dot{x}_1 = x_2$$

$$\dot{x}_2 = -a_1(t)x_1 - a_2(t)x_2 - b(t)u + b(t)f \qquad (3.2)$$

where the control $u = a_M(t)$ and disturbance $f = a_T(t)$,

$$a_1(t) = \frac{\ddot{r}(t)}{r(t)} \qquad (3.3)$$

$$a_2(t) = \frac{2\dot{r}(t)}{r(t)} \qquad (3.4)$$

and

$$b(t) = \frac{1}{r(t)} \qquad (3.5)$$

[see equations (2.31)–(2.33)].

Analogous to the approach in Chapter 2, the guidance problem can be formulated as the problem of choosing control u to guarantee the asymp-totic stability of the system (3.2) with respect to x_2.

For the Lyapunov function:

$$Q = \frac{1}{2}cx_2^2 \qquad (3.6)$$

where c is a positive coefficient, its derivative along any trajectory of equa-tion (3.2) equals:

$$\dot{Q} = cx_2(-a_1(t)x_1 - a_2(t)x_2 - b(t)u + b(t)f) \qquad (3.7)$$

The PN guidance law (2.44) is called admissible if it guarantees intercept for a finite time t_F.

We consider the PN class of guidance laws that have the form (2.44) or contain (2.44) as a component. Despite the fact that even the PN laws of the form (2.44) with various N were compared by experiments, we will introduce a criterion of comparison that has a certain physical justification. Because proportional navigation is the guidance law that implements parallel navigation ($\dot{\lambda}(t) = 0$), we will compare the laws belonging to the PN class by their closeness to parallel navigation.

Of course, the most reliable performance index should evaluate the guidance law during the entire engagement time. However, this time is unknown and, in its turn, depends on the guidance law implemented. To avoid this "catch-22" situation, we assume that the guidance law with $\dot{\lambda}(t)$ tending to zero faster (closer to parallel navigation) at each t is preferable.

The module of the Lyapunov function derivative $|\dot{Q}(t)|$ [see equation (3.7)] will be the performance index for comparing the PN laws and creating the new ones. Proceeding in this way, we change the finite interval engagement problem to a specific infinite interval partial stability problem. The Lyapunov approach will be used to compare and design controls—guidance laws.

3.3 LYAPUNOV APPROACH TO CONTROL LAW DESIGN

The Lyapunov approach to control law design can be explained in the following way (more rigorous formulations and theorems can be found, e.g., in [8,14]): If there exist positive definite functions $Q(x,t)$ and $R(x,t)$ so that the derivative $\dot{Q}$ with respect to t along any trajectory of the system of equations that describes the control system under consideration (x and u are its state vector and control, respectively) satisfy the inequality:

$$\dot{Q} = \dot{Q}(x,u,t) \le -R(x,t) \tag{3.8}$$

then the system is stabilized by control u that can be determined from this inequality.

To apply this sufficient condition in practice, the above-indicated positive definite forms must be found. Unfortunately, there are no universal recommendations how to find these forms. The relation between $Q(x,t)$ and $R(x,t)$ was established for the so-called linear quadratic optimal control problems (Riccati type equations) [6,13]. Based on this, the design procedure was expanded on a certain class of nonlinear system [13].

However, for special types of equations, it is not difficult to find $Q(x,t)$ and $R(x,t)$ satisfying the inequality (3.8). Below the control law design procedure based on the Lyapunov approach is demonstrated for the guidance problem [see equations (3.2) and (3.7); for simplicity, we consider in equation (3.2) $f = 0$].

By choosing $Q(x,t)$ in the form (3.6) and $R(x,t) = c_1 x_2^2$, where c_1 is a positive coefficient, the inequality (3.8) can be written as [see also equation (3.7)]:

$$\dot{Q} = cx_2(-a_1(t)x_1 - a_2(t)x_2 - b(t)u) \le -c_1 x_2^2 \tag{3.9}$$

or

$$(-a_2(t) + c_1/c)x_2^2 - a_1(t)x_1 x_2 - b(t)x_2 u \le 0 \tag{3.10}$$

It follows from (3.10) that for $a_1(t) = 0$ and $c_1 << c$, the control $u = kx_2$ [see equations (2.42) and (2.44)] stabilizes the system (3.2) if k satisfies equation (2.43).

For $R(x,t) = c_1 x_2^2 + c_2 x_2^4$, where c_2 is a positive coefficient, instead of equation (3.9) we have:

$$\dot{Q} = cx_2(-a_1(t)x_1 - a_2(t)x_2 - b(t)u) \le -c_1 x_2^2 - c_2 x_2^4 \tag{3.11}$$

or

$$(-a_2(t) + c_1/c)x_2^2 - a_1(t)x_1 x_2 + \frac{c_2}{c}x_2^4 - b(t)x_2 u \le 0 \tag{3.12}$$

It is easy to conclude that for $a_1(t) = 0$, $c_1 << c$ and the control $u = kx_2 + N_1 x_2^3$, where k satisfies equation (2.43) and $N_1 > 0$, the left part of inequality (3.12) is negative definite, so that this control stabilizes the system (3.2) with respect to x_2.

By including the additional term in $R(x,t)$ we imposed "harder" requirements on the rate of decreasing Q. Despite $|\dot{Q}|$ being used as a system estimate in some applications of the Lyaponov method (see, e.g., [13]), it cannot be applied as a reliable criterion of quality of control systems. It serves only as an instantaneous criterion. The quality estimate of control system includes (directly or indirectly) time of control. For example, an oscillatory long transient even with a small amplitude in many cases is unacceptable. However, when choosing the guidance laws implementing parallel navigation, the only requirement is to be closer, as soon as possible, to zero LOS rate. The $|\dot{Q}|$ criterion reflects this requirement.

Let us assume that there exists a capture range domain over which the control (guidance law) $u(t)$ guarantees engagement ($x_2(t) \to 0$). Then based on the above mentioned, it is easy to conclude that the guidance law:

$$u = N v_{cl} \dot{\lambda}(t) + N_1 \dot{\lambda}^3(t), \quad N > 2, \quad N_1 > 0 \tag{3.13}$$

is better than the PN law (2.44).

The PN law reacts almost identically on various changes of LOS rate (assuming that the closing velocity does not vary drastically), i.e., small and fast changes of LOS result in proportional changes of acceleration. According to equations (3.11) and (3.13), by increasing N in the PN law, we can decrease the LOS rate faster. But this will increase the level of noise when the LOS rate becomes small and, hence, the accuracy of guidance is decreased. Moreover, big gains can make the whole guidance system unrobust. From a purely physical consideration, we can assume that the system with a variable gain that is bigger when the LOS rate is big and smaller when the LOS rate is small will act better than the traditional PN system. The second component of equation (3.13) (the "cubic" term) with a properly chosen N_1 serves this purpose.

It was shown that the PN law (2.44) can be improved by using a complex exponential type function of time $N(t)$ instead of a constant N [2,16]. This function is obtained as the result of the solution of an optimal guidance problem and depends on the predicted time-to-go. Its calculation presents certain difficulties for utilization of such laws in practice [5].

The guidance law (3.13) can be written in the form (2.44):

$$u(t) = \left(N + \frac{N_1}{v_{cl}} \dot{\lambda}^2(t) \right) v_{cl} \dot{\lambda}(t) = N(t) v_{cl} \dot{\lambda}(t) \tag{3.14}$$

with a time varying coefficient $N(t)$ that formally is an exponential type function (asymptotic with respect to x_2 solution of equation (3.2), under the assumption mentioned above, is an exponential type function). The form (3.14) looks similar to the guidance laws considered in [2]. In contrast to the law with variable $N(t)$, obtained as an optimal guidance in [2], the guidance law (3.13) does not require special complex computations.

If in equation (3.7) $f \neq 0$, instead of equation (3.11) we have:

$$\dot{Q} = cx_2(-a_1(t)x_1 - a_2(t)x_2 - b(t)u + b(t)f) \leq -c_1 x_2^2 - c_2 x_2^4 \tag{3.15}$$

An additional component $a_T(t) = f$ in control "compensates" in equation (3.15) the $b(t) f$ term, so that the control $u = kx_2 + N_1 x_2^3 + a_T(t)$ stabilizes the system (3.2) with respect to x_2 [see also equation (2.52)].

The Lyapunov approach is demonstrated here in details for the system (3.2), when $a_1(t) = 0$. In an analogous way, it is easy to establish the negative definiteness of the function (3.7) for $a_1(t) \neq 0$ and $a_2(t) \leq 0$ if the control u is:

$$u(t) = Nv_{cl}\dot{\lambda}(t) + N_1\dot{\lambda}^3(t) - N_2\ddot{r}(t)\lambda(t) + N_3 a_T(t) \quad N > 0, \quad N_1 > 0$$
(3.16)

$$N_2 \begin{matrix} \geq 1 \\ \leq 1 \end{matrix} \quad \text{if} \quad sign(\ddot{r}(t)\dot{\lambda}(t)\lambda(t)) \begin{matrix} \leq 0 \\ \geq 0 \end{matrix}$$

$$N_3 \begin{matrix} \leq 1 \\ \geq 1 \end{matrix} \quad \text{if} \quad sign(a_T(t)\dot{\lambda}(t)) \begin{matrix} \leq 0 \\ \geq 0 \end{matrix}$$

The term $N_3 a_T(t)$ differs from the corresponding term in the augmented proportional navigation (APN) law because the parameter N_3 is time-varying. The term $N_2\ddot{r}(t)\lambda(t)$, the shaping term, acts along the line-of-sight, changes the shape of the missile trajectory, and also influences the terminal velocity of a missile.

3.4 BELLMAN–LYAPUNOV APPROACH. OPTIMAL GUIDANCE PARAMETERS

As shown in Chapter 2, under the assumption (2.54) the proportional navigation law minimizes the functional (2.55), and the optimal value of the effective navigation ratio $N = 3$ guarantees zero miss distance assuming unbounded control resources. This assumption significantly undermines the reason of choosing $N = 3$ for practical applications.

Below the choice of the effective navigation ratio $N = 3$ will be justified based on the consideration of a different optimal problem. The offered approach enables us to choose more argumentative parameters for a wide class of guidance laws implementing parallel navigation.

3.4.1 OPTIMAL GUIDANCE FOR NONMANEUVERING TARGETS

For the planar model of engagement, the missile-target engagement is described by the system of the differential equations (2.37). Assuming a constant closing velocity, i.e., $a_1(t) = 0$, and $a_T(t) = 0$, we consider the problem of determining the guidance law that minimizes the functional:

$$I = \int_0^{t_F} (cx_2^2 + u^2(t))dt$$
(3.17)

where c is a constant coefficient, subject to equation (2.37), which can be rewritten as:

$$\dot{x}_2 = -a_2(t)x_2 - b(t)u$$
(3.18)

The Bellman functional equation for this optimal problem is (see Appendix A.2):

$$\min_u \left\{ cx_2^2 + u^2 + \frac{\partial \varphi}{\partial x_2} \left(\frac{2v_{cl}}{r(t)} x_2 - \frac{1}{r(t)} u \right) + \frac{\partial \varphi}{\partial t} \right\} = 0 \qquad (3.19)$$

or

$$u = \frac{1}{2r(t)} \frac{\partial \varphi}{\partial x_2} \qquad (3.20)$$

where $\varphi(x_2, t)$ is the minimal value of the functional (3.17).

Seeking the minimum of the functional (3.17) in the form:

$$\varphi(x_2,t) = w(x_2,t)x_2^2 = Nv_{cl}r(t)x_2^2 \qquad (3.21)$$

where

$$r(t) = r(0) - v_{cl}t \qquad (3.22)$$

we have

$$\frac{\partial \varphi}{\partial x_2} = 2w(x_2,t)x_2 = 2Nv_{cl}r(t)x_2 \qquad (3.23)$$

$$\frac{\partial \varphi}{\partial t} = \frac{\partial w}{\partial t}x_2^2 = -Nv_{cl}^2x_2^2 \qquad (3.24)$$

and

$$u = \frac{w(x_2,t)}{r(t)}x_2 = Nv_{cl}x_2 \qquad (3.25)$$

and the Bellman equation (3.19) can be presented in the well-known form:

$$\frac{\partial w}{\partial t} + w\frac{4v_{cl}}{r(t)} - w^2\frac{1}{r^2(t)} + c = 0, \quad w(x_2,t_F) = 0 \qquad (3.26)$$

which can be reduced, based on equations (3.21) and (3.24), to the algebraic Riccati equation:

$$4Nv_{cl}^2 - N^2v_{cl}^2 - Nv_{cl}^2 + c = 0 \qquad (3.27)$$

The optimal navigation ratio:

$$N = 3/2 + \sqrt{9/4 + c/v_{cl}^2} > 3 \qquad (3.28)$$

$(N = 3/2 - \sqrt{9/4 + c/v_{cl}^2} < 0$ corresponds to the unstable system (3.26)).

Formally, in the equation (3.26) $r(t)$ is a time-varying parameter, and the terminal condition $w(x_2, t_F) = 0$ (which follows from $\varphi(x_2, t_F) = 0$) can be interpreted as the existence of such $t_F \in (0, \infty)$ that $r(t_F) = 0$ (i.e., t_F corresponds to the time of intercept). If $t_F \to \infty$ then $x_2 \to 0$, i.e., the system (3.18) under control (3.25) is asymptotically stable and the guidance law (3.25) implements parallel navigation. The asymptotical stability requirement is important to exclude the trivial solution $u = 0$ for the case $c = 0$ corresponding to the minimum of energy spent while executing the guidance law.

It follows from equation (3.28) that the optimal solution for $c = 0$ corresponds to $N = 3$ (i.e., the PN law with the effective navigation ratio $N = 3$ is the most energy efficient); it implements parallel navigation $\left(\lim_{t_F \to \infty} x_2 \to 0\right)$ with minimal energy resources. For a finite t_F, the minimal energy guidance with $N = 3$ is possible under the assumption of the existence of intercept, i.e., $r(t_F) = 0$.

The values of $N > 3$ decrease the LOS rate faster than in the case $N = 3$ (i.e., the guidance is closer to parallel navigation). However, the increase of the effective navigation ratio is bounded because of existing control limits and the increase of noise in the guidance and control system.

According to [1], the PN guidance with $N = 3$ can guarantee zero miss distance for any t_F assuming boundless control resources; the value $N = 3$ is obtained as the most energy efficient for the modification of the PN guidance law using the predicted time-to-go (predicted intercept point). The above-given justification of values of the effective navigation ratio takes into account the restriction on controls actions, constrains on the lateral acceleration. Based on the results of this section, it is reasonable to recommend for the radar-seeker system the value of N about 3, whereas higher N values can be taken in optical-seeker systems.

Now we consider the guidance law in the form [see equation (3.16)]:

$$a_M(t) = u(t) = N v_{cl} \dot{\lambda}(t) + N_1 v_{cl} \dot{\lambda}^3(t) \qquad (3.29)$$

and link the choice of the coefficients N and N_1 with the solution of a certain optimal problem.

Considering the inverse optimal problem, we can determine the functional whose minimal value subject to equation (3.18) corresponds to the control law $u(t)$ described by equation (3.29). The functional has the form:

$$I = \int_0^{t_F} (c(x_2)x_2^2 + u^2(t))dt \tag{3.30}$$

where

$$c(x_2) = c_0 + c_1 x_2^2 + c_2 x_2^4 \tag{3.31}$$

and for $N = 3$ the coefficients in equation (3.31) equal:

$$c_0 = 0, \quad c_1 = 2.5 \, N_1 v_{cl}^2 \quad \text{and} \quad c_2 = N_1^2 v_{cl}^2 \tag{3.32}$$

To prove the above statement, we will seek the solution of the Bellman functional equation (3.19), where c is changed to $c(x_2)$. The solution of the Bellman equation is sought in the form:

$$\varphi(x_2, t) = w(x_2, t)x_2^2 + w_1(x_2, t)x_2^4 \tag{3.33}$$

Substituting $\varphi(x_2, t)$ in equation (3.19) with c changed to $c(x_2)$ and taking into account that:

$$\frac{\partial \varphi}{\partial x_2} = 2w(x_2, t)x_2 + 4w_1(x_2, t)x_2^3 \tag{3.34}$$

$$\frac{\partial \varphi}{\partial t} = \frac{\partial w}{\partial t}x_2^2 + \frac{\partial w_1}{\partial t}x_2^4 \tag{3.35}$$

and

$$u = \frac{w(x_2, t)}{r(t)}x_2 + \frac{2w_1(x_2, t)}{r(t)}x_2^3 \tag{3.36}$$

the Bellman equation (3.19) can be presented as:

$$\left(\frac{\partial w}{\partial t} + w\frac{4v_{cl}}{r(t)} - \frac{w^2}{r^2(t)} + c_0\right)x_2^2$$

$$+ \left(\frac{\partial w_1}{\partial t} + w_1\frac{8v_{cl}}{r(t)} - \frac{4ww_1}{r^2(t)} + c_1\right)x_2^4 - \left(\frac{4w_1^2}{r^2(t)} - c_2\right)x_2^6 = 0 \tag{3.37}$$

$$w(x_2, t_F) = 0, \quad w_1(x_2, t_F) = 0$$

Seeking the solution of equation (3.37) in the form:

$$w(x_2,t) = Nv_{cl}r(t), \quad w_1(x_2,t) = \frac{1}{2}N_1v_{cl}r(t) \tag{3.38}$$

and equating to zero the components of second, fourth, and sixth power terms in equation (3.37), we obtain:

$$4Nv_{cl}^2 - N^2v_{cl}^2 - Nv_{cl}^2 + c_0 = 0$$

$$4N_1v_{cl}^2 - 0.5N_1^2v_{cl}^2 - 2NN_1v_{cl}^2 + c_1 = 0 \tag{3.39}$$

$$-N_1^2v_{cl}^2 + c_2 = 0$$

For $N = 3$, it immediately follows from equation (3.39) that the coefficients c_0, c_1, and c_2 of the functional (3.30) should satisfy equation (3.32).

Comparing the functionals (3.17) and (3.30) we can conclude that the cubic term in equation (3.29) decreases the LOS derivative x_2 but this requires additional control resources. However, as we mentioned earlier, its efficiency decreases with the decrease of x_2.

3.4.2 Optimal Augmented Guidance Laws

Determine the guidance law that minimizes the functional (3.17) subject to:

$$\dot{x}_2 = -a_2(t)x_2 - b(t)u + b(t)f \tag{3.40}$$

Instead of equation (3.19) we have:

$$\min_u \left\{ cx_2^2 + u^2 + \frac{\partial \varphi}{\partial x_2} \left(\frac{2v_{cl}}{r(t)}x_2 - \frac{1}{r(t)}u + \frac{1}{r(t)}f \right) + \frac{\partial \varphi}{\partial t} \right\} = 0 \tag{3.41}$$

where $\varphi(x_2, t)$ is the minimal value of the functional (3.17).

Seeking the minimum of the functional (3.17) in the form:

$$\varphi(x_2,t) = w(x_2,t)x_2^2 + L(t)x_2 + L_0(t) = Nv_{cl}r(t)x_2^2 + L(t)x_2 + L_0(t) \tag{3.42}$$

we have the guidance law:

$$u = \frac{1}{2r(t)}\frac{\partial \varphi}{\partial x_2} = Nv_{cl}x_2 + \frac{L(t)}{2r(t)} \tag{3.43}$$

and by grouping the terms containing x_2^2 and x_2 the Bellman equation (3.41) can be presented as:

$$\frac{\partial w}{\partial t} + w\frac{4v_{cl}}{r(t)} - w^2\frac{1}{r^2(t)} + c = 0, \quad w(x_2,t_F) = 0 \tag{3.44}$$

and

$$\dot{L}(t) - \frac{N-2}{r(t)}v_{cl}L(t) + 2Nv_{cl}f(t) = 0, \quad L(t_F) = 0 \tag{3.45}$$

where $w(x_2,t)$ coincides with the solution of equation (3.26), the function $L(t)$ should satisfy (3.45), and $L_0(t)$ should be equal to the free terms of equation (3.41) (this expression is not given since $L_0(t)$ is not present in the guidance law).

The solution of equation (3.45) is

$$L(t) = \frac{1}{(r(0)-v_{cl}t)^{N-2}}\int_t^{t_F} 2Nv_{cl}(r(0)-v_{cl}t)^{N-2}f(t)dt \tag{3.46}$$

so that using equation (3.43) we obtain the guidance law:

$$a_M(t) = Nv_{cl}\dot{\lambda}(t) + \frac{1}{(r(0)-v_{cl}t)^{N-1}}\int_t^{t_F} Nv_{cl}(r(0)-v_{cl}t)^{N-2}a_T(t)dt \tag{3.47}$$

For a step maneuver $a_T(t) = a_T$ we obtain the expression:

$$a_M(t) = Nv_{cl}\dot{\lambda}(t) + \frac{N}{(N-1)}a_T \tag{3.48}$$

which is different from the APN guidance law given in many papers.

In the general case, it follows from equation (3.47) that the optimal missile acceleration increases near t_F. For $N = 3$ and $a_T(t) = a_T \sin \omega_T t$ (assuming $r(t_F)$ is very small and $a_T(t_F) = 0$) we can obtain:

$$a_M(t) = Nv_{cl}\dot{\lambda}(t) + \frac{Nv_{cl}}{\omega_T^2 r(t)}\dot{a}_T(t) + \frac{Nv_{cl}^2}{\omega_T^2 r^2(t)}a_T(t) \qquad (3.49)$$

As mentioned earlier, the augmented proportional navigation guidance law was offered without rigorous justification. The above expressions fill this gap.

3.5   MODIFIED LINEAR PLANAR MODEL OF ENGAGEMENT

The majority of guidance laws have one objective: to reduce to zero the miss distance between the missile and target. However, this is not always sufficient. The direction from which the missile approaches the target is also important. In certain scenarios, the mission requirements call for the payload to impact the target from a specific direction. Final impact angle requirements are very important for hitting ground targets. There exist specific angles to hit a target most effectively.

When we guide a missile with a seeker, the impact point (point of the missile warhead detonation and/or hitting a target) is heavily dependent upon the target information provided by the seeker. In the case of IR seekers, the most probable impact point lies near the heat source of the target, if the conventional guidance law is used. However, the heat source of many targets is located near the tail of fuselage, and therefore the kill probability could be significantly low if the missile simply follows the heat source. The concern about low kill probability due to the impact point can be partially resolved by choosing the impact angle properly.

The above-considered linear planar model of engagement can be enhanced by specifying a missile-target impact achieved at a fixed LOS angle λ_0.

By introducing the state variables:

$$z_1 = \lambda - \lambda_0 = x_1 - \lambda_0, \quad x_2 = \dot{z}_1 \qquad (3.50)$$

and acting analogously to the described above, instead of equation (3.2) we obtain:

$$\dot{z}_1 = x_2$$

$$\dot{x}_2 = -a_1(t)(z_1 + \lambda_0) - a_2(t)x_2 - b(t)u + b(t)a_T \qquad (3.51)$$

In contrast to equation (3.6), the Lyapunov function Q is chosen as:

$$2Q = cx_2^2 + c_0 z_1^2 \qquad (3.52)$$

and the guidance law is obtained from the stability conditions of the whole system (3.51) rather than the partial stability with respect to the x_2 coordinate.

From the condition of negative definiteness of the derivative of the Lyapunov function (3.52):

$$\dot{Q} = cx_2(-a_1(t)(z_1 + \lambda_0) - a_2(t)x_2 - (b(t)u - b(t)a_T)) + c_0 z_1 x_2 \quad (3.53)$$

we can derive the following guidance law:

$$u(t) = Nv_{cl}\dot{\lambda}(t) + N_1\dot{\lambda}^3(t) - N_2\left(\ddot{r}(t) - \frac{c_0}{c}r(t)\right)(\lambda(t) - \lambda_0) \quad (3.54)$$
$$- \ddot{r}(t)\lambda_0 + N_3 a_T(t)$$

where

$$N_2 \begin{matrix} \geq 1 \\ \leq 1 \end{matrix} \quad \text{if} \quad sign\left(\left(\ddot{r}(t) - \frac{c_0}{c}r(t)\right)\dot{\lambda}(t)(\lambda(t) - \lambda_0)\right) \begin{matrix} \leq 0 \\ \geq 0 \end{matrix}$$

Comparing the guidance laws and equations (3.16) and (3.54), we can see that the specified missile-target impact LOS angle λ_0 influences only the shaping term:

$$N_2\left(\ddot{r}(t) - \frac{c_0}{c}r(t)\right)(\lambda(t) - \lambda_0) - \ddot{r}(t)\lambda_0$$

3.6   GENERAL PLANAR CASE

Instead of using a small linear approximation (2.9), we will consider a general nonlinear case. The expressions for the LOS angle [see equation (2.8)] and its derivatives can be presented in the following form:

$$\sin(\lambda(t)) = \frac{y(t)}{r(t)} \quad (3.55)$$

$$\ddot{\lambda}(t)\cos(\lambda(t)) - \dot{\lambda}^2(t)\sin(\lambda(t))$$
$$= -a_1(t)\sin(\lambda(t)) - a_2(t)\cos(\lambda(t))\dot{\lambda}(t) + b_1\ddot{y}(t) \quad (3.56)$$

or

$$\dot{x}_1 = x_2$$

$$\dot{x}_2 = x_2^2 \tan x_1 - a_1(t) \tan x_1 - a_2(t)x_2 - \frac{b(t)}{\cos x_1} u + \frac{b(t)}{\cos x_1} f \quad (3.57)$$

where $u(t)$ is a commanded missile acceleration, $x_1 = \lambda(t)$, $x_2 = \dot{\lambda}(t)$, and the coefficients $a_1(t)$, $a_2(t)$, and $b(t)$ are determined by equations (3.3)–(3.5).

It is of importance to mention that when considering a linear approximation of the trigonometric functions in equation (3.57) we will not obtain the linear system (3.2). There will be the additional nonlinear term $x_2^2 x_1$ (for small x_1, $x_2^2 \tan x_1 \approx x_2^2 x_1$). The linearization for small LOS angles at this stage is more rigorous than linearization (2.9) and sequential differentiation, as it was done in many publications. The main reason for using equations (3.1) and (3.2) is the difficulty in dealing with nonlinear differential equations.

The derivative of the Lyapunov function (3.6) along any trajectory of equation (3.57) is:

$$\dot{Q} = cx_2(x_2^2 \tan x_1 - a_1(t)x_1 \tan x_1 - a_2(t)x_2 - (b(t)u - b(t)f)/\cos x_1)$$

or

$$\dot{Q} = c(x_2^3 \tan x_1 - a_1(t)x_1 x_2 \tan x_1 - a_2(t)x_2^2 - (b(t)x_2 u - b(t)x_2 f)/\cos x_1)$$
$$(3.58)$$

The negative definiteness of the derivative (3.58) can be guaranteed by the control-guidance law:

$$u(t) = Nv_{cl} \cos(\lambda(t))\dot{\lambda}(t) + N_1 \cos(\lambda(t))\dot{\lambda}^3(t)$$
$$- N_2 \sin(\lambda(t))\ddot{r}(t) - N_0 r(t)\dot{\lambda}^2(t)\sin(\lambda(t)) + N_3 a_T(t) \quad (3.59)$$

$$N > 2, \quad N_1 > 0,$$

$$N_0 \begin{array}{c} \geq 1 \\ \leq 1 \end{array} \quad \text{if} \quad sign(\dot{\lambda}(t)\lambda(t)) \begin{array}{c} \leq 0 \\ \geq 0 \end{array}$$

$$N_2 \begin{array}{c} \geq 1 \\ \leq 1 \end{array} \quad \text{if} \quad sign(\ddot{r}(t)\dot{\lambda}(t)\lambda(t)) \begin{array}{c} \leq 0 \\ \geq 0 \end{array}$$

and

$$N_3 \begin{array}{c} \leq 1 \\ \geq 1 \end{array} \quad \text{if} \quad sign(a_T(t)\dot{\lambda}(t)) \begin{array}{c} \leq 0 \\ \geq 0 \end{array}$$

The guidance law (3.59) can be presented as the sum of the main PN law and additional correcting controls:

$$u = N v_{cl} \cos(\lambda(t)) \dot{\lambda}(t) + \sum_{k=0}^{3} u_k \qquad (3.60)$$

where

$$u_0 = -N_0 r(t) \dot{\lambda}^2(t) \sin(\lambda(t)) \qquad (3.61)$$

$$u_1 = N_1 \cos(\lambda(t)) \dot{\lambda}^3(t) \qquad (3.62)$$

$$u_2 = -N_2 \ddot{r}(t) \sin(\lambda(t)) \qquad (3.63)$$

and

$$u_3 = N_3 a_T(t) \qquad (3.64)$$

For small LOS angles and short homing ranges (the case that was discussed mostly in the guidance literature), the term $x_2^3 \tan x_1$ in equation (3.58) is smaller than a dominant x_2^2 component. That is why the analysis of the linear system (3.2) is justified if such conditions are satisfied. For a larger spectrum of LOS angles the u_0 component is needed.

The effectiveness of the u_1 correction was discussed earlier for the linearized model. The u_2 correction is needed for maneuvering targets and when the second derivative of range is not small enough. The augmented proportional navigation term (3.64) differs from the well-known one [see also equation (2.53)] that was obtained for step maneuvers but was recommended to be used for all types of maneuvers. The $sign(a_T(t)\dot{\lambda}(t))$ factor reflects the dependence of the correction on the target behavior. Each of the controls u_k ($k = 0, 1, 2, 3$) increases the effectiveness of the PN navigation law with respect to the criterion chosen. The number of the controls applied in practice should depend on the problem under consideration (target distances, LOS angles, maneuvering or nonmaneuvering targets, etc., as well as the system's ability to realize the correction control in practice).

Taking into account that we consider a class of the modified proportional navigation laws, the control-corrections (3.61)–(3.64) are considered as the means of improving the PN law (2.49), extending its area of applicability. The coefficients $N_0 - N_3$ (constant or time-varying) can be determined based on simulation results of the whole missile system taking into account the autopilot limits on a missile acceleration, airframe dynamics, and some other factors, i.e., the same way as the most appropriate values of $N = 3 - 4$ were established.

The considered nonlinear planar model of engagement can be enhanced by specifying a missile-target impact achieved at a fixed LOS angle λ_0, as in the case of the linearized planar model, by introducing the state variables:

$$z_1 = \sin(\lambda(t)) - \sin\lambda_0, \quad x_2 = \dot{z}_1 = \dot{\lambda}(t)\cos(\lambda(t)) \tag{3.65}$$

and considering the Lyapunov function (3.52).

Analogous to equations (2.29) and (2.30) we can obtain:

$$\dot{z}_1 = \frac{\dot{y}(t)}{r(t)} - \frac{\sin(\lambda(t))\dot{r}(t)}{r(t)} \tag{3.66}$$

$$\dot{x}_2 = \frac{\ddot{y}(t)}{r(t)} - \frac{\dot{z}_1\dot{r}(t)}{r(t)} - \frac{\dot{y}(t)\dot{r}(t)}{r^2(t)} - \frac{\ddot{r}(t)\sin(\lambda(t))}{r(t)}$$

$$+ \frac{\dot{r}^2(t)\sin(\lambda(t))}{r^2(t)} - \frac{\dot{r}(t)\dot{\lambda}(t)\cos(\lambda(t))}{r(t)} \tag{3.67}$$

$$= \frac{\ddot{y}(t)}{r(t)} - \frac{2\dot{z}_1\dot{r}(t)}{r(t)} - \frac{z_1\ddot{r}(t)}{r(t)} - \sin\lambda_0\frac{\ddot{r}(t)}{r(t)}$$

so that instead of equation (3.51) we have:

$$\dot{z}_1 = x_2$$

$$\dot{x}_2 = -a_1(t)z_1 - a_2(t)x_2 - a_1(t)\sin\lambda_0 - b(t)u + b(t)a_T \tag{3.68}$$

Acting analogous to (3.53), we can derive the following guidance law:

$$u(t) = Nv_{cl}\cos(\lambda(t))\dot{\lambda}(t) + \sum_{k=1}^{3} u_k \tag{3.69}$$

where u_k ($k = 1, 3$) coincide with equations (3.62) and (3.64), respectively, and

$$u_2 = N_2\left(\ddot{r}(t) - \frac{c_0}{c}r(t)\right)(\sin(\lambda(t)) - \sin\lambda_0) - \ddot{r}(t)\sin\lambda_0 \tag{3.70}$$

The modified Lyapunov function, which contains trigonometric functions of the LOS, enables us to reduce the guidance problem with an impact angle constraint for the nonlinear planar model to the analogous one for the linearized model and the Lyapunov function (3.52). As in the case of the linearized planar model, the modified Lyapunov function "eliminates" the quadratic term u_0 [see equation (3.61)] in the guidance law of the enhanced nonlinear model of engagement.

The guidance laws with the impact angle constraints (3.54) and (3.69) differ insignificantly from the corresponding laws obtained without the impact angle constraints. Their realization does not present any difficulties.

3.7 THREE-DIMENSIONAL ENGAGEMENT MODEL

For the three-dimensional case and the Earth-based coordinate system we rewrite equation (2.45):

$$\ddot{\lambda}_s(t) = -a_1(t)\lambda_s(t) - a_2(t)\dot{\lambda}_s(t) + b(t)(a_{Ts}(t) - u_s) \qquad (3.71)$$

where $a_{Ts}(t)$ are the coordinates of the target acceleration vector and $u_s(t)$ ($s = 1, 2, 3$) are the coordinates of the missile acceleration vector, which are considered as controls.

As in equation (2.46), the Lyapunov function is chosen as the sum of squares of the LOS derivative components that corresponds to the nature of parallel navigation, i.e.,

$$Q = \frac{1}{2}\sum_{s=1}^{3} d_s \dot{\lambda}_s^2 \qquad (3.72)$$

where d_s are positive coefficients.

Its derivative can be presented in the following form:

$$2\dot{Q} = \sum_{s=1}^{3} d_s(-a_1(t)\lambda_s\dot{\lambda}_s - a_2(t)\dot{\lambda}_s^2 + b(t)\dot{\lambda}_s(a_{Ts}(t) - u_s)) \qquad (3.73)$$

so that the three-dimensional guidance problem is similar to the linearized planar guidance problem.

Analogous to equation (3.16), the controls $u_s(t)$ that guarantee $\lim\|\dot{\lambda}\| \to 0$, $t \to \infty$ can be presented as:

$$u_s = Nv_{cl}\dot{\lambda}_s + \sum_{k=1}^{3} u_{sk} \qquad (3.74)$$

where

$$u_{s1}(t) = N_1 \dot{\lambda}_s^3(t), \quad N_1 > 0 \tag{3.75}$$

$$u_{s2}(t) = -N_{2s}\lambda_s(t)\ddot{r}(t) \quad N_{2s} \begin{matrix} \geq 1 \\ \leq 1 \end{matrix} \quad \text{if} \quad sign(\ddot{r}(t)\lambda_s(t)\lambda_s(t)) \begin{matrix} \leq 0 \\ \geq 0 \end{matrix} \tag{3.76}$$

and

$$u_{s3}(t) = N_{3s}a_{Ts}(t) \quad N_{3s} \begin{matrix} \leq 1 \\ \geq 1 \end{matrix} \quad \text{if} \quad sign(a_{Ts}(t)\dot{\lambda}_s(t)) \begin{matrix} \leq 0 \\ \geq 0 \end{matrix} \quad (s = 1, 2, 3) \tag{3.77}$$

The expressions (3.74)–(3.77) can be obtained similar to the linear planar case [see equation (3.16)]. However, for the three-dimensional engagement model, in the case $d_s = 1$, the term $\sum_{s=1}^{3} -a_1(t)\lambda_s\lambda_s$ in equation (3.73) equals zero. This means that controls $u_{s2}(t) = -N_{2s}\lambda_s(t)\ddot{r}(t)$ are not needed to guarantee $\lim \|\dot{\lambda}\| \to 0$, $t \to \infty$. Nevertheless, the above-mentioned controls are also important parts of the guidance law.

The commanded acceleration can be considered as consisting of two components—radial (along the line-of-sight, also called longitudinal) and tangential (perpendicular to the line-of-sight, also called lateral), [see equations (1.21) and (1.22)]. As it follows from equation (3.76), the components $u_{s2}(t)$ belong to the radial acceleration (i.e., they influence the closing velocity).

Usually, during a missile flight, only two LOS rate components are dominant, so that the case of equal d_s is not typical and then $u_{s2}(t)$ ($s = 1, 2, 3$) also influence the tangential acceleration. However, the radial component is dominant.

As indicated above, controls $u_{s2}(t) = -N_{2s}\lambda_s(t)\ddot{r}(t)$ ($s = 1–3$) do not influence the tangential component of a missile acceleration in the case of equal d_s; they change the radial acceleration component, which is important to guarantee an appropriate acceleration (force) at the moment of intercept.

It is important to mention that for many types of existing missiles (e.g., without throttleable engines) radial acceleration cannot be used as a control action. Such missiles are not able to use thrust control as a part of a guidance law. Controls u_{s2} can influence a missile trajectory only by decelerating its motion.

As seen from equations (3.40)–(3.43), the multidimensional PN law follows immediately as one of the possible solutions and as a component of a more complicated law with nonlinear terms. The described guidance laws can be used for the midcourse and terminal guidance. During the midcourse stage the components of the LOS are obtained from equation (1.8).

For the terminal stage these components are usually calculated based on measurements of azimuth and elevation angles. The vectors $\lambda(t)$ and $\dot{\lambda}(t)$ can be presented as [see equations (1.9) and (1.16)]:

$$\lambda(t) = \begin{bmatrix} \cos\alpha\cos\beta \\ \cos\alpha\sin\beta \\ \sin\alpha \end{bmatrix}, \quad \dot{\lambda}(t) = \begin{bmatrix} -\sin\alpha\cos\beta \\ -\sin\alpha\sin\beta \\ \cos\alpha \end{bmatrix}\dot{\alpha} + \begin{bmatrix} -\cos\alpha\sin\beta \\ \cos\alpha\cos\beta \\ 0 \end{bmatrix}\dot{\beta} \quad (3.78)$$

where α and β are elevation and azimuth angles.

By comparing the guidance law (3.74) using equation (3.78) with the guidance law for the nonlinear planar model (3.27), we can conclude that the guidance law (3.60) can be used to analyze the coordinate u_3 in the three-dimensional case, i.e., the three-dimensional, Lyapunov-based guidance law embeds the Lyapunov-based guidance laws obtained for the planar case.

The established similarity between the equations determining the guidance law for the three-dimensional and linearized planar engagement models enables us to present the guidance law for the three-dimensional enhanced model of engagement with the specified missile-target impact achieved at a fixed LOS angle $\lambda_0 = (\lambda_{01}, \lambda_{02}, \lambda_{03})$ in the form (3.40) with the modified term $u_{s2}(t)$ [see equation (3.76)].

Acting analogously to the above described, instead of equation (3.51) we obtain:

$$\dot{z}_{1s} = x_{2s}$$

$$\dot{x}_{2s} = -a_1(t)(z_{1s} + \lambda_{0s}) - a_2(t)x_{2s} - b(t)u_s + b(t)a_{Ts} \quad (3.79)$$

$$(s = 1, 2, 3)$$

In contrast to equation (3.52), the Lyapunov function Q is chosen as:

$$2Q = \sum_{s=1}^{3} (d_s x_{2s}^2 + c_{0s}z_{s1}^2) \quad (3.80)$$

Repeating the earlier considered procedure we obtain the guidance law in the form (3.74) with:

$$u_{s2}(t) = -N_{2s}\left(\ddot{r}(t) - \frac{c_{0s}}{d_s}r(t)\right)(\lambda_s(t) - \lambda_{0s}) - \ddot{r}(t)\lambda_{0s} \quad (3.81)$$

where

$$N_{2s} \overset{\geq 1}{\leq 1} \quad if \quad sign\left((\ddot{r}(t) - \frac{c_{0s}}{d_s} r(t))\dot{\lambda}_s(t)(\lambda_s(t) - \lambda_{0s}) \right) \overset{\leq 0}{\geq 0} \qquad (s = 1, 2, 3)$$

Many interceptors use a lethality enhancement device to improve their hit-to-kill capabilities. For example, endoatmospheric guided missiles typically employ a fusing system and fragmentation warhead to accomplish this. The performance of these lethality enhancement systems can be sensitive to endgame conditions. Suitably controlling the terminal interceptor body rates and interceptor-threat approach angles can help to maximize the performance and effectiveness of the lethality enhancement device.

3.8   GENERALIZED GUIDANCE LAWS

As indicated earlier, the guidance laws considered in this chapter were obtained under assumption that both guidance components, radial and tangential, can be realized in practice. However, this is possible only for certain types of missiles. For missiles without throttleable engines, only a tangential component of the developed guidance laws can be implemented.

Unlike missiles without throttleable engines, missiles with axial control can employ thrust control as a part of guidance. Because of their superior guidance ability, for the purpose of the detailed analysis of this type of missiles, as well as other types of unmanned aerial vehicles, we consider the guidance laws based on the analysis of the longitudinal and lateral motions.

According to equations (1.21) and (1.22), the dynamic equations of the three-dimensional engagement can be presented in the form:

$$\ddot{r}(t) = a_T(t) - a_M(t) = a_{Tr}(t) + a_{Tt}(t) - a_{Mr}(t) - a_{Mt}(t) \qquad (3.82)$$

Combining equations (1.19), (1.20), and (3.82), we obtain the following system of equations describing the three-dimensional engagement:

$$\ddot{\lambda}_s(t)r(t) + 2\dot{r}(t)\dot{\lambda}_s(t) + \ddot{r}(t)\lambda_s(t) = a_{Ts}(t) - a_{Ms}(t) \qquad (s = 1, 2, 3) \quad (3.83)$$

where $a_{Ts}(t)$ and $a_{Ms}(t)$ are the coordinates of $a_T(t)$ and $a_M(t)$, respectively.

The last term of the left part of equation (3.83) corresponds to the vector directed along the LOS. The components $q\lambda_s$ of $h_s = \ddot{\lambda}_s(t)r(t) + 2\dot{r}(t)\dot{\lambda}_s(t)$ $(s = 1, 2, 3)$ that correspond to the vector directed along the LOS are determined from the orthogonality of radial and tangential vectors, i.e.,

$$\sum_{s=1}^{3}(h_s - q\lambda_s)q\lambda_s = 0$$

Using the equalities obtained from the sequential differentiation of

$$\sum_{s=1}^{3} \lambda_s^2 = 1$$

the following expression for the factor q can be obtained:

$$q = r(t)\sum_{s=1}^{3} \ddot{\lambda}_s(t)\lambda_s(t) = -r(t)\sum_{s=1}^{3} \dot{\lambda}_s^2(t) \tag{3.84}$$

The expressions for the missile longitudinal and lateral motions follow from equations (3.83) and (3.50). We analyze these motions in the Cartesian frame of coordinates of an inertial reference coordinate system, in contrast to the well-known presentation of the three-dimensional kinematics of guidance (see, e.g., [9]) describing the longitudinal and lateral motions using a rotating frame of coordinates with axes along the unit vectors 1_r directed along $\boldsymbol{r}$, 1_w directed along $\boldsymbol{r} \times \dot{\boldsymbol{r}}$, and $1_t = 1_r \times 1_w$.

For the longitudinal motion we have:

$$\ddot{r}(t)\lambda_s(t) - r(t)\sum_{s=1}^{3} \dot{\lambda}_s^2(t)\lambda_s = a_{Trs}(t) - a_{Mrs}(t) \qquad (s = 1, 2, 3) \tag{3.85}$$

By presenting the radial vectors $\boldsymbol{a}_{Tr}(t)$ and $\boldsymbol{a}_{Mr}(t)$ in the form:

$$a_{Trs}(t) = a_{Tr}(t)\lambda_s(t), \quad a_{Mrs}(t) = a_{Mr}(t)\lambda_s(t) \qquad (s = 1, 2, 3) \tag{3.86}$$

where $\boldsymbol{a}_{Tr}(t)$ and $\boldsymbol{a}_{Mr}(t)$ are the target and missile radial accelerations, respectively, equation (3.85) can be reduced to:

$$\ddot{r}(t) - r(t)\sum_{s=1}^{3} \dot{\lambda}_s^2(t) = a_{Tr}(t) - a_{Mr}(t) \tag{3.87}$$

For the lateral motion we have:

$$\ddot{\lambda}_s(t)r(t) + 2\dot{r}(t)\dot{\lambda}_s(t) + r(t)\sum_{s=1}^{3} \dot{\lambda}_s^2(t)\lambda_s(t) = a_{Tts}(t) - a_{Mts}(t) \qquad (s = 1, 2, 3) \tag{3.88}$$

The system of equations (3.87) and (3.88) is equivalent to the system (3.83). The analysis of their specifics enables us to simplify the analysis of the original system (3.83).

As mentioned earlier, missiles without axial control are able to control only the lateral motion using information about missile thrust, drag, and target acceleration and considering them as external factors with respect to control actions. The basic widespread philosophy behind controlling the lateral motion is that the lateral missile acceleration should nullify the LOS rate, (i.e., the lateral acceleration as control is aimed at implementing parallel navigation). In the ideal case $\dot{\lambda}_s(t) = 0$ ($s = 1, 2, 3$) the system (3.87) and (3.88) is reduced to:

$$\ddot{r}(t) = a_{Tr}(t) - a_{Mr}(t) \tag{3.89}$$

It can be easily observed from equations (3.87) and (3.88) that the dynamics of longitudinal and lateral motions can be decoupled by using a pseudo-acceleration $a_{Mr1}(t)$ in the radial direction:

$$a_{Mr1}(t) = a_{Mr}(t) - r(t) \sum_{s=1}^{3} \dot{\lambda}_s^2(t) \tag{3.90}$$

so that instead of (3.87) we can analyze (3.88), where $a_{Mr}(t)$ is changed to $a_{Mr1}(t)$.

The terms *lateral acceleration* and *lateral motion* were used above to characterize the motion in a plane orthogonal to the LOS. The TPN (true proportional navigation) law $a_{Mts}(t) = -N\dot{r}(t)\dot{\lambda}_s(t)$, $N > 2$ characterizes the motion belonging to this plane. However, the class of guidance laws implementing parallel navigation does not necessarily satisfy (3.88) because the acceleration vector required by the guidance law does not lie in this plane. For example, in the PPN (pure PN) law the commanded acceleration is applied normally to the missile velocity vector; in the GPN (generalized PN) the commanded acceleration forms a constant angle with the normal to the LOS [9]. We use the term lateral acceleration to characterize the motion satisfying equation (3.88) and the term longitudinal (radial) acceleration to characterize the motion satisfying equation (3.89).

In accordance with the Lyapunov-based control design approach, used earlier, the guidance problem can be formulated as the problem of choosing controls $a_{Mr}(t)$ and $a_{Mts}(t)$ ($s = 1, 2, 3$) to guarantee $\dot{r}(t) < 0$ and the asymptotic stability of the system (3.88) with respect to $\dot{\lambda}_s(t)$ ($s = 1, 2, 3$). Because in reality we deal with a finite problem, for simplicity and a more rigorous utilization of the term "asymptotic stability" we assume, as earlier, disturbance (target acceleration) to be a vanishing function (i.e., contains a factor $e^{-\varepsilon t}$), where ε is an infinitely small positive number; moreover, if t_F is the time of intercept then $\lim_{t \to t_F} r(t) \to 0$ and $a_T(t) = 0$ for $t > t_F$.

From equation (3.89), the conditions $\dot{r}(t) < 0$ and $\lim_{t \to t_F} r(t) \to 0$ can be achieved by choosing $a_{Mr1}(t) > a_{Tr}(t)$ for $t \le t_F$ and $a_{Mr1}(t) = 0$ for $t > t_F$, i.e.,

$$a_{Mr1}(t) = k_1(t)a_{Tr}(t), \quad k_1(t) \ge 1 \tag{3.91}$$

This follows from the condition of negative definiteness of the derivative of the Lyapunov function $r^2(t)$ along any trajectory of equation (3.89), i.e., $r(t)\dot{r}(t) < 0$ where

$$\dot{r}(t) = \dot{r}(t_0) + \int_{t_0}^{t} (a_{Tr}(t) - a_{Mr1}(t))dt < 0;$$

t_0 is the initial moment of guidance.

The system described by equation (3.89) has been examined thoroughly in the literature; various optimal problems have been considered and solved (see, e.g., [1,6]). Without considering concrete optimal problems here (their practical application is limited because of the lack of information about future values of a target acceleration), we indicate only that a pseudoacceleration $a_{Mr1}(t)$ in the radial direction should exceed the radial target acceleration, so that the larger their difference the faster the decrease in range.

The asymptotic stability of equation (3.88) with respect to $\lambda_s(t)$ ($s = 1, 2, 3$) is guaranteed by the guidance law [see equations (3.74)–(3.77)]:

$$a_{Mts}(t) = Nv_{cl}\dot{\lambda}_s(t) + \sum_{k=1}^{3} u_{sk}(t) \tag{3.92}$$

where $v_{cl} = -\dot{r}(t)$,

$$u_{sl}(t) = N_{1s}\dot{\lambda}_s^3(t), \quad N_{1s} > 0 \tag{3.93}$$

$$u_{s2}(t) = -N_2 r(t) \sum_{s=1}^{3} \dot{\lambda}_s^2(t)\lambda_s(t), \quad N_{2s} \begin{matrix} \ge 1 \\ \le 1 \end{matrix} \quad \text{if} \quad sign(\dot{\lambda}_s(t)\lambda_s(t)) \begin{matrix} \le 0 \\ \ge 0 \end{matrix} \tag{3.94}$$

and

$$u_{3s}(t) = N_{3s}a_{Tts}(t), \quad N_{3s} \begin{matrix} \le 1 \\ \ge 1 \end{matrix} \quad \text{if} \quad sign(a_{Tts}(t)\dot{\lambda}_s(t)) \begin{matrix} \le 0 \\ \ge 0 \end{matrix} \quad (s = 1, 2, 3) \tag{3.95}$$

The above expressions (3.93)–(3.95) immediately follow from the procedure based on the Lyapunov approach described in the previous sections of this chapter, if the Lyapunov function has the form (3.72).

Based on (3.88)–(3.95), the guidance law can be presented in the following form:

$$a_{Ms}(t) = Nv_{cl}\dot{\lambda}_s(t) + N_1\dot{\lambda}_s^3(t)$$

$$+ (1 - N_{2s})r(t)\sum_{s=1}^{3}\dot{\lambda}_s^2(t)\lambda_s(t) + k_1(t)a_{Trs}(t) + N_{3s}a_{Trs}(t) \tag{3.96}$$

$$(s = 1, 2, 3)$$

where $N_1,$ N_{2s}, and N_{3s} $(s = 1,$ 2, 3) are the same as in equations (3.93)–(3.95).

The first term of equation (3.96) corresponds to the traditional PN law. The second component of equation (3.96), with a properly chosen N_1, as discussed earlier, reacts to significant values of the LOS rate and should not influence the missile acceleration when the LOS rate is small. The coefficient $k_1(t)$ is chosen to guarantee a fast decrease of $r(t)$. It can be constant or time-varying depending on available information about a target. The term $N_{3s}a_{Trs}(t)$ is different from the corresponding term in the augmented proportional navigation law because the parameter N_{3s} is time-varying and of a bang-bang type. The $sign(a_{Trs}(t)\lambda_s(t))$ factor reflects the dependence of the correction on the target behavior. The case $N_{3s} = 1$ corresponds to the cancellation of the effect of the target maneuver on the Lyapunov function derivative $\dot{Q}$ (3.73) by forward-compensating the component of $a_{Tr}(t)$ normal to the LOS. The coefficients N_1, N_2, N_3, and k_1 (constant or time-varying) can be determined based on simulation results of the whole missile system taking into account the autopilot limits on a missile acceleration, airframe dynamics, and some other factors, i.e., the same way as the most appropriate values $N = 3$–4 were established.

The guidance law (3.96) assumes that a missile is able to control all three-dimensional space. For missiles enabled to control only their lateral acceleration, instead of equations (3.88) and (3.89) the initial equation (3.83) should be examined. This equation is analogous to equation (3.71), and the guidance law corresponding to this case is presented by equations (3.74)–(3.77). Comparison of equations (3.96) and (3.74) shows that in the case of missiles with uncontrollable thrust the $u_{3s}(t)$ terms depend upon the total target acceleration rather than its tangential component and that instead of the radial components:

$$k_1(t)a_{Trs}(t) + (1 - N_{2s})r(t)\sum_{s=1}^{3}\dot{\lambda}_s^2(t)\lambda_s(t)$$

the guidance law contains the radial component $u_{2s}(t)$ ($s = 1, 2, 3$). As mentioned earlier, $u_{2s}(t)$ in equation (3.74) influence the derivative $\dot{Q}$ (3.73) only in the case of unequal coefficients d_s ($s = 1, 2, 3$). Moreover, only negative $u_{2s}(t)$ ($s = 1, 2, 3$), i.e., deceleration can be realized in practice.

In our simplified model of engagement we assumed that the missile and target are point masses and considered the radial acceleration acting along the LOS. In reality, the radial acceleration acts along a missile's body and the tangential acceleration acts in the orthogonal direction, so that the real tangential acceleration obtained by projecting the acceleration (3.74) on the axis perpendicular to a missile's body axis may reflect the influence of the $u_{2s}(t)$ components ($s = 1, 2, 3$).

The obtained guidance laws assume that current information about a target acceleration is available. Usually, we operate only with the estimated target acceleration, so that a result worse than in the ideal estimation case can be expected. Many missiles are unable to measure a target acceleration and use it in a guidance law. In this case, the components $u_{3s}(t)$ ($s = 1, 2, 3$) are not present in the guidance law, and its performance is worse compared to the case when a target acceleration can be measured.

3.9   MODIFIED GENERALIZED GUIDANCE LAWS

The ability to control the longitudinal motion enable us, potentially, to speed up the intercept process [see the term with $k_1(t)$ in equation (3.96)]. As to the lateral motion, it is possible to impose additional requirements to engagement by specifying missile-target impact achieved at a fixed LOS angle $\lambda_0 = (\lambda_{01}, \lambda_{02}, \lambda_{03})$.

Presenting equation (3.88) as the system of first-order differential equation and comparing it with the equivalent presentation of the system (3.83) [see also equations (2.36), (2.37), and (2.71)], we can see that instead of the term $a_1(t)$ there will be the term $\sum_{s=1}^{3} \lambda_s^2(t)$. By introducing the variables $z_{1s} = \lambda_s - \lambda_{0s}$ ($s = 1, 2, 3$), the Lyapunov function (3.80) and repeating the discussed earlier procedure, we can derive the following lateral component of the guidance law:

$$a_{Mts}(t) = Nv_{cl}\dot{\lambda}_s(t) + \sum_{k=1}^{3} u_{sk}(t) \tag{3.97}$$

where the components $u_{sk}(t)$ ($k = 1, 3$) are the same as in equation 3.92 [see equations (3.93) and (3.95)] and the term $u_{s2}(t)$ equals:

$$u_{s2}(t) = -N_{2s}\left(r(t)\sum_{s=1}^{3}\dot{\lambda}_s^2 - \frac{c_{0s}}{d_s}r(t)\right)(\lambda_s(t) - \lambda_{0s}) - r(t)\sum_{s=1}^{3}\dot{\lambda}_s^2\lambda_{0s} \quad (3.98)$$

$$N_{2s} \begin{array}{c}\geq 1\\\leq 1\end{array} \quad \text{if} \quad sign\left((r(t)\sum_{s=1}^{3}\dot{\lambda}_s^2 - \frac{c_{0s}}{d_s}r(t))\dot{\lambda}_s(t)(\lambda_s(t) - \lambda_{0s})\right) \begin{array}{c}\leq 0\\\geq 0\end{array}$$

$$(s = 1, 2, 3)$$

Since the longitudinal acceleration component remains unchanged, instead of the guidance law (3.96) we can write:

$$a_{Ms}(t) = Nv_{cl}\dot{\lambda}_s(t) + N_1\dot{\lambda}_s^3(t) - N_{2s}\left(r(t)\sum_{s=1}^{3}\dot{\lambda}_s^2(t) - \frac{c_{0s}}{d_s}r(t)\right)(\lambda_s(t) - \lambda_{0s})$$

$$+ r(t)\sum_{s=1}^{3}\dot{\lambda}_s^2(t)(\lambda_s(t) - \lambda_{0s}) + k_1(t)a_{Trs}(t) + N_{3s}a_{Tts}(t)$$

or

$$a_{Ms}(t) = Nv_{cl}\dot{\lambda}_s(t) + N_1\dot{\lambda}_s^3(t) + (1 - N_{2s})r(t)\sum_{s=1}^{3}\dot{\lambda}_s^2(t)(\lambda_s(t) - \lambda_{0s})$$

$$+ N_{2s}\frac{c_{0s}}{d_s}r(t)(\lambda_s(t) - \lambda_{0s}) + k_1(t)a_{Trs}(t) + N_{3s}a_{Tts}(t) \quad (s = 1, 2, 3)$$

$$(3.99)$$

where N_1, N_{2s}, and N_{3s} ($s = 1$, 2, 3) are the same as in equations (3.93), (3.95), and (3.98).

3.10 EXAMPLES

First, we consider a realistic example of a tail-controlled aerodynamic missile operating at high altitude to illustrate the effectiveness of the described guidance laws and compare it to PN guidance results.

The flight control dynamics are assumed to be presented by a third-order transfer function:

$$W(s) = \frac{1 - \dfrac{s^2}{\omega_z^2}}{(\tau s + 1)\left(\dfrac{s^2}{\omega_M^2} + \dfrac{2\zeta}{\omega_M}s + 1\right)} \quad (3.100)$$

with damping ζ and natural frequency ω similar to [18] ($\zeta = 0.7$ and $\omega_M = 20$ *rad/s*), the flight control system time constant $\tau = 0.5$ *s* and the right-half plane zero $\omega_z = 5$ *rad/s*.

As it was mentioned in [18], at high altitudes where the airframe zero frequency ω_z can be low, optimal guidance, similar to [2] for the single-lag model, has no advantage when compared to proportional navigation and can produce even worse results. Miss distance, when using optimal guidance, increases as the airframe zero frequency decreases. A new optimal guidance law that accounts for the presence of airframe zero was developed and tested in [18]. It works better than a proportional navigation law but cannot be presented as a closed-form solution and is developed numerically and stored as a tabulated function of time depending on several factors.

The performance of the guidance laws (3.16) is compared to proportional navigation. We assume that the effective navigation ratio $N = 4$ and the closing velocity $v_{cl} = 1219.2$ *m/s* and consider the homing stage when LOS angle is relatively small, so that the expression (3.16) can be used. As in [18], two error sources are considered: a 3-*g* constant target maneuver and 1 *mr* of range-independent angle measurement noise. The acceleration limit is 10-g.

A simplified model of the missile engagement is presented in Figure 3.1. Here R_{TM} is the range r between a missile and a target and $\hat{R}_{TM}$ is its estimate. The measurement of the LOS angle λ_k^* is corrupted by noise. A pseudomeasurement of relative position y_k^* is created by a multiplication of λ_k^* by $\hat{R}_{TM}$. The Kalman filter then provides optimal estimates of relative position, relative velocity, and a target acceleration [17,18]. Three guidance laws are considered: proportional navigation; nonlinear guidance, discussed in the previous sections, without measurements of target acceleration; and nonlinear guidance utilizing measurements of target acceleration.

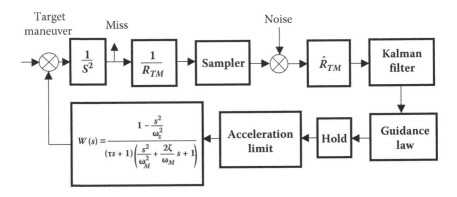

FIGURE 3.1 Missile guidance model.

The nonlinear guidance law has the form:

$$u(t) = 4v_{cl}\dot{\lambda}(t) + N_1\dot{\lambda}^3(t) + N_3 a_T(t) \qquad (3.101)$$

For the linearized engagement model in Figure 3.1, we assume a constant closing velocity, so that the term with the second derivative $\ddot{r}(t)$ in equation (3.16) equals zero.

We use the discrete form of the nonlinear guidance law and the estimates of $\hat{\lambda}$ in the form (symbol "∧" denotes estimates):

$$\hat{\dot{\lambda}}_k = \frac{\hat{y}_k + \hat{\dot{y}}_k t_{go}}{v_{cl}t_{go}^2} \qquad (3.102)$$

The results of a Monte Carlo simulation for a step target maneuver and zero initial conditions are presented in Figure 3.2. The mean absolute value of the resultant miss distances is given based on 50 simulation trials. The nonlinear term with gain $N_1 = 40{,}000v_{cl}$ (data with symbol "×") significantly improves the performance of a missile when compared to the PN guidance law (data with symbol "---"). A further improvement is reached by measuring target acceleration: by using a constant gain $N_3 = 1$ (solid curve) or a time dependent gain $N_3 = 0.75; 1.25$ [data with symbol "ooo"; see also equation (3.16)]. Each component of the guidance law

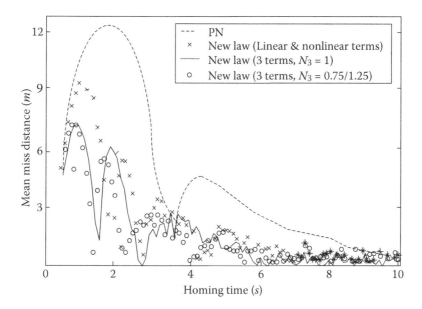

FIGURE 3.2 Comparative analysis of guidance laws performance.

(3.16) increases $|\dot{Q}(t)|$. This fact enables us to choose the gains N_i ($i = 1 - 3$) sequentially.

In the above example, N_1 was chosen based on value of the LOS angle rate estimates for the case of the PN guidance law (i.e., when only the first component of equation (3.101) was used), which was about 0.006 *rad/s* at the beginning of the homing stage and significantly less (3 times and more) at the end of the homing stage. For the given N_1 value, we indirectly increased N in the PN law at the beginning of the homing stage (according to equation (3.14), about 30%). However, the "cubic term" has a negligible influence at the end of the homing stage [see equation (3.14)].

Now we consider an example of a tail-controlled aerodynamic missile guided by the guidance laws discussed earlier and compare their effectiveness against a weaving maneuvering target with the acceleration $a_T(t) = 5g\ \sin(1.75\ t)$. The flight control dynamics are assumed to be presented by equation (3.100) with damping $\zeta = 0.65$ and natural frequency $\omega = 5$ *rad/s*, the flight control system time constant $\tau = 0.1$ *s* and the right-half plane zero $\omega_z = 30$ *rad/s*.

The following guidance laws are analyzed (the effective navigation ratio $N = 3$, the closing velocity $v_{cl} = 7000$ *m/s*):

1. The PN law: $u(t) = 3v_{cl}\dot{\lambda}(t)$
2. The APN law: $u(t) = 3v_{cl}\dot{\lambda}(t) + 1.5\,a_T(t)$
3. $u(t) = 3v_{cl}\dot{\lambda}(t) + N_1\dot{\lambda}^3(t)$
4. $u(t) = 3v_{cl}\cos(\lambda(t))\dot{\lambda}(t) + N_1\cos(\lambda(t))\dot{\lambda}^3(t) + N_3 a_T(t)$

$$\left(N_1 = 30{,}000\ v_{cl},\ N_3 = \begin{cases} 0.75 \\ 1.75 \end{cases} \text{if } sign(a_T(t)\dot{\lambda}(t)) \begin{matrix} \leq 0 \\ \geq 0 \end{matrix}\right)$$

Simulation results are shown in Figure 3.3. Miss distances for the PN and APN guidance are shown by dashed and dash-dot lines, respectively. The effectiveness of the "cubic" term for the linearized planar model, together with the PN law, is shown by the dotted line. This term, as well as other additional terms, in the guidance law for the nonlinear planar model (solid line) decreases the miss distance significantly.

In conclusion, the guidance laws are tested on an example of the engagement model with the parameters close to those considered in [7]: the effective navigation ratio $N = 3$; target initial conditions $R_{T1} = 4500\,m$, $R_{T2} = 2500\ m$, $R_{T3} = 0$; $V_{T1} = -350$ *m/s*, $V_{T2} = 30$ *m/s*; $V_{T3} = 0$; missile initial conditions $R_{M1} = R_{M2} = R_{T3} = 0$, $V_{M1} = -165$ *m/s*, $V_{M2} = 475$ *m/s*, $V_{T3} = 0$; target acceleration $a_{T1} = 0$, $a_{T2} = 3g\ \sin(1.31t)$, $a_{T3} = 0$; missile

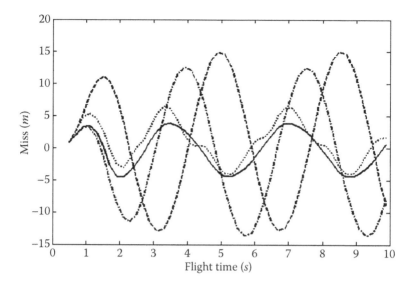

FIGURE 3.3 Miss distance comparison.

acceleration limit 5-g. In contrast to [7], the missile dynamics are taken into consideration: the missile flight control system right half-plane airframe zero frequency $\omega_z = 30$ *rad/s*, damping $\zeta = 0.7$, natural frequency $\omega_M = 20$ *rad/s*, and time constant $\tau = 0.5$ *s*. A target weaving frequency is chosen according to [12].

Figure 3.4 corresponds to the guidance law (3.96) and the case when the missile dynamics are ignored. It shows the trajectories of the target (cross solid line) and missile for the PN law and the laws considered in this chapter. The time of intercept for the APN and PN laws equals 8 *s*. The APN does not improve the PN guidance in this case. However, the additional terms in equation (3.96) enable us to improve the PN performance. The symbol "ATN" indicates the components N_{3s} $a_{Tts}(t)$ of equation (3.96) ($N_{31} = N_{32} = \{0.5; 3.5\}$, $N_{33} = 0$). The "cubic" term corresponds to $u_{s1}(t)$ components with gains $N_{11} = 20{,}000 v_{cl}$, $N_{12} = 2000 v_{cl}$, and $N_{13} = 0$. The guidance law with all terms of equation (3.96) ($k_1(t) = 7$) gives the best results. The time of intercept equals 7.35 *s*. As expected, inability to measure the target acceleration and absence of axial control decreases the missile performance. The time of intercept equals 7.8 *s* for the guidance law (3.74), where $u_{s2}(t) = 0$, $N_{31} = N_{32} = \{1; 3.5\}$, $N_{33} = 0$.

Figure 3.5 repeats the numerical simulations of Figure 3.4 taking into account the missile dynamics. In Figure 3.5 the miss distance and the time of intercept correspond to the moment of time when the closing velocity became positive. In the case of "PN + ATN" $N_{31} = \{0; 1.5\}$, $N_{32} = 1$, $N_{33} = 0$.

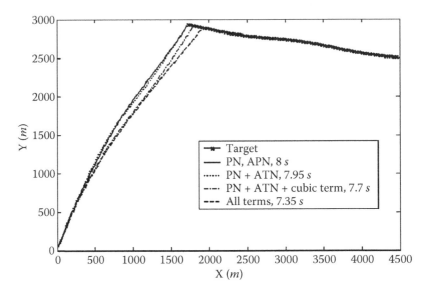

FIGURE 3.4 Comparison of the new guidance laws with PN and APN guidance (engagement model without missile dynamics).

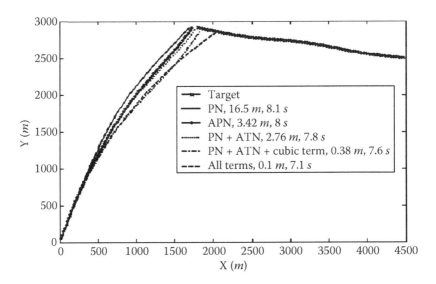

FIGURE 3.5 Comparison of the new guidance laws with PN and APN guidance (engagement model includes missile dynamics).

In the case of "PN + ATN + cubic term" $N_{11} = 28{,}000v_{cl}$, $N_{12} = 40{,}000v_{cl}$, $N_{13} = 0$. As in the case when the missile dynamics were ignored, the guidance law with all terms of equation (3.96) gives the best results. The parameters of the guidance law are: $k_1(t) = 2.8$; $N_{11} = 400{,}000v_{cl}$, $N_{12} = 19{,}400v_{cl}$, $N_{13} = 0$; $N_{31} = \{0;1.5\}$, $N_{32} = 1$, $N_{33} = 0$. The time of intercept and miss distance are significantly better than obtained under the PN and APN guidance laws.

In this chapter, the examples show the effectiveness of the considered guidance laws against maneuvering targets, their superiority to PN and APN guidance. In addition to showing better performance, these laws can be easily implemented in practice because they use the same parameters as the PN and APN laws.

REFERENCES

1. Bryson, A. E. Linear Feedback Solution for Minimal Effort Intercept Rendezvous, and Soft Landing, *AIAA Journal* 3, no. 8 (1965): 1542–48.
2. Ben-Asher, J. Z., and Yaesh, I. *Advances in Missile Guidance Theory, Progress in Astronautics and Aeronautics.* Vol. 180. Washington, DC: American Institute of Astronautics and Aeronautics, Inc., 1998.
3. Ghose, D. True Proportional Navigation with Maneuvering Target, *IEEE Transactions on Aerospace and Electronic Systems* 30, no. 1 (1993): 229–37.
4. Guelman, M. A Qualitative Study of Proportional Navigation, *IEEE Transactions on Aerospace and Electronic Systems*, 7, no. 4 (1971): 637–43.
5. Kim, K. B., Kim, M. J., and Kwon, W. H. Receding Horizon Guidance Laws with No Information on the Time-To-Go, *Journal of Guidance, Control, and Dynamics* 23, no. 2 (2000): 193–99.
6. Lee, E. B., and Markus, L. *Foundations of Optimal Control Theory.* New York, NY: John Willey & Sons, Inc., 1986.
7. Moon, J., Kim, K., and Kim, Y. Design of Missile Guidance Law via Variable Structure Control, *Journal of Guidance, Control, and Dynamics* 24, no. 4 (2001): 659–64.
8. Rumyantsev, V.V. On Asymptotic Stability and Instability of Motion with Respect to a Part of the Variables, *Journal of Applied Mathematics and Mechanics* 35, no. 1 (1971): 19–30.
9. Shneydor, N. A. *Missile Guidance and Pursuit.* Chichester, UK: Horwood Publishing, 1998.
10. Yanushevsky, R., and Boord, W. New Approach to Guidance Law Design, *Journal of Guidance, Control, and Dynamics* 28, no. 1 (2003): 162–66.
11. Yanushevsky, R. Concerning Lyapunov-Based Guidance, *Journal of Guidance, Control, and Dynamics* 29, no. 2 (2006): 509–11.
12. Yanushevsky, R. Analysis of Optimal Weaving Frequency of Maneuvering Targets, *Journal of Spacecraft and Rockets* 41, no. 3 (2003): 477–79.
13. Yanushevsky, R. An Approach to Design on Control Systems with Parametric-Coordinate Feedback, *IEEE Transactions on Automatic Control* 36, no. 11 (1991): 1293–95.
14. Yanushevsky, R. Lyapunov Approach to Guidance Laws Design. In *Proceedings of the WCNA 2004*, Orlando, Florida, June 29–July 7, 2003.
15. Yanushevsky, R. Generalized Missile Guidance Laws Against Manoeuvring Targets, *Proceedings of the Institution of Mechanical Engineers, Part I: Journal of Systems and Control Engineering* 221, no. I 3, 2007.

16. Yanushevsky, R. *Missile Guidance*, Lecture Notes, AIAA Guidance, Navigation, and control conference, Chicago, IL, 2009.
17. Zarchan, P. *Tactical and Strategic Missile Guidance, Progress in Astronautics and Aeronautics*. Vol. 176. Washington, DC: American Institute of Astronautics and Aeronautics, Inc., 1997.
18. Zarchan, P., Greenberg, E., and Alpert, J. Improving the High Altitude Performance of Tail-Controlled Endoatmospheric Missiles, *AIAA Guidance, Navigation and Control Conference*, AIAA Paper 2002-4770, August 2002.

4 Analysis of Proportional Navigation Guided Systems in the Time Domain

4.1 INTRODUCTION

It is well known that an investigation of processes and phenomena is linked, first of all, with the construction of mathematical models describing these processes and phenomena using mathematical language. The model is characterized by some parameters. These parameters include input variables or control actions as they are called, or simply controls, output variables or output coordinates, or controlled variables, and also intermediate variables, the so-called state variables. In most cases processes are not considered in isolation but in direct connection with other processes and phenomena. The influence of external conditions—the environment—is characterized by the so-called disturbing influences or, simply, disturbances.

As a matter of fact, the mathematical model is nothing but the analytical expression of an interconnection of the specified parameters. The parameters chosen are determined by the problem under consideration.

In Chapter 2, the control theory approach was used to obtain the proportional navigation (PN) guidance law. The line-of-sight (LOS) rate was considered as the system output; the PN law, the commanded missile acceleration, was considered as control or input; and the target acceleration was considered as disturbance. The Lyapunov approach, the pivot of control theory, was used in the previous chapter to obtain a wide class of guidance laws implementing parallel navigation. Since the PN term is the main component of the considered guidance laws and, separately, the PN law produces the lateral motion of unmanned aerial vehicles, we will pay special attention to this guidance law. Below we will build and analyze the models of PN guided missile systems. Similar models can be built for the PN guided UAV systems. The miss distance, the parameter that characterizes the missile guidance system performance, is the system output. The missile and target accelerations are control and disturbance, respectively.

For simplicity, we will analyze the planar model with one control action. The planar model itself is widely used on the preliminary design stage. As seen from Chapter 3, the results obtained for this model can be easily generalized for the three-dimensional engagement model.

In control theory, analytical tools were developed for describing the characteristics of control systems based on the concept of the system error. The goal of control is to reduce the error to the smallest feasible amount. The ability to adjust the transient and steady-state response of a control system to meet certain performance requirements is the main goal of its design. To analyze systems their performance criterion should be defined. Then, based on the desired performance, the parameters of the system or/and its structure should be adjusted to provide the desired response. Because the actual input signals are usually unknown, a standard test input signal is normally chosen. The time-domain analysis is usually based on the so-called step input.

The miss distance in the guidance system analysis and design is, at a certain degree, analogous to the error in conventional control systems. The goal of guidance is to reduce the miss distance to the smallest feasible amount. Target maneuver plays a major role in determining missile system performance. The miss distance, due to a step-target maneuver, is the miss step response similar to the well-known time-domain characteristic in control theory. Below we obtain analytical expressions of miss distance for simple models of PN guidance systems. Unfortunately, in the time domain the closed-form solutions cannot be obtained for high-order models realistically reflecting autopilot and airframe dynamics. Nevertheless, the models under consideration enable us to establish some properties of linear models of PN guided missile systems.

4.2   INERTIALESS PN GUIDANCE SYSTEM

Although PN guidance presents a nonlinear control problem, to apply a known technique of analysis and design, the system equations are linearized yielding a linear time-varying system. The linearization is valid when it is assumed that the missile and target approach the so-called collision course. The results of simulation of linear and nonlinear models show that the linearized model faithfully represents the guidance system dynamics, i.e., the linearization is valid close to the interception where the closing velocity can be considered constant so that the range can be approximated by a linear function of time [1,2,5].

In Chapter 2 we considered equation (2.35) to obtain the expression for the PN guidance law. We will use this equation to analyze the performance of the idealized linearized inertialess model of the PN missile guidance system (2.35). Substituting equation (2.7) into (2.35) we have:

$$\ddot{y}(t) = -Nv_{cl}\dot{\lambda}(t) + a_T(t) \tag{4.1}$$

After integration of both parts of equation (4.1) and taking into account equations (2.9) and (2.11), it becomes:

$$\dot{y}(t) = -Nv_{cl}\lambda(t) + \int a_T(t)dt = -\frac{N}{t_F - t}y(t) + V_T(t) \tag{4.2}$$

The solution of equation (4.2) is presented in the following form:

$$y(t) = \frac{1}{M(t)}\left(\int V_T(t)M(t)dt + C\right) \tag{4.3}$$

where C is a constant of integration and

$$M(t) = \exp\left(\int \frac{N}{t_F - t}dt\right) = (t_F - t)^{-N} \tag{4.4}$$

It can be simplified as:

$$y(t) = C(t_F - t)^N + \frac{t_F - t}{N - 1}V_T(t) - \frac{(t_F - t)^N}{N - 1}\int (t_F - t)^{-N+1}a_T(t)dt \tag{4.5}$$

In the case of a step target maneuver, i.e., $a_T(t) = a_T$ the last expression has the form:

$$y(t) = C(t_F - t)^N + \left(\frac{(t_F - t)t}{N - 1} - \frac{(t_F - t)^2}{(N - 1)(N - 2)}\right)a_T \tag{4.6}$$

where the constant C is determined based on the initial conditions for $y(t)$.

The analysis of equation (4.6) enables us to conclude that the miss distance $y(t_F)$ is zero, i.e., proportional navigation with the effective navigation ratio $N > 2$ is an effective way to hit a target. To be more rigorous, we should mention that the expression (4.6) indicates that only values $N = 1$ and $N = 2$ are dangerous. But by choosing $N > 2$ we guarantee zero miss.

The model of the missile guidance system considered above is too simple to make immediate optimistic estimates of the PN law performance. Even a slightly more complicated linear model of the missile guidance system (e.g., by presenting autopilot inertia by the first-order dynamic unit) makes the problem of the miss distance analysis very complicated.

The miss distance model for a missile with a first-order acceleration lag τ_1, against a target undergoing a step acceleration maneuver, is described by the following equations:

$$\dot{y}_1 = y_2$$

$$\dot{y}_2 = a_T - a_M$$

$$\dot{a}_M = (a_c - a_M)/\tau_1$$

$$\dot{a}_T = \delta(t) \tag{4.7}$$

$$y(t) = y_1$$

$$a_c(t) = N\left(\frac{y_1}{(t_F - t)^2} + \frac{y_2}{t_F - t}\right)$$

In this case the commanded acceleration $a_c(t)$ does not coincide with the real missile acceleration $a_M(t)$ and is presented in the form (2.14). A step maneuver $a_T(t)$ at $t = 0$ is described by a differential equation with the delta-function $\delta(t)$ in its right part.

It is impossible to obtain a visible analytical solution $y(t)$ of the above linear equation with time-varying coefficients and the singularity at $t = t_F$. The general approach to analysis of this type of equation is the use of simulation tools. Since our main interest lies in analyzing the miss distance $y(t_F)$, it means that we should simulate the system (4.7) for various t_F. To avoid multiple simulation trials and obtain $y(t_F)$ in one computer run, the method of adjoints is used [4,5]. Moreover, a specific structure of equation (4.7) enables us, based on the method of adjoints, to obtain the analytical solution of equation (4.7) with respect to $y(t_F)$.

4.3 METHOD OF ADJOINTS

The method of adjoints, which is a useful tool to simulate the impulse response $P(\sigma, t)$ of time-varying linear systems for the fixed observation time $\sigma = t_F$ with respect to the impulse application time t, has been widely used in missile guidance system design and analysis, especially for linearized engagement models. An approach to obtaining the adjoint system is based on a structural representation of the guidance system model.

The method of adjoint will be explained on the example of a linear time-varying system described by the system of the differential equation presented in the vector-matrix form:

$$\dot{y} = A(t_F - t)y + f, \quad 0 \leq t \leq t_F \tag{4.8}$$

where y and f are n-dimensional vectors, A is a matrix with coefficients depending on $t_F - t$.

By introducing the system adjoint to the system (4.8) as

$$\dot{x} = -A^T(t_F - t)x \tag{4.9}$$

it is easy to check that the adjoint vector x satisfies the condition:

$$\frac{d(x^T y)}{dt} = x^T f \tag{4.10}$$

or

$$x^T(t_F)y(t_F) - x^T(0)y(0) = \int_0^{t_F} x^T(\sigma)f(\sigma)d\sigma \tag{4.11}$$

where the upper symbol "T" denotes transposition.

To present the miss distance $y(t_F)$ due a constant target maneuver, we should put $x^T(t_F) = (1, 0,\ldots,0)$ and $f(t) = \delta(t)$, so that:

$$y(t_F) = x^T(0)y(0) \tag{4.12}$$

It is easy to verify that the transition matrix of the adjoint system (4.9) $\Phi_a(t, t_0) = \Phi^T(t, t_0)$, where $\Phi(t, t_0)$ is the transition matrix of equation (4.8).

For the class of guidance problems under investigation, we should present disturbances (target acceleration and other external factors) as the result of the solution of a system of differential equations. As seen from equation (4.7), for the case of a step acceleration maneuver (see the condition $f(t) = \delta(t)$) it reduces to a simple operation of differentiation.

The initial conditions $x(0)$ of the adjoint system (4.9) can be obtained by integrating equation (4.9) backward in time, or by considering the modified adjoint system with respect to time $\tau = t_F - t$, i.e., the miss distance $y(t_F)$ can be obtained in one run by simulating the system:

$$\dot{z} = A^T(\tau)z, \quad y(t_F) = z^T(t_F)y(0) \tag{4.13}$$

where $z^T(0) = (1, 0,\ldots,0)$.

For equation (4.7) the modified adjoint system has the following form:

$$\dot{z}_1 = \frac{N}{\tau_1 \tau^2} z_3$$

$$\dot{z}_2 = z_1 + \frac{N}{\tau_1 \tau} z_3$$

$$\dot{z}_3 = -z_2 - \frac{1}{\tau_1} z_3 \tag{4.14}$$

$$\dot{z}_4 = z_2$$

with the initial condition $z^T(0) = (1, 0,\ldots,0)$.

The matrix of coefficients of equation (4.13) is transposed with respect to the matrix $A(t_F - t)$ of the initial system (4.8). Hence, the adjoint modified system can be modeled by changing inputs by outputs and vice versa in all elements of the initial system (4.8) and by changing time t in the arguments of all time-varying coefficients by $t_F - t$. Changing inputs by outputs is equivalent to the following structural changes: nodes of the original system become summation units of the modified adjoint system, summation units of the original system become nodes of the modified adjoint system, and the direction of all signal flow is reversed. In addition, as mentioned above, the structural changes of the original system may be needed to convert its actual input to the equivalent impulsive input.

By differentiating the second equation of the system (4.14), it can be transformed to:

$$\ddot{z}_2 = \frac{N}{\tau_1 \tau} \dot{z}_3 \tag{4.15}$$

Using the Laplace transform and substituting z_3 from the third equation of (4.14), equation (4.15) can be presented as:

$$\frac{d}{ds}(s^2 Z_2(s)) = \frac{N s^2}{s(\tau_1 s + 1)} Z_2(s)$$

or

$$\frac{d}{ds} X(s) = N H(s) X(s) \tag{4.16}$$

where s is the symbol of the Laplace transform:

$$X(s) = s^2 Z_2(s) \tag{4.17}$$

and

$$H(s) = \frac{1}{s(\tau_1 s + 1)} = \frac{W(s)}{s} \tag{4.18}$$

The solution of equation (4.16) can be presented as:

$$X(s) = s^2 Z_2(s) = C \exp\left(\int NH(s)ds\right) = C\left(\frac{s}{s+1/\tau_1}\right)^N \tag{4.19}$$

where C is a constant determined by the initial conditions.
    From the last equation of the system (4.14) we obtain:

$$Z_4(s) = s^{-3}C\left(\frac{s}{s+1/\tau_1}\right)^N = s^{-1}X(s) \tag{4.20}$$

The constant $C = 1$ is determined from the condition $z_1(0) = \dot{z}_2(0) = 1$, which follows from the first two equations of (4.14), so that $\lim_{s \to \infty} s^2 Z_2(s) = \lim_{s \to \infty} X(s) = 1$ and equation (4.20) becomes:

$$Z_4(s) = s^{-3}\left(\frac{s}{s+1/\tau_1}\right)^N \tag{4.21}$$

Taking into account equations (4.13) and (4.21), for the effective navigation ratio $N = 4$ the miss distance due to the unit step target acceleration can be presented as:

$$y(t_F) = t_F^2 \exp(-t_F/\tau_1)(0.5 - t_F/6\tau_1) \tag{4.22}$$

The block diagram of the original missile guidance system (4.7) is given in Figure 4.1 (D denotes the operator of differentiation).
    The adjoint system (4.14) structure is presented in Figure 4.2. The above-given block diagrams of the original (4.7) and adjoint (4.14) systems for the unit step-target acceleration $a_T = 1$ can be simplified. Their simplified form, based on equations (2.10)–(2.14), is presented in Figure 4.3 and Figure 4.4, respectively.
    The modified systems are more convenient for the analysis. The modified original system operates directly with the line-of-sight $\lambda = y/(v_{cl}t)$ and its derivative. The modified adjoint system corresponds to the transformation (4.16)–(4.18); the closing velocity v_{cl} is shown in Figure 4.4 to correspond fully to Figure 4.3.

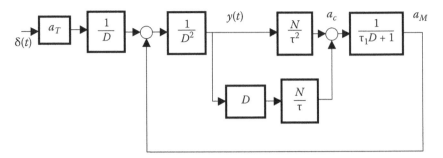

FIGURE 4.1 Block diagram of original guidance system.

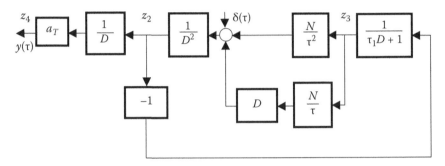

FIGURE 4.2 Block diagram of adjoint system.

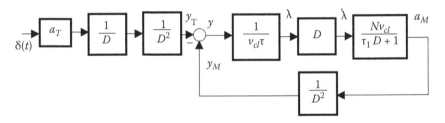

FIGURE 4.3 Modified block diagram of original guidance system.

The result of simulation of the adjoint system shown in Figure 4.5 for $\tau_1 = 0.5s$ and $a_T = 1g$ (the acceleration of gravity $g = 9.81$ m/s^2) presents the miss step response of the missile guidance system.

As seen in Figure 4.5, in contrast to the idealized linearized inertialess model (4.1), the miss distance of the inertial missile guidance system (4.7) is not zero. The acceleration time lag τ_1 significantly influences the miss step characteristic. It is obvious [see, e.g., equation (4.22)] that the miss distance is smaller for a smaller τ_1.

The analysis of the analytical expressions (4.6) and (4.22) for the miss distance due to a step target maneuver allows us to conclude that, despite

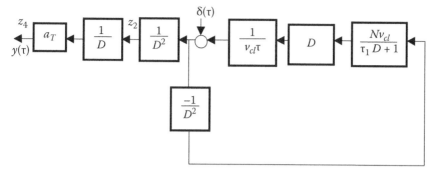

FIGURE 4.4 Modified block diagram of adjoint system.

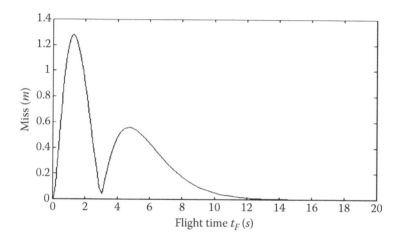

FIGURE 4.5 Miss distance for step target maneuver.

the commanded acceleration of the linearized PN guidance system model [see equation (4.7)] tends to infinity when t tends to t_F, it does not influence significantly the miss distance $y(t_F)$; it does not influence at all in the case of the idealized inertialess model.

The miss step response is one of the most used estimates of missile system performance. It is an important time-domain characteristic of missile guidance systems that allows designers to choose appropriate parameters of the missile guidance system to minimize the miss step. The method of adjoints was developed to simplify the simulation procedure. However, the necessity to simulate the system response for each impulse application time by using the model of the original system or using the adjoint system to simplify this procedure was stipulated by the inability to obtain an analytical expression for the miss step that can be used for analysis and design of missile guidance systems.

The zero- and single-lag guidance systems are convenient analytical models; but they did not quite match reality. The binomial representation $1/(1 + sT/n)^n$, where T is the effective guidance system time constant and n the system order, used for more accurate high-order guidance system models in [5], still does not accurately reflect flight control system dynamics. The binomial units are usually used to approximate delay units [3] and, therefore, they cannot be considered a reliable tool for guidance system design.

Because of an inability to obtain analytical expressions for the miss step for the high-order planar models, the simulation process using the method of adjoints still remains a very useful tool of the time-domain analysis. The analytical difficulties do not allow researchers to build and analyze more complicated models that also include the dynamics of a maneuvering target. The frequency approach to analysis and design of missile guidance systems, described in the next chapter, enables up to overcome the above-mentioned difficulties.

REFERENCES

1. Yanushevsky, R., and Boord, W. New Approach to Guidance Law Design, *Journal of Guidance, Control, and Dynamics* 28, no. 1 (2005): 162–66.
2. Yanushevsky, R. Concerning Lyapunov-Based Guidance, *Journal of Guidance, Control, and Dynamics* 29, no. 2 (2006): 509–11.
3. Yanushevsky, R. Optimal Control of Differential-Difference Systems of Neutral Type, *International Journal of Control*, 1 (1989): 1835–50.
4. Zadeh, L., and Desoer, C. *Linear System Theory*, New York, NY: McGraw Hill, 1964.
5. Zarchan, P. *Tactical and Strategic Missile Guidance*, *Progress in Astronautics and Aeronautics*. Vol. 176. Washington, DC: American Institute of Astronautics and Aeronautics, Inc., 1997.

5 Analysis of Proportional Navigation Guided Systems in the Frequency Domain

5.1 INTRODUCTION

The difficulty in analysis of differential equations with time-varying coefficients describing linearized models of guidance systems does not allow researchers to obtain analytical expressions for miss distance that can be effectively used in practice. As mentioned in [10], "the disadvantage of the single time constant representation of a missile guidance system is that the miss distance can be seriously underestimated." The same remark can also be applied to the binomial representation [6,10].

The method of adjoints used in the previous chapter can be presented in the integral form as the impulse response of the adjoint system (i.e., its reaction to a unit impulse function). In the case of a single time constant guidance system, the transfer function between the target acceleration and the guidance system miss distance can be obtained from equation (4.21). In control theory, the method of transfer functions, as input-output characteristics of linear systems, is a foundation of frequency methods (i.e., analysis of systems in the frequency domain). Analogous to the unit step signal in the time domain, the unit sinusoidal input signal is the standard test signal in the frequency domain. The response of the system to a changing frequency is considered. The frequency response is defined as the steady-state response of the system to a sinusoidal input signal. The frequency approach is very popular among engineers because the design of a system in the frequency domain provides the designer with control of the bandwidth of a system. It is very physical and enables researchers and designers to build realistic models and make justifiable simplifications.

The below analysis of the linearized proportional navigation (PN) guidance system models is based on the frequency response of the linearized guidance system that corresponds to the miss distance due to a weaving target (i.e., due to a sinusoidal acceleration). As shown later, the miss due to a step target maneuver (i.e., the miss step response) can also be obtained from the frequency characteristics of the system under consideration.

The block diagram of an interceptor's main subsystems is given in Figure 5.1. The seeker provides a guidance system with target information, which together with information from onboard sensors is necessary to generate a guidance law. The guidance system generates acceleration commands for the autopilot channels to control the motion of the missile. The warhead subsystem receives a burst-hit command from the guidance system. Performance of the guided missile systems is assessed by their terminal effect. The generation and intelligent control of this "terminal effect" is one of the key requirements to missile systems.

The above-mentioned subsystems are interconnected. The performance of a separate subsystem dictates requirements to the interconnected ones. For example, the missile airframe parameters determine the airframe poles ω_z that significantly influence missile dynamics and, as a result, influence autopilot system τ_1 characteristic requirements. Higher accuracy guidance and autopilot systems can employ smaller warheads. The seeker dynamic parameters τ_2 and τ_3 influence the guidance system accuracy. The traditional approach to designing missile guidance and autopilot systems usually neglects interaction between these systems and treats individual missile subsystems separately. The subsystems are designed separately and then integrated before verifying their performance. The quantification of the impact of missile parameters on the miss distance is the first important step toward integrated design of missile guidance and autopilot systems. The main factors that influence the miss distance in homing missiles are the seeker errors, aero frame characteristics, autopilot lag τ_1, and target maneuvers. An appropriate choice of the estimation system parameters (in Figure 5.1 it is combined with a seeker and presented by τ_2 and τ_3) can reduce requirements to a seeker's accuracy and a guidance law effective navigation ratio N.

The performance of guided UAVs flying according to the predetermined flight pass, which is represented by a sequence of waypoints—dummy

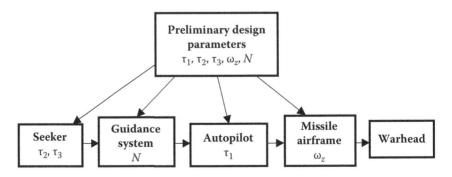

FIGURE 5.1   Block diagram of an interceptor's main subsystems.

targets, can be judged by the accuracy (miss distance) of reaching these waypoints.

The above-mentioned shows the importance of establishing an analytical relationship between the miss distance and main parameters of the guidance system.

Below, the analytical expressions for the miss distance (frequency response) and related expressions for missile system performance are obtained for the proportional navigation law and the guidance control system that reflects the most important characteristics of the flight control system that combines airframe and autopilot dynamics (damping, natural frequency, time constant, and airframe zero frequency). These analytical expressions can also be used to evaluate UAV system performance. They need to employ significantly simpler computational programs than by using the method of adjoints to analyze the influence of the basic guidance system parameters on the miss distance for step and weave maneuvers.

5.2 ADJOINT METHOD. GENERALIZED MODEL

As shown earlier, the method of adjoints is a useful tool to simulate the impulse response of the system (i.e., its response of an impulse function). The analysis in the time domain was bounded by the consideration of a step target acceleration signal. Because the frequency-domain analysis uses a different test signal, it is convenient to consider a more general model of the adjoint system commonly used in modern control theory.

In contrast to equations (4.8), (4.9), and (4.13), here we present the equations of the system in a more general form than in equation (4.8). We will consider the so-called canonical form used in modern control theory:

$$\dot{x}(t) = A(t)x(t) + B(t)u(t) \quad y(t) = C(t)x(t) \tag{5.1}$$

where the state equation similar to equation (4.8) is accompanied by the output equation; $x(t)$ is the state vector, $u(t)$ and $y(t)$ are the input (control) and output, respectively; $A(t)$, $B(t)$, and $C(t)$ are matrices of appropriate dimensions.

The adjoint system of equations has the form:

$$\dot{z}(t) = -A^T(t)z(t) + C^T(t)v(t) \quad w(t) = B^T(t)z(t) \tag{5.2}$$

where $z(t)$, $v(t)$, and $w(t)$ are the state vector, input and output, respectively.

Similar to the earlier indicated relationship between the transition matrices of the original and adjoint systems, the impulse response matrices $P(t, \sigma)$ of equation (5.1) and $P_a(t, \sigma)$ of equation (5.2) are connected by:

$$P_a(t, \sigma) = P^T(\sigma, t) \tag{5.3}$$

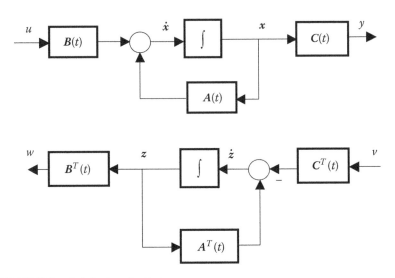

FIGURE 5.2   Original and adjoint systems.

The structure of the original and the adjoint system is shown in Figure 5.2 (note the reversal of flow direction of signals). For the single input-output systems and for $\sigma = t_F$ we have $P(t_F, t) = P_a(t, t_F)$, i.e., $P(t_F, t)$ corresponds to the reaction of the adjoint system to the delta function $\delta(t - t_F)$, i.e., to the δ-function applied at time t_F. In contrast to the physically realizable initial system (5.1) with $P(t_F, t) = 0$ for $t_F < t$ (it is also called *causal* or *nonanticipative*, as the system output does not anticipate future values of the input) the adjoint system (5.2) with $P_a(t, t_F) = 0$ for $t > t_F$ is physically unrealizable (it is also called *pure anticipative* [9]). To operate with the physically realizable adjoint system we will consider the dynamics of (5.2) with respect to time $\tau = t_F - t$. The modified adjoint system has the impulse response $P_{ma}(t_F - \tau, 0)$, $0 \le \tau \le t_F$, and is described by the following equation:

$$\dot{z}(\tau) = A^T(t_F - \tau)z(\tau) + C^T(t_F - \tau)v(\tau) \quad w(\tau) = B^T(t_F - \tau)z(\tau) \quad (5.4)$$

The structure of the modified adjoint system is shown in Figure 5.3. As it follows from the comparison of Figures 5.1 and 5.3, to build the modified adjoint system we should:

i. Replace t by $t_F - \tau$ in all arguments of all time-varying coefficients
ii. Reverse all signal flow, redefining nodes as summing junctions and vice versa

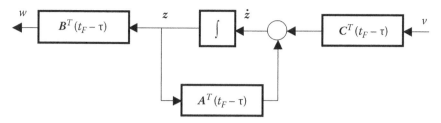

FIGURE 5.3   Modified adjoint system.

Instead of equation (5.3) we can write:

$$P_{ma}^{T}(t_F - \tau, 0) = P(t_F, \tau) \tag{5.5}$$

i.e., the impulse response $P(t_F, t)$ can be obtained by applying the delta function $\delta(\tau)$ in the modified adjoint system.

The state, input, and output matrices of equation (4.7) presented in the form (5.1) are:

$$A = \begin{bmatrix} 0 & 1 & 0 \\ 0 & 0 & -1 \\ \dfrac{N}{\tau_1 \tau^2} & \dfrac{N}{\tau_1 \tau} & \dfrac{-1}{\tau_1} \end{bmatrix} \quad B = \begin{bmatrix} 0 \\ 1 \\ 0 \end{bmatrix} \quad C = \begin{bmatrix} 1 & 0 & 0 \end{bmatrix} \tag{5.6}$$

As it follows from equation (5.3), the output of the adjoint system is $w(\tau) = z_2 = P(t_F, \tau)$, where $P(t_F, t)$ is the impulse response to $a_T(t)$ at time t_F. Specifics of the linearized models of missile guidance systems that contain a time-varying coefficient depending on $t_F - t$ [see, e.g., equation (4.8)] enable us to use the method of adjoints not only as a simulation tool but also to use it to obtain analytical expressions for the impulse response. As seen from equations (5.1) and (5.3), for a class of linear time-varying systems with the state matrix $A(t) = A(t_F - t)$ the state matrix of the modified adjoint system $A(t_F - \tau) = A(\tau)$, i.e., it depends on adjoint time τ, rather than directly on t_F. In this case, the impulse response of the modified adjoint system does not depend directly on t_F, and the adjoint time $0 \le \tau \le t_F$ can be interpreted as time of flight t_F. For this class of system we will denote the impulse response as $P(t_F, t)$.

The block diagram of the system with the state, input, and output matrices (5.6) and $W(s) = (\tau_1 s + 1)^{-1}$ shown in Figure 5.4 is similar to Figure 4.3. The input-output relationships in the frequency domain are characterized by the transfer functions of the corresponding units, which will be

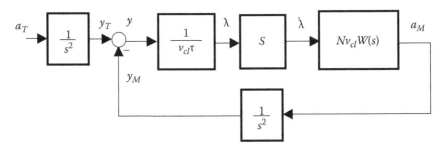

FIGURE 5.4 Modified block diagram of original system.

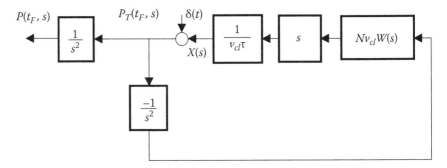

FIGURE 5.5 Modified block diagram of adjoint system.

analyzed in details later. Based on the above mentioned, the relationship between the target trajectory $y_T(t)$, the target acceleration $a_T(t)$, and the miss distance $y(t_F)$ can be obtained from the analysis of the block diagram of the modified adjoint system in Figure 5.5, which for $W(s) = (\tau_1 s + 1)^{-1}$ is similar to Figure 4.4.

As mentioned earlier, a simple first-order transfer function representation with time constant τ_1 does not describe accurately the relationship between the line-of-sight (LOS) rate and the missile acceleration. The transfer function $W(s)$ should reflect the dynamic responses of the airframe, autopilot, guidance filters, and seeker. However, the more complicated $W(s)$ does not change the structure of Figure 5.5. Analyzing this structure we will obtain the analytical expression for the transfer function $P(t_F, s)$ corresponding to the impulse response $P(t_F, t)$, and the transfer function $P_T(t_F, s)$ corresponding to the impulse response $P_T(t_F, t)$ to $y_T(t)$.

Let $X(\tau)$ be the impulse response of the closed loop of the structure in Figure 5.5. Then taking into account equation (4.18), the closed-loop dynamics can be presented as:

$$\frac{1}{\tau}\int_0^\tau NH(\tau-\sigma)(\delta(\sigma)-X(\sigma))d\sigma = X(\tau) \tag{5.7}$$

or using the Laplace transform:

$$\int_s^\infty NH(q)(1-X(q))dq = X(s) \tag{5.8}$$

Differentiating equation (5.8) we obtain:

$$-\frac{d(1-X(s))}{ds} = NH(s)(1-X(s)) \tag{5.9}$$

which is similar to equation (4.16). Taking into account that $P_T(t_F, s) = 1 - X(s)$, we can write similar to equation (4.19):

$$P_T(t_F,s) = \exp\left(\int_\infty^s NH(\sigma)d\sigma\right) \tag{5.10}$$

and, correspondingly:

$$P(t_F,s) = \frac{1}{s^2}\exp\left(\int_\infty^s NH(\sigma)d\sigma\right) = \frac{1}{s^2}P_T(t_F,s) \tag{5.11}$$

where the lower infinite limit in equations (5.10) and (5.11) follows from the condition $\lim_{s\to\infty} P_T(t_F,s) = 0$.

The step miss equals the integral of the impulse response, i.e., in the frequency domain it corresponds to $s^{-1} P(t_F, s)$. It is easy to check that for the first-order system it coincides with the expression for $Z_4(s)$ obtained from the previous chapter [see equation (4.20)]. However, the above-described approach is not limited to determining the miss step only, the miss distance due to a step target maneuver. It will be used to determine the miss distance for a wide class of target maneuvers.

5.3 FREQUENCY DOMAIN ANALYSIS

First we consider the fourth-order flight control system, which is widely used at the initial stage of analysis and design. Then the obtained expressions will be generalized for the n-th order system.

A block diagram of the guidance system under consideration is given in Figure 5.6. Here, missile acceleration a_M is subtracted from target acceleration a_T, and the result is integrated to obtain relative separation between a missile and target y, which at the end of flight t_F is the miss distance $y(t_F)$. Division by range (closing velocity v_{cl} multiplied by time-to-go t_{go} until intercept) yields the geometric line-of-sight (LOS) angle λ, where the time-to-go is defined as $t_{go} = t_F - t$. The missile seeker is presented formally as a perfect differentiator that effectively provides a measurement

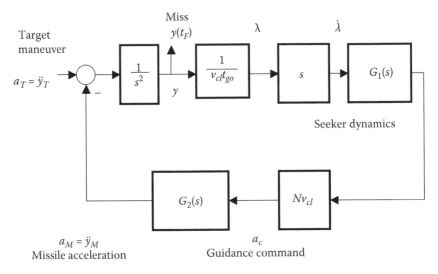

FIGURE 5.6   Missile guidance model.

of the rotation rate of LOS from the interceptor to the target. The filter and seeker dynamics are represented by a transfer function

$$G_1(s) = \frac{\tau_z s + 1}{\tau_2 s + 1}$$

where τ_z and τ_2 are constant coefficients. An estimation of the LOS rate generates a guidance command a_c based on the proportional navigation law with the effective navigation ratio $N > 2$.

The flight control system guides the missile to follow this acceleration command.

The flight control system dynamics, which combine its airframe and autopilot dynamics, are represented by the following transfer function:

$$G_2(s) = \frac{a(s)}{(\tau_1 s + 1)\left(\dfrac{s^2}{\omega_M^2} + \dfrac{2\zeta}{\omega_M} s + 1\right)} \tag{5.12}$$

where

$$a(s) = 1 - \frac{s^2}{\omega_z^2}$$

for tail-controlled missiles and $a(s)$ is a first-order polynomial for nontail-controlled missiles; the flight control system damping ζ, natural frequency ω_M, time constant τ_1, and airframe zero ω_z are the flight control system parameters.

According to equation (5.10), in the complex domain, the miss distance at time t_F can be presented as:

$$Y(t_F, s) = \exp\left(N \int_{\infty}^{s} H(\sigma) d\sigma \right) Y_T(s) \tag{5.13}$$

where $Y_T(s)$ is the Laplace transform of a target vertical position $y_T(t)$, $Y(t_F, s)$ is the Laplace transform of $y(t_F)$ and similar to equation (4.18):

$$H(s) = \frac{W(s)}{s} \tag{5.14}$$

and

$$W(s) = G_1(s) * G_2(s) = \frac{1 + r_1 s + r_2 s^2 + r_3 s^3}{(1 + \tau_1 s)(1 + \tau_2 s)\left(1 + \dfrac{2\zeta}{\omega_M} s + \dfrac{s^2}{\omega_M^2} \right)} \tag{5.15}$$

where r_k ($k = 1,...,3$) are constant coefficients.

The integral $\int_{\infty}^{i\omega} H(\sigma) d\sigma$ can be calculated by writing $H(s)$ in the form:

$$H(s) = \frac{A}{s} + \frac{B_1/\tau_1}{s + 1/\tau_1} + \frac{B_2/\tau_2}{s + 1/\tau_2} + \frac{Cs + D}{\dfrac{s^2}{\omega_M^2} + \dfrac{2\zeta}{\omega_M} s + 1} \tag{5.16}$$

where the coefficients A, B_1, B_2, C, and D can be calculated as:

$$A = 1$$

$$B_1 = \frac{\tau_1^2 - r_1 \tau_1 + r_2 - \dfrac{r_3}{\tau_1}}{\left(1 - \dfrac{\tau_2}{\tau_1} \right)\left(\dfrac{2\zeta}{\omega_M} - \tau_1 - \dfrac{1}{\tau_1 \omega_M^2} \right)}$$

$$B_2 = \frac{\tau_2^2 - r_1 \tau_2 + r_2 - \dfrac{r_3}{\tau_2}}{\left(1 - \dfrac{\tau_1}{\tau_2} \right)\left(\dfrac{2\zeta}{\omega_M} - \tau_2 - \dfrac{1}{\tau_2 \omega_M^2} \right)} \tag{5.17}$$

$$C = -\frac{1}{\omega_M^2} - \frac{B_1}{\tau_1 \omega_M^2} - \frac{B_2}{\tau_2 \omega_M^2}$$

$$D = r_1 - B_1 - B_2 - (\tau_1 + \tau_2) - \frac{2\zeta}{\omega_M}$$

For $\tau_2 = 0$:

$$B_2 = 0 \quad \text{and} \quad \lim_{\tau_2 \to 0} \frac{B_2}{\tau_2 \omega_M^2} = -\frac{r_3}{\tau_1} \tag{5.18}$$

If also $\tau_1 = 0$ and $r_3 = 0$, then:

$$B_1 = 0 \quad \text{and} \quad \lim_{\tau_1 \to 0} \frac{B_1}{\tau_1 \omega_M^2} = -r_2 \tag{5.19}$$

To obtain the transfer function $P(t_f, s)$ of the guidance system with respect to a target acceleration, the components (5.16) of the integral $\int_\infty^{i\omega} H(\sigma)d\sigma$ should be calculated.

The upper limit of integration of the first three terms of equation (5.16) gives, respectively:

$$\ln s, \quad \frac{B_1}{\tau_1} \ln(s + 1/\tau_1) \quad \text{and} \quad \frac{B_2}{\tau_2} \ln(s + 1/\tau_2)$$

The integral of the last term of equation (5.16) can be presented as:

$$
\int_\infty^s \frac{Cs + D}{\dfrac{s^2}{\omega_M^2} + \dfrac{2\zeta}{\omega_M} s + 1} \, ds
$$

$$
= \int_\infty^s \frac{C\omega_M^2 s + D\omega_M^2}{s^2 + 2\omega_M \zeta s + \omega_M^2} \, ds
$$

$$
= \frac{C\omega_M^2}{2} \ln(s^2 + 2\omega_M \zeta s + \omega_M^2) - \int_\infty^s \frac{\omega_M^2(\zeta\omega_M C - D)}{s^2 + 2\omega_M \zeta s + \omega_M^2} \, ds
$$

$$
= \frac{C\omega_M^2}{2} \ln(s^2 + 2\omega_M \zeta s + \omega_M^2)
$$

$$\tag{5.20}$$

$$
+ \omega_M^2 (D - \zeta\omega_M C) \frac{1}{\omega_M \sqrt{1-\zeta^2}} Arc\tan \frac{s + \zeta\omega_M}{\omega_M \sqrt{1-\zeta^2}}
$$

$$
= \frac{C\omega_M^2}{2} \ln(s^2 + 2\omega_M \zeta s + \omega_M^2)
$$

$$
+ \omega_M^2 (D - \zeta\omega_M C) \frac{1}{\omega_M \sqrt{1-\zeta^2}} \frac{1}{2i} \ln \frac{i\omega_M \sqrt{1-\zeta^2} - (s + \zeta\omega_M)}{i\omega_M \sqrt{1-\zeta^2} + (s + \zeta\omega_M)}
$$

Based on equations (5.13), (5.15), and (5.20), the upper limit of equation (5.13) for

$$Y_T(s) = \frac{1}{s^2} a_T(s) \quad \text{and} \quad a_T(s) = g$$

can be presented as:

$$P(t_F, s) = g s^{N-2} \prod_{k=1}^{2} \left(s + \frac{1}{\tau_k} \right)^{B_k N / \tau_k}$$

$$(s^2 + 2\omega_M \zeta s + \omega_M^2)^{C N \omega_M^2 / 2} \left(\frac{-s - \zeta \omega_M + i\omega_M \sqrt{1-\zeta^2}}{s + \zeta \omega_M + i\omega_M \sqrt{1-\zeta^2}} \right)^{\frac{N\omega_M (D - \zeta \omega_M C)}{2i\sqrt{1-\zeta^2}}} \quad (5.21)$$

The lower infinite limit of integration of equation (5.13) for the transfer function (5.15) with the degree of its numerator less than the degree of its denominator equals zero (it will be explained later in details (see also [5]), and the above equation represents the transfer function characterizing the relationship between the miss distance and target acceleration.

The frequency response of the guidance system follows from equation (5.21) when $s = i\omega$.

For $s = i\omega$, the last factor of equation (5.21) can be written as:

$$-i \frac{\omega_M (D - \zeta \omega_M C)}{2\sqrt{1-\zeta^2}} \ln \frac{i(-\omega + \omega_M \sqrt{1-\zeta^2}) - \zeta \omega_M}{i(\omega + \omega_M \sqrt{1-\zeta^2}) + \zeta \omega_M} = \text{Re}(.) + i \, \text{Im}(.) \quad (5.22)$$

where

$$\text{Re}(.) = \frac{\omega_M (D - \zeta \omega_M C)}{2\sqrt{1-\zeta^2}} \left(\tan^{-1} \frac{\omega - \omega_M \sqrt{1-\zeta^2}}{\zeta \omega_M} - \tan^{-1} \frac{\omega + \omega_M \sqrt{1-\zeta^2}}{\zeta \omega_M} \right)$$

$$(5.23)$$

and

$$\text{Im}(.) = -\frac{\omega_M (D - \zeta \omega_M C)}{4\sqrt{1-\zeta^2}} \ln \left(\frac{\omega_M^2 + \omega^2 - 2\omega\omega_M \sqrt{1-\zeta^2}}{\omega_M^2 + \omega^2 + 2\omega\omega_M \sqrt{1-\zeta^2}} \right) \quad (5.24)$$

Here the symbol "Acrtan" is used to denote the inverse tangent function of the complex variable and the symbol "tan^{-1}" denotes the inverse tangent function of the real variable that characterizes the argument of the complex variable of the logarithmic function.

Based on equations (5.22)–(5.24) we can present the last factor of equation (5.21) for $s = i\omega$ in the following form (it follows directly from equation (5.21) based on the definition of complex exponents [2]):

$$\left(\frac{-i\omega - \zeta\omega_M + i\omega_M\sqrt{1-\zeta^2}}{i\omega + \zeta\omega_M + i\omega_M\sqrt{1-\zeta^2}}\right)^{\frac{N\omega_M(D-\zeta\omega_M C)}{2i\sqrt{1-\zeta^2}}} = \exp(N\,\mathrm{Re}(.))\exp(iN\,\mathrm{Im}(.)) \tag{5.25}$$

The amplitude and frequency characteristics of the guidance system follow immediately from equations (5.21)–(5.25).

The amplitude characteristic $|P(t_F, i\omega)|$ has the following form:

$$|P(t_F, i\omega)|$$
$$= g\omega^{N-2}\prod_{k=1}^{2}(\omega^2 + 1/\tau_k^2)^{B_k N/2\tau_k}((\omega_M^2 - \omega^2)^2 + 4\omega_M^2\omega^2\zeta^2)^{CN\omega_M^2/4}\exp(.) \tag{5.26}$$

where

$$\exp(.)$$
$$= \exp\left(N\frac{\omega_M(D-\zeta\omega_M C)}{2\sqrt{1-\zeta^2}}\left(\tan^{-1}\frac{\omega - \omega_M\sqrt{1-\zeta^2}}{\zeta\omega_M} - \tan^{-1}\frac{\omega + \omega_M\sqrt{1-\zeta^2}}{\zeta\omega_M}\right)\right) \tag{5.27}$$

The phase characteristic $\varphi(t_F, i\omega)$ has the following form:

$$\varphi(t_F, i\omega) = -\pi + N\frac{\pi}{2} + N\frac{B_1}{\tau_1}\tan^{-1}(\omega\tau_1) + N\frac{B_2}{\tau_2}\tan^{-1}(\omega\tau_2)$$

$$+ N\frac{C}{2}\omega_M^2\tan^{-1}\left(\frac{2\omega\omega_M\zeta}{\omega_M^2 - \omega^2}\right) \tag{5.28}$$

$$- \frac{\omega_M(D-\zeta\omega_M C)}{4\sqrt{1-\zeta^2}}\ln\left(\frac{\omega_M^2 + \omega^2 - 2\omega\omega_M\sqrt{1-\zeta^2}}{\omega_M^2 + \omega^2 + 2\omega\omega_M\sqrt{1-\zeta^2}}\right)$$

The first factor of equation (5.25) corresponds to exp(.) in equation (5.27), and the second factor corresponds to the last term of the phase characteristic.

The above expressions were obtained for the fourth-order model. Below we generalize them assuming that the flight control system has an arbitrary n-th order. Instead of equation (5.15) we have:

$$W(s) = G_1(s) * G_2(s) = \frac{1 + \sum_{k=1}^{n-1} r_k s^k}{\prod_{q=1}^{l}(1 + \tau_q s) \prod_{j=1}^{m}\left(1 + \frac{2\varsigma_j s}{\omega_j} + \frac{s^2}{\omega_j^2}\right)} \quad (5.29)$$

where $l + 2m = n$; $l = 2$ and $m = 1$ correspond to the denominator of equation (5.15); r_k ($k = 1,...,n-1$) are constant coefficients.

The partial-fraction expansion of $H(s)$ has the form:

$$H(s) = \frac{A}{s} + \sum_{q=1}^{l} \frac{B_q/\tau_q}{s + 1/\tau_q} + \sum_{j=1}^{m} \frac{C_j s + D_j}{\frac{s^2}{\omega_j^2} + \frac{2\varsigma_j s}{\omega_j} + 1} = \frac{A}{s} + \sum_{p=1}^{n} \frac{K_p}{s - \alpha_p} \quad (5.30)$$

where α_p ($p = 1,..,n$) are the poles of $W(s)$—for simplicity they are assumed to be distinct—and the coefficients A, B_q, C_j, D_j, and K_p can be calculated as:

$$K_p = \lim_{s \to \alpha_p}(s - \alpha_p)H(s), \quad A = \lim_{s \to 0} sH(s) = W(0) = 1 \quad (5.31)$$

for the real poles $\alpha_p = -1/\tau_p$:

$$B_q = K_q \tau_q \quad (5.32)$$

for the pair of complex-conjugated poles $\alpha_{p,p+1} = -\varsigma_p \omega_p \pm i\omega_p \sqrt{1 - \varsigma_p^2}$ the coefficients K_p and K_{p+1} are also complex conjugated, so that from equation (5.30) and

$$\frac{\operatorname{Re}K_p + i\operatorname{Im}K_p}{s + \varsigma_p \omega_p + i\omega_p \sqrt{1 - \varsigma_p^2}} + \frac{\operatorname{Re}K_p - i\operatorname{Im}K_p}{s + \varsigma_p \omega_p - i\omega_p \sqrt{1 - \varsigma_p^2}}$$

$$= 2\frac{\operatorname{Re}K_p s + \varsigma_p \omega_p \operatorname{Re}K_p + \omega_p \sqrt{1 - \varsigma_p^2}\operatorname{Im}K_p}{s^2 + 2\varsigma_p \omega_p s + \omega_p^2}$$

we have:

$$C_j = \frac{2\operatorname{Re}K_p}{\omega_p^2}, \quad D_j = \frac{2(\zeta_p \operatorname{Re}K_p + \sqrt{1-\zeta_p^2}\,\operatorname{Im}K_p)}{\omega_p}, \quad p = 2j-1 \quad (5.33)$$

The transfer function $P(t_F, s)$ for the n-dimensional flight control system is:

$$P(t_F, s) = gs^{N-2} \prod_{k=1}^{l} \left(s + \frac{1}{\tau_k}\right)^{\frac{B_k N}{\tau_k}}$$

$$\cdot \prod_{j=1}^{m} (s^2 + 2\omega_j\zeta_j s + \omega_j^2)^{\frac{C_j N\omega_j^2}{2}} \left(\frac{-s - \zeta_j\omega_j + i\omega_j\sqrt{1-\zeta_j^2}}{s + \zeta_j\omega_j + i\omega_j\sqrt{1-\zeta_j^2}}\right)^{\frac{N\omega_j(D_j - \zeta_j\omega_j C_j)}{2i\sqrt{1-\zeta_j^2}}}$$

$$(5.34)$$

The frequency response of the guidance system follows from equation (5.34) when $s = i\omega$. The amplitude characteristic $|P(t_F, i\omega)|$ has the following form:

$$|P(t_F, i\omega)| = g\omega^{N-2} \prod_{k=1}^{l} (\omega^2 + 1/\tau_k^2)^{\frac{B_k N}{2\tau_k}}$$

$$(5.35)$$

$$\cdot \prod_{j=1}^{m} ((\omega_j^2 - \omega^2)^2 + 4\omega_j^2\omega^2\zeta_j^2)^{C_j N\omega_j^2/4} \exp(.)$$

where

$$\exp(.) = \exp\left(N \sum_{j=1}^{m} \frac{\omega_j(D_j - \zeta_j\omega_j C_j)}{2\sqrt{1-\zeta_j^2}} \left(\tan^{-1}\frac{\omega - \omega_j\sqrt{1-\zeta_j^2}}{\zeta_j\omega_j}\right.\right.$$

$$(5.36)$$

$$\left.\left. - \tan^{-1}\frac{\omega + \omega_j\sqrt{1-\zeta_j^2}}{\zeta_j\omega_j}\right)\right)$$

The phase characteristic $\varphi(t_F, i\omega)$ has the following form:

$$\varphi(t_F, i\omega) = -\pi + N\frac{\pi}{2} + N\sum_{k=1}^{l}\frac{B_k}{\tau_k}\tan^{-1}(\omega\tau_k)$$

$$+ N\sum_{j=1}^{m}\frac{C_j}{2}\omega_j^2\tan^{-1}\left(\frac{2\omega\omega_j\zeta_j}{\omega_j^2 - \omega^2}\right) \tag{5.37}$$

$$- \frac{\omega_j(D_j - \zeta_j\omega_jC_j)}{4\sqrt{1-\zeta_j^2}}\ln\left(\frac{\omega_j^2 + \omega^2 - 2\omega\omega_j\sqrt{1-\zeta_j^2}}{\omega_j^2 + \omega^2 + 2\omega\omega_j\sqrt{1-\zeta_j^2}}\right)$$

The equations (5.35)–(5.37) follow immediately from the below expressions that present the generalization of equations (5.22)–(5.27):

$$\sum_{j=1}^{m} -i\frac{\omega_j(D_j - \zeta_j\omega_jC_j)}{2\sqrt{1-\zeta_j^2}}\ln\frac{i(-\omega + \omega_j\sqrt{1-\zeta_j^2}) - \zeta_j\omega_j}{i(\omega + \omega_j\sqrt{1-\zeta_j^2}) + \zeta_j\omega_j} = \text{Re}(.) + i\,\text{Im}(.) \tag{5.38}$$

where:

$$\text{Re}(.) = \sum_{j=1}^{m}\frac{\omega_j(D_j - \zeta_j\omega_jC_j)}{2\sqrt{1-\zeta_j^2}}$$

$$\left(\tan^{-1}\frac{\omega - \omega_j\sqrt{1-\zeta_j^2}}{\zeta_j\omega_j} - \tan^{-1}\frac{\omega + \omega_j\sqrt{1-\zeta_j^2}}{\zeta_j\omega_j}\right) \tag{5.39}$$

$$\text{Im}(.) = \sum_{j=1}^{m} -\frac{\omega_j(D_j - \zeta_j\omega_jC_j)}{4\sqrt{1-\zeta_j^2}}\ln\left(\frac{\omega_j^2 + \omega^2 - 2\omega\omega_j\sqrt{1-\zeta_j^2}}{\omega_j^2 + \omega^2 + 2\omega\omega_j\sqrt{1-\zeta_j^2}}\right) \tag{5.40}$$

and

$$\prod_{j=1}^{m}\left(\frac{-i\omega - \zeta_j\omega_j + i\omega_j\sqrt{1-\zeta_j^2}}{i\omega + \zeta_j\omega_j + i\omega_j\sqrt{1-\zeta_j^2}}\right)^{\frac{N\omega_M(D-\zeta\omega_MC)}{2i\sqrt{1-\zeta^2}}} = \exp(N\,\text{Re}(.))\exp(iN\,\text{Im}(.)) \tag{5.41}$$

Based on equation (5.21), analogous to equations (5.35) and (5.37), for the amplitude $|P_T(t_F, i\omega)|$ and phase $\varphi_T(t_F, i\omega)$ characteristics of $P_T(t_F, i\omega)$ we have:

$$|P_T(t_F, i\omega)| = \omega^N \prod_{k=1}^{l} (\omega^2 + 1/\tau_k^2)^{\frac{B_k N}{2\tau_k}}$$

$$\cdot \prod_{j=1}^{m} ((\omega_j^2 - \omega^2)^2 + 4\omega_j^2 \omega^2 \zeta_j^2)^{C_j N \omega_j^2 / 4} \exp(.) \tag{5.42}$$

and

$$\varphi_T(t_F, i\omega) = N\frac{\pi}{2} + N\sum_{k=1}^{l} \frac{B_k}{\tau_k} \tan^{-1}(\omega\tau_k) + N\sum_{j=1}^{m} \frac{C_j}{2} \omega_j^2 \tan^{-1}\left(\frac{2\omega\omega_j \zeta_j}{\omega_j^2 - \omega^2}\right)$$

$$- \frac{\omega_j(D_j - \zeta_j \omega_j C_j)}{4\sqrt{1 - \zeta_j^2}} \ln\left(\frac{\omega_j^2 + \omega^2 - 2\omega\omega_j \sqrt{1 - \zeta_j^2}}{\omega_j^2 + \omega^2 + 2\omega\omega_j \sqrt{1 - \zeta_j^2}}\right) \tag{5.43}$$

The real and imaginary parts of $P(t_F, i\omega)$ and $P_T(t_F, i\omega)$ are:

$$\text{Re}[P(t_F, i\omega)] = |P(t_F, i\omega)| \cos(\varphi(t_F, i\omega))$$
$$\text{Re}[P_T(t_F, i\omega)] = |P_T(t_F, i\omega)| \cos(\varphi_T(t_F, i\omega)) \tag{5.44}$$

and

$$\text{Im}[P(t_F, i\omega)] = |P(t_F, i\omega)| \sin(\varphi(t_F, i\omega))$$
$$\text{Im}[P_T(t_F, i\omega)] = |P_T(t_F, i\omega)| \sin(\varphi_T(t_F, i\omega)) \tag{5.45}$$

The obtained analytical expressions for the missile guidance system transfer function and its frequency characteristics enable us to analyze the missile system performance without resorting to simulation utilizing adjoint models in the time domain.

5.4   STEADY-STATE MISS ANALYSIS

The frequency approach also enables us to analyze the steady-state miss for various types of maneuvers analogous to the analysis of the steady-state mode in control theory [3]. As known, the steady-state solution may be a good approximation for sufficiently large values of time.

For the step maneuver the steady-state miss $Miss_s$ is determined as:

$$Miss_s = P(t_F, s)|_{s=0} \qquad (5.46)$$

It follows from equations (5.21) and (5.34) that $Miss_s = 0$ if $N \geq 2$.
For the ramp maneuver:

$$Miss_s = \frac{d}{ds}(P(t_F, s))|_{s=0} \qquad (5.47)$$

It follows from equations (5.21) and (5.34) that $Miss_s = 0$ if $N \geq 3$.
For the parabolic maneuver:

$$Miss_s = \frac{d^2}{ds^2}(P(t_F, s))|_{s=0} \qquad (5.48)$$

It follows from (5.21) and (5.34) that $Miss_s = 0$ if $N \geq 4$.

For weave maneuvers, the miss steady-state response is determined directly from the frequency response (5.26)–(5.28), (5.35)–(5.37), (5.44), and (5.45).

5.5 WEAVE MANEUVER ANALYSIS

Maneuvers present the best strategy for missiles to achieve their goals. Evasive maneuvers are one of the most effective defense penetration features used on offensive missiles. The evasive maneuver causes the interceptor to expend additional energy, so that it becomes unable to reach the necessary point of engagement. As a result, interceptor miss distances are inevitable and subsequently intolerable, especially for hit-to-kill missiles. If designed properly, the maneuver can render the entire defense system useless. As indicated in [7,8], sinusoidal or weave maneuvers of a target can make it particularly difficult for a pursuing missile to engage the threat. Targets with very low weaving frequency appear as targets with "near-constant" maneuvers and in many cases will cause no problems for a proportional navigation guidance system. Targets with very high weaving frequencies also cause minimal problems for a missile guidance system because there is very little resultant target displacement as a result of the maneuver. The miss distance increases between these target weaving frequency extremes. The existence of the optimal maneuvering frequency, that is, the frequency that maximizes the steady-state miss distance amplitude, was established in [5].

The obtained above closed-form solution for the miss distance as a function of the effective navigation ratio, guidance system time constant, natural frequency, and damping ratio makes it possible to use the frequency analysis in practice. The established existence of the optimal evasive weave frequency and the procedure for determining it lead to the optimization approach for the design of attacking maneuvering missiles, as well as to the evaluation of the worst-case scenario when developing defensive missiles to defeat maneuvering targets.

First, we consider the simplest dynamic model of the missile guidance system with $W(s) = (\tau_1 s + 1)^{-1}$. As it follows from equations (5.26) and (5.28), the amplitude and phase characteristics $|P(t_F, i\omega)|$ are:

$$|P(t_F, i\omega)| = g\omega^{N-2}(\omega^2 + 1/\tau_1^2)^{B_1 N/2\tau_1} \tag{5.49}$$

and

$$\varphi(t_F, i\omega) = -\pi + N\frac{\pi}{2} + N\frac{B_1}{\tau_1}\tan^{-1}(\omega\tau_1) \tag{5.50}$$

where $B_1 = -\tau_1$.

The steady-state miss distance due to a weaving target with a frequency ω can be presented in the time domain as:

$$y(t_F) = |P(t_F, i\omega)|\sin(\omega t_F + \varphi(t_F, i\omega)) \tag{5.51}$$

For example, for $N = 3$ we have:

$$y(t_F) = \frac{g\omega}{(\omega^2 + 1/\tau_1^2)^{1.5}}\sin(\omega t_F + \pi/2 - 3\tan^{-1}(\omega\tau_1)) \tag{5.52}$$

For $\tau_1 = 0.5$ s the peak miss as a function of a target frequency is shown in Figure 5.7.

The maximum magnitude of the steady-state miss distance 0.93 m corresponds to a target frequency 1.5 rad/s (i.e., the weaving period is 4.2 s).

5.6   EXAMPLE

To illustrate the effectiveness of the approach described above we consider a realistic example of a tail-controlled aerodynamic missile operating at high altitude [5,12].

The flight control dynamics are assumed to be presented by a third-order transfer function with damping $\zeta = 0.7$ and natural frequency $\omega_M = 20$ rad/s; the flight control system time constant $\tau_1 = \tau = 0.5$ s; and the right-half plane zero $\omega_z = 5$ rad/s corresponds to high altitudes of missile flight. The filter and seeker dynamics are neglected ($G_1(s) = 1$) and

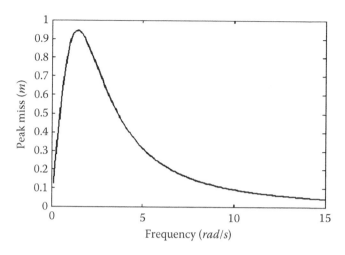

FIGURE 5.7　Peak miss distance for 1-*g* target maneuver amplitude.

perfect estimation of the LOS rate is assumed to generate a guidance command a_c based on the proportional navigation law, so that in equation (5.15) $\tau_2 = 0$, $r_1 = 0$, $r_3 = 0$, and $r_2 = -1/\omega_z^2$.

The flight control system of a tail-controlled endoatmospheric missile with the indicated parameters was considered in [12]. It was shown that at high altitude the performance of a tail-controlled aerodynamic missile can deteriorate because of the existence of low frequency right half-plane zeroes ω_z.

Table 5.1 shows the influence of the flight control system parameters on the miss amplitude and the optimal weaving frequency ω_{opt}; deviations were considered with respect to the values used in Figure 5.7. As it is seen from Table 5.1, the peak miss decreases drastically when $\omega_z \geq 10$ *rad/s*, but right half-plane zeroes do not significantly influence the optimal weaving frequency. The increase of the amplitude miss for smaller values of time constant and larger values of damping and natural frequency is stipulated by the unsatisfactory dynamic properties of the flight control system (peak overshoot, settling time, etc.).

The frequency analysis enables us to evaluate the miss without simulation of the guidance system. Moreover, the miss for weave maneuvers, which are more realistic than step maneuvers, can be analyzed directly from the analytical expressions for the frequency response given above. They allow us to examine the influence of the guidance system parameters on the missile system performance.

The model of the flight control system (4.1) is more precise than the binomial model considered in [10,11]. The results obtained based on this model are more reliable.

TABLE 5.1

Influence of Flight Control System Parameters on Optimal Weaving Frequency and Peak Miss Distance

Case Number	ω_z rad/s	τ s	ζ	ω_M rad/s	Peak miss m	ω_{opt} rad/s
1	5	0.5	0.7	20	234	1.4
2	10	0.5	0.7	20	7.9	1.3
3	20	0.5	0.7	20	4.4	1.3
4	100	0.5	0.7	20	2.8	1.3
5	5	0.2	0.7	20	22100	5.5
6	5	0.6	0.7	20	151	1.2
7	5	0.7	0.7	20	115	1.0
8	5	0.5	0.6	20	265	1.4
9	5	0.5	0.8	20	208	1.4
10	5	0.5	0.7	10	34.8	1.3
11	5	0.5	0.7	30	2325	1.5

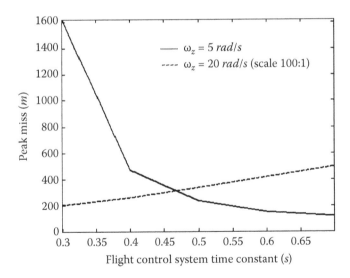

FIGURE 5.8 Relationship between the maximum miss and the flight control system time constant.

For example, the larger effective guidance time constant in binomial models gives the larger miss [10,11]. However, for the given example, because of the significant influence of the right-half plane zero $\omega_z = 5 \ rad/s$, the decrease of the flight control system time constant τ increases the miss. The solid line in Figure 5.8 presents the relationship between the maximum miss and τ obtained from equation (5.26) for $\omega_z = 5$ *rad/s* and demonstrates tail-controlled missiles performance problems at very high altitudes. For lower altitudes (i.e., for higher values of ω_z), the maximum miss decreases with the decrease of τ (see dashed line in Figure 5.8 for $\omega_z = 20$ *rad/s*; for this case the maximum miss scale is 100:1).

5.7 FREQUENCY ANALYSIS AND MISS STEP RESPONSE

There exists a relationship between the frequency response and the step response [3] that enables us to use frequency analysis to build the miss step response based on the frequency response of the missile guidance system.

The relationship between the transfer function and the impulse response of the missile guidance system is described by:

$$P(t_F, s) = \int_0^\infty P(t_F, t) e^{-st} dt \tag{5.53}$$

Assuming $s = i\omega$, we obtain the expressions that relate the frequency and impulse response:

$$P(t_F, i\omega) = \int_0^\infty P(t_F, t) e^{-i\omega t} dt \tag{5.54}$$

and

$$P(t_F, t) = \frac{1}{2\pi} \int_{-\infty}^\infty P(t_F, i\omega) e^{\omega t} d\omega \tag{5.55}$$

which are valid only for a stable $P(t_F, s)$; otherwise the integral of the right part of (5.54) would diverge.

By presenting

$$P(t_F, i\omega) = \mathrm{Re}[P(t_F, i\omega)] + i \, \mathrm{Im}[P(t_F, i\omega)] \tag{5.56}$$

and taking into account that

$$e^{i\omega t} = \cos \omega t + i \sin \omega t$$

the expression for the impulse response can be written as:

$$P(t_F, t) = \frac{1}{2\pi} \int_{-\infty}^\infty (\mathrm{Re}[P(t_F, i\omega)] \cos \omega t - \mathrm{Im}[P(t_F, i\omega)] \sin \omega t) d\omega$$

$$+ \frac{1}{2\pi} \int_{-\infty}^\infty (\mathrm{Re}[P(t_F, i\omega)] \sin \omega t + \mathrm{Im}[P(t_F, i\omega)] \cos \omega t) d\omega \tag{5.57}$$

The integrand of the second integral is an odd function of frequency ω, so that this integral equals zero. The integrand of the first integral is an even function of frequency ω, so that this integral can be changed by the double value of the integral with limits 0 and ∞, i.e.,

$$P(t_F,t) = \frac{1}{\pi} \int_0^\infty (\text{Re}[P(t_F,i\omega)]\cos\omega t - \text{Im}[P(t_F,i\omega)]\sin\omega t)d\omega \quad (5.58)$$

Taking into account the condition of physical realization:

$$P(t_F,t) \equiv 0 \quad \text{for} \quad t \leq 0$$

i.e.,

$$P(-t_F,t) = \frac{1}{\pi} \int_0^\infty (\text{Re}[P(t_F,i\omega)]\cos(-\omega t) - \text{Im}[P(t_F,i\omega)]\sin(-\omega t))d\omega = 0$$

or

$$P(-t_F,t) = \frac{1}{\pi} \int_0^\infty (\text{Re}[P(t_F,i\omega)]\cos\omega t + \text{Im}[P(t_F,i\omega)]\sin\omega t)d\omega = 0 \quad (5.59)$$

and adding equations (5.58) and (5.59) we obtain:

$$P(t_F,t) = \frac{2}{\pi} \int_0^\infty \text{Re}[P(t_F,i\omega)]\cos\omega t d\omega \quad (5.60)$$

The miss step response equals the integral of the impulse response $P(t_F, t)$, i.e.,

$$\text{Miss} = \int_0^{t_F} P(t_F,\sigma)d\sigma \quad (5.61)$$

Substituting equations (5.60) in (5.61) and changing the order of integration we have:

$$\text{Miss} = \frac{2}{\pi} \int_0^\infty \text{Re}[P(t_F,i\omega)] \int_0^{t_F} \cos\omega\sigma d\sigma d\omega$$

or

$$\text{Miss} = \frac{2}{\pi} \int_0^\infty \frac{\text{Re}[P(t_F,i\omega)]}{\omega} \sin\omega t_F d\omega \quad (5.62)$$

As established in control theory (see, e.g., [3,4]), if ω_s is the frequency that characterizes a system bandwidth, then the time of the transient response satisfies the inequality

$$\frac{\pi}{\omega_s} \leq t \leq \frac{4\pi}{\omega_s},$$

which for guidance systems can be reformulated in the following way: If determined from the real part of the frequency response ω_s characterizes the guidance system bandwidth, then the step miss is small for the flight time:

$$t_F \geq \frac{4\pi}{\omega_s} \tag{5.63}$$

The procedure of obtaining the miss step response based on equation (5.62) is demonstrated on the example of the guidance system analyzed above.

The frequency response of the guidance system [see equations (5.44) and (5.45)] for $N = 3$ is given in Figure 5.9. Figure 5.10 presents the real part of the frequency response [see equation (5.44)]. As indicated above, based on this characteristic it is possible to evaluate the time of flight t_F, when the miss becomes small enough. Substituting in equation (5.63) $\omega_s \approx 3rad/s$ (see Figure 5.10), we obtain the estimate $t_F \geq 4.2$ s.

The miss due to the step maneuver is calculated based on equation (5.62). The miss values are shown in Figure 5.11 by the "*" symbol. The miss due to the unit step maneuver, obtained by simulation of the guidance system in Figure 5.6, is shown in Figure 5.11 by the solid line.

As seen from Figure 5.11, the frequency analysis enables us to evaluate the miss step without simulation of the missile guidance system.

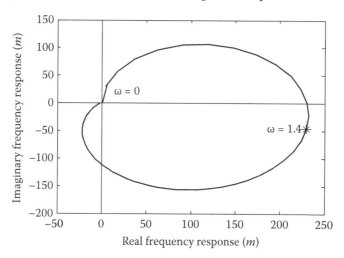

FIGURE 5.9 Frequency response of guidance system.

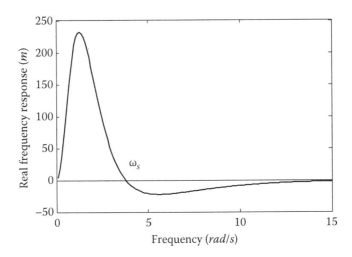

FIGURE 5.10   Real frequency response of guidance system.

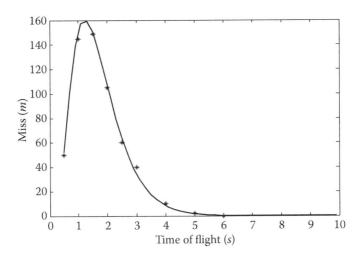

FIGURE 5.11   Miss due to step maneuver.

5.8   BOUNDED INPUT—BOUNDED OUTPUT STABILITY

The expression for the frequency response of the missile guidance system is obtained assuming that the Fourier transform (5.54) and (5.55) exists or, in other words, the system is stable with respect to $y(t_F)$, $t_F \in [0, \infty)$. However, the stability conditions present the most difficult part of analysis and synthesis of guidance systems. Since guidance systems operate on a finite time interval, their stability is determined as finite-time stability and is called in [1] the Lyapunov stability. The known conditions are sufficient and are based on results related to the stability of nonlinear systems.

In contrast to finite-time stability that analyzes $y(t)$, $t \in [0, t_F]$, the input-output relation between the miss distance $y(t_F)$ and target acceleration [see equations (5.13), (5.14), and (5.34)] enables us to analyze $y(t_F)$, $t_F \in [0, \infty)$ and formulate the stability of the proportional navigation guidance systems as BIBO (bounded input – bounded output) stability.

DEFINITION: The proportional navigation guidance system is BIBO stable, if for any bounded target acceleration its miss $y(t_F)$ is bounded for all times of flight $t_F \in [0, \infty)$.

It is obvious that $y(t_F)$ is bounded on a finite interval. We can expect that $y(t_F)$ is bounded, when $t_F \to \infty$ because $1/t_{go} \to 0$.

Using the expression for the transfer function (5.34), the stability condition can be written similar to the BIBO stability condition of linear systems, i.e., $L^{-1}(P(t_F, s))$ should be absolutely integrable on $[0, \infty)$ (L is the symbol of the Laplace transform). This condition is equivalent to the requirement for the transfer function $P(t_F, s)$ to be analytical in the right half-plane of the complex variable (including the imaginary axis) and $\lim_{s \to \infty} P(t_F, s) = 0$.

THEOREM: The proportional navigation guidance system with the transfer function $W(s)$ [see equation (5.29)] is BIBO stable if and only if the following condition is satisfied:

$$N - 2 + N \sum_{k=1}^{l} B_k / \tau_k + N \sum_{j=1}^{m} C_j \omega_j^2 < 0 \qquad (5.64)$$

where τ_k and ω_j are parameters of the missile guidance system; N is the effective navigation ratio; B_k, and C_j are coefficients of the partial fraction expansion (5.30) of $W(s)/s$.

PROOF. NECESSITY: If the condition (5.64) does not hold, then $\lim_{s \to \infty} P(t_F, s) \neq 0$. This contradicts the condition of the existence of an inverse Laplace transform.

SUFFICIENCY: Let the condition (5.64) hold. The function $P(t_F, s)$ defined by equation (5.34) is analytic in the region $C_v = \{s : \mathrm{Re}\, s > -\sigma\}$, where $\sigma = \min(1/\tau_k, \zeta_j \omega_j)$, $k = 1,..,l$, $j = 1,...,m$, i.e., it is analytic in the right-half plane ($\mathrm{Re}\, s \geq 0$) so that $L^{-1}(P(t_F, s))$ is absolutely integrable on $[0, \infty)$. The last statement needs additional clarification taking into account that $P(t_F, s)$ is a multiple-valued function of the complex variable s. Appendix B contains the rigorous proof of this statement.

COROLLARY: Proportional navigation missile guidance systems with the transfer function (5.29) are BIBO stable for all r_i $(i = 1,...,n-1)$.

Since $H(s)$ is a proper rational function and the degree of its numerator equals n-1, presenting equation (5.30) in the form (5.29) and equating to zero the term of the numerator of power n, we have:

$$1 + N \sum_{k=1}^{l} B_k / \tau_k + N \sum_{j=1}^{m} C_j \omega_j^2 = 0$$

so that the inequality (5.64) is always satisfied.

The established property of the structures in Figure 5.6 serves as a justification of the described procedure that can be used for analysis and synthesis of proportional navigation guidance systems in the frequency domain.

5.9 FREQUENCY RESPONSE OF THE GENERALIZED GUIDANCE MODEL

The missile guidance models widely used in the literature do not take into account target dynamics. A target acceleration considered in most publications is, in essence, a commanded target acceleration rather than a real target acceleration. Nevertheless, this acceleration is compared with a missile acceleration that is presented as a result of the transformation of a missile commanded acceleration by a certain dynamic unit (first-order or higher) that reflects dynamic features of a missile flight control system.

Ignoring target missile dynamics can bring inaccuracies when evaluating engagement performance. The generalized missile guidance model presented in Figure 5.12 can be used to obtain more accurate results.

Analogous to equation (5.12), the flight control dynamics of a target are presented by a third-order transfer function (below we consider a tail-controlled missile):

$$W_T(s) = \frac{1 - \dfrac{s^2}{\omega_{Tz}^2}}{(1 + \tau_T s)\left(1 + \dfrac{2\zeta_T}{\omega_T} s + \dfrac{s^2}{\omega_T^2}\right)} \qquad (5.65)$$

with damping ζ_T, natural frequency ω_T, the flight control system time constant τ_T, and the right-half plane zero ω_{Tz}.

The transfer function $P_G(t_F, s)$ of the generalized model of the guidance system with respect to a commanded target acceleration can be presented

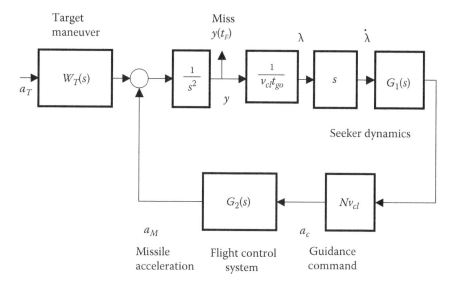

FIGURE 5.12 Generalized missile guidance model.

as the product of $P(t_F, s)$ and $W_T(s)$. Taking into account that the frequency response of the target flight control system $W_T(i\omega)$ is

$$W_T(i\omega) = |P_T(i\omega)| \exp(i\varphi_T(i\omega)) \tag{5.66}$$

where

$$|P_T(i\omega)| = \omega_T^2(1+\omega^2/\omega_{Tz}^2)(\tau_T^2\omega^2+1)^{-2}((\omega_T^2-\omega^2)^2+4\zeta_T^2\omega_T^2\omega^2)^{-2} \tag{5.67}$$

and

$$\varphi_T(i\omega) = -\tan^{-1}(\omega\tau_T) - \tan^{-1}\left(\frac{2\omega\omega_T\zeta_T}{\omega_T^2-\omega^2}\right) \tag{5.68}$$

the amplitude $|P_G(t_f, i\omega)|$ and phase $\varphi_G(t_f, i\omega)$ characteristics of the generalized model of the missile guidance system have the following form:

$$|P_G(t_F, i\omega)| = |P(t_F, i\omega)| \cdot |P_T(i\omega)| \tag{5.69}$$

and

$$\varphi_G(t_F, i\omega) = \varphi(t_F, i\omega) + \varphi_T(i\omega) \tag{5.70}$$

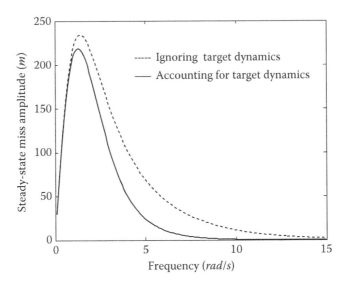

FIGURE 5.13 Amplitude characteristics of frequency response.

The other frequency characteristics of the generalized model and the estimates of the miss distance can be obtained from equations (5.44)–(5.48), and (5.62) by changing $|P(t_F, i\omega)|$ and $\varphi(t_F, i\omega)$ to $|P_G(t_F, i\omega)|$ and $\varphi_G(t_F, i\omega)$, respectively.

As mentioned, the generalized missile guidance model gives more accurate results than the model that does not take into account target dynamics. Figure 5.13 presents the amplitude characteristics of the missile guidance system considered above (dashed line) and the generalized missile guidance model (solid line) for a target with $\zeta_T = 0.8$, $\omega_T = 3.5$ *rad/s*, $\tau_T = 0.15$ *s* and $\omega_{T_z} = 15$ *rad/s*. The miss for the generalized model is less than for the model that ignores target dynamics.

The above discussion and examples were focused mainly on tail-controlled missiles. As shown in Figure 5.8, the airframe zeroes can significantly decrease missile performance. The tail configuration is also known as the nonminimum phase, due to the location of a zero in the right half *s*-plane in the corresponding transfer function of the linear model representation. The long moment arm between the tail controls and the forward position of the center of gravity after burnout requires smaller control forces for constructing the angle of attack, resulting in a lower drag configuration. These control forces are exerted in a direction opposite to the required maneuver, thus, generating a delay in the missile response in the correct direction.

The delayed response of the tail controls can be compensated by employing an additional forward control device, using either divert thrusters or aerodynamic canard fins. Missiles with forward control fins, or canards, have been used for many years. However, this type of missile can suffer from adverse induced rolling moments. The use of grid fins, or "lattice controls," for the tail control surfaces—instead of conventional planar fins— was recently proposed as a possible remedy for the roll control problems. Studies have shown that when compared to conventional planar fins, grid fins have certain advantages, such as effective aerodynamic control at high angles of attack and high Mach number, attenuated body-vortex interference, and improved roll control. The primary disadvantage of the grid fin concept is a higher drag than conventional planar fins. The canard fins, located in the front part of the fuselage, generate an aerodynamic force that is in the same direction as the required maneuvering force, thus generating an immediate response in the correct direction. Canard missiles have forward and aft control systems.

In contrast to equation (5.12) and Figure 5.6, the flight control system dynamics of this type of missile can be represented by two transfer functions: a minimum phase transfer function for the forward control and a nonminimum phase transfer function for the aft control. The use of two control systems offers new capabilities, for example, the ability to generate a very high angle of attack for fast and large turns. The additional degree of freedom offered by the dual control system requires special consideration in the guidance and control design. The appropriate blending of the two controls can significantly improve performance of canard missiles. The material of this chapter allows readers to obtain the analogous equations for this type of missiles.

REFERENCES

1. Bhat, S., and Bernstein, D. Finite-Time Stability of Continuous Autonomous Systems, *SIAM Journal of Control and Optimization* 38, no. 3 (2000): 751–66.
2. Churchill, R. V. *Complex Variables and Applications*. New York, NY: McGraw-Hill, 1960.
3. Dorf, R. C. *Modern Control Systems*. New York, NY: Addison-Wesley, Inc., 1989
4. Solodovnikov, V. V. *Introduction to the Dynamics of Automatic Control Systems*. New York, NY: Dover, 1960.
5. Yanushevsky, R. Analysis of Optimal Weaving Frequency of Maneuvering Targets, *Journal of Spacecraft and Rockets* 41, no. 3 (2005): 477–79.
6. Yanushevsky, R. Optimal Control of Differential-Difference Systems of Neutral Type, *International Journal of Control* 1 (1989): 1835–50.
7. Yanushevsky, R. Analysis and Design of Missile Guidance Systems in Frequency Domain, *44th AIAA Space Sciences Meeting*, Paper AIAA 2006-825, Reno, Nevada, 2006.
8. Yanushevsky, R. Frequency Domain Approach to Guidance System Design, *IEEE Transactions on Aerospace and Electronic Systems* 43 (2007).

9. Zadeh, L., and Desoer, C. *Linear System Theory.* New York, NY: McGraw Hill, 1964.

10. Zarchan, P. *Tactical and Strategic Missile Guidance, Progress in Astronautics and Aeronautics.* Vol. 176. Washington, DC: American Institute of Astronautics and Aeronautics, Inc., 1997.

11. Zarchan, P. Proportional Navigation and Weaving Targets, *Journal of Guidance, Control, and Dynamics* 18, no. 5 (1995): 969–74.

12. Zarchan, P., Greenberg, E., and Alpert, J. Improving the High Altitude Performance of Tail-Controlled Endoatmospheric Missiles, *AIAA Guidance, Navigation and Control Conference*, AIAA Paper 2002-4770, August 2002.

6 Design of Guidance Laws Implementing Parallel Navigation. Frequency–Domain Approach

6.1 INTRODUCTION

The classical approach to missile guidance is usually based on applying a guidance law obtained from certain line-of-sight (LOS) geometrical rules. The guidance law is the algorithm by which the desired geometrical rule is implemented. According to the well-known proportional navigation (PN) law, widely used in military applications, the missile acceleration is proportional to the measured LOS rate. However, acting as the commanded missile acceleration, this law produces the missile real acceleration, which differs from the desired commanded acceleration. Usually, kinematics of PN are analyzed without taking into account missile dynamics, and most recommendations concerning guidance law parameters are made based on this analysis. In the previous chapters, we acted the same way. As shown in Chapter 4 [see equation (4.6)], the miss distance due to a step target maneuver is exactly zero for an idealized linearized inertialess two-dimensional PN missile-target engagement model. The influence of missile dynamics was examined analytically for single-lag models of guidance systems by using the method of adjoints [see equation (4.22)]. As indicated, the single-lag models, as well as the binomial models, do not quite match reality and do not accurately reflect flight control system dynamics. The analytical approach to analysis of effectiveness of PN for more realistic models of guidance systems, reflecting airframe and autopilot dynamics against weaving targets was considered in Chapter 5.

It is known that PN demonstrates good performance for nonmaneuvering or moderately maneuvering targets. For highly maneuvering targets the so-called optimal guidance laws (based on optimal control or game theory) can theoretically get significantly better results. However, as indicated earlier, these laws require complete and detailed information about missile dynamics and future behavior of a target. They are too complicated and the closed-form solution is obtained only for simple guidance system models [6,12–14].

As mentioned above, the actual missile acceleration differs from the commanded acceleration because of the flight control system dynamics. On the one hand, its transient response may make the difference significant. For weaving targets, the frequency response of the flight control system determines the steady-state amplitude and phase shift of the real acceleration compared to the commanded acceleration [7,8]. On the other hand, external disturbances usually ignored in many engagement models (e.g., drag) contribute to the difference between the actual and commanded accelerations and increase the miss distance.

The PN guidance law acts as a simple proportional controller that was used at the initial stage of control systems development [3–5]. Now the PID (proportional-integral-differential) controllers are widely used in practice. Usually, the instruction of utilizing these controllers contains the following: a proportional controller will have the effect of reducing the rise time and will reduce, but never eliminate, the steady-state error; an integral control will have the effect of eliminating the steady-state error, but it may make the transient response worse; a derivative control will have the effect of increasing the stability of the system, reducing the overshoot, and improving the transient response. Can these recommendations be applied to the PN guidance law?

Over the years, control theory has made enormous progress, and various types of control laws have been developed and used in practice. Nevertheless, the guidance laws used in the aerospace field have not changed significantly and the PN continues to dominate research and development. The so-called neoclassical approach that, at a certain degree, is similar to the utilization of proportional-differential controllers (that have been used since the 1940s) was considered only in 2001 [1].

Taking into account that both transient and frequency responses can be improved by utilizing feedback/feedforward control signals, we will consider how to use the classical control theory approach to improve the performance of missile systems with PN guidance. The approach to decrease significantly the miss distance by modifying the PN guidance and using in the guidance law the actual missile acceleration signals is discussed. New guidance laws and the conditions for choosing their parameters are considered.

Although the below discussion focuses on missile guidance, it is applied also to a wide class of unmanned aerial vehicles, including UAVs.

6.2 NEOCLASSICAL MISSILE GUIDANCE

Let us again consider the missile guidance system discussed in Chapter 5 (see Figure 5.6 and 6.1). The relative separation $y(t)$ between a missile and

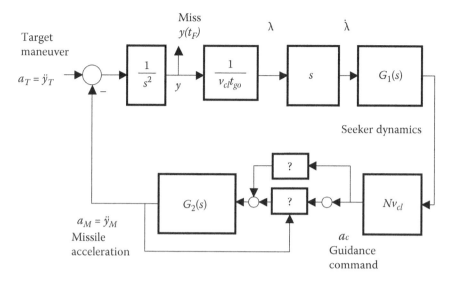

FIGURE 6.1   Modified missile guidance model.

a target is obtained by integrating the missile acceleration a_M subtracted from the target acceleration a_T. A division by range (closing velocity v_{cl} multiplied by time-to-go t_{go} until intercept) yields the geometric line-of-sight (LOS) angle λ, where the time-to-go is defined as $t_{go} = t_F - t$. Analogous to Figure 5.6, the missile seeker is presented formally as a perfect differentiator; the filter and seeker dynamics are represented by a transfer function

$$G_1(s) = \frac{\tau_z s + 1}{\tau_2 s + 1},$$

where τ_z and τ_2 are constant coefficients. An estimation of the LOS rate generates a guidance command a_c based on the proportional navigation law with the effective navigation ratio N. The flight control system dynamics, which combine its airframe and autopilot dynamics, are represented by the transfer function (5.12).

Here we rewrite the main relationships (5.13)–(5.15) in Chapter 4 (for simplicity we do not consider here $G_1(s)$ of higher order and do not use equation (5.29) for $W(s)$)

$$Y(t_F, s) = \exp\left(N \int_\infty^s H(\sigma)d\sigma \right) Y_T(s) \qquad (6.1)$$

where $Y_T(s)$ is the Laplace transform of a target vertical position $y_T(t)$, $Y(t_F, s)$ is the Laplace transform of $y(t_F)$:

$$H(s) = \frac{W(s)}{s} \tag{6.2}$$

and

$$W(s) = G_1(s) \cdot G_2(s) = \frac{1 + r_1 s + r_2 s^2 + r_3 s^3}{(1 + \tau_1 s)(1 + \tau_2 s)\left(1 + \dfrac{2\zeta}{\omega_M} s + \dfrac{s^2}{\omega_M^2}\right)} \tag{6.3}$$

where r_k ($k = 1, \ldots, 3$) are constant coefficients.

In contrast to Figure 5.6, Figure 6.1 contains the feedforward and feedback units, which should be determined to improve the performance of the PN guidance law.

The problem of obtaining small miss distance is similar to the problem of reaching high accuracy of conventional feedback systems. It is known that high accuracy can be achieved by increasing the controller gain in conventional feedback systems. However, the controversy between accuracy and stability makes the problem of designing high accuracy systems difficult.

A special class of linear systems admitting infinite gains was considered in [4], and the link between this class of linear structures and a class of linear optimal systems was discussed in [9]. The class of linear structures examined in [4,9], described by the n-order differential equations, requires n-1 "pure" differentiators. As shown in [10,11], their practical realization can make the system unrobust.

For idealized linearized inertialess two-dimensional PN missile-target engagement model, discussed in Chapter 4 (see (4.6)), $W(s) = 1$, so that an intuitive approach to achieve zero-miss-distance for the PN guided missile systems with the transfer function $W(s)$ described by equation (6.3) consists in utilizing the feedforward sequential unit with the transfer function $1/W(s)$, so that the transfer function (6.3) of the modified system (see Figure 6.1) would be equal to 1. Such a "naïve" approach was used in the inverse operator method applied to control systems admitting infinitely high gain [9]. Offered in the 1960s, this method suffers significant drawbacks. First, it ignores the transient response that cannot be eliminated. As the result, it cannot be applied to the systems with unstable zeroes, because the whole system with the inverse operator is unrobust; it becomes unstable [9]. Finally, its realization usually requires multiple differentiation operations that makes the real system susceptible to noise.

It looks like the similar idea to decrease the miss distance in the PN guided systems by including in the "acceleration channel" additional differentiating units was considered in [1]. The approach of achieving zero-miss-distance (ZMD) was called the neoclassical guidance. The main result is stated by the following theorem [1].

THEOREM: Consider a strictly proper rational function $H(s) = W(s)/s$ of the form:

$$H(s) = \frac{b(s)}{a(s)} = \frac{b_1 s^{n-1} + b_2 s^{n-2} + \cdots + b_n}{s^n + a_1 s^{n-1} + \cdots + a_{n-1} s}, \quad b_1 \geq 0$$

where $a(s)$ and $b(s)$ are coprime polynomials.

Denote by r the relative order of $H(s)$, i.e., $r = \deg[a(s)] - \deg[b(s)]$. Under these conditions, if $s = \mathrm{Re}s + i\omega$:

$$F(\infty) = \lim_{\omega \to \infty} F(s) = \lim_{\omega \to \infty} \left[\int H(s) ds \right] \to \begin{cases} 0 & \text{iff} \quad r \geq 2 \\ \infty & \text{iff} \quad r = 1 \end{cases}$$

PROOF: By presenting $H(s)$ in the form

$$H(s) = \frac{b_1}{s} + \sum_{j=2}^{\infty} h_j s^{-j},$$

where h_j are coefficients of this series, and integrating the above expression, we obtain for $b_1 \neq 0$ (when $r = 1$) the integral tends to infinity since $\ln(\infty) \to \infty$. For $b_1 = 0$ (when $r \geq 2$), the result of integration gives the components $h_j s^{1-j}/(1-j), j > 1$, which tend to zero.

Since the infinite value of the above integral corresponds to the lower limit of integration in equation (6.1), i.e., the corresponding exponent factor equals zero, the condition $r = 1$ gives zero $Y(t_F, s)$.

As stated in [1], if the guidance system is linear and the degree of the numerator of $W(s)$ equals the degree of the denominator (the biproper transfer function $W(s)$) and $b_1 > 0$, then the zero miss distance (ZMD) is obtained for any bounded target maneuver.

It is important to indicate that the expression (6.1) was obtained based on the impulse response presentation by using the method of adjoints, so that it is assumed that the planar model of the PN guided missile system has zero initial conditions. Hence, ZMD can be achieved only for the mentioned zero initial conditions.

Although the guidance systems operate on a finite interval of time so that $y(t_F)$ is limited, to conclude only based on equation (6.1) that the ZMD property is attainable for the linearized models and satisfying the condition of the above-given theorem means to ignore the missile system dynamics. This is inadmissible and, as the result, the neoclassical approach can significantly worsen, rather than improve, the missile system performance.

The biproper transfer function can be obtained only if a compensator contains a "pure" differential operator. In practice, operations of differentiation can be performed only approximately, so that instead of the case $r = 1$ we have the case $r = 2$.

Using the analytical expressions (5.16)–(5.21), we obtain the expression for the miss distance for these two cases and compare the results for the "ideal" guidance system and the system with very close characteristics (with a real differentiation operator). The miss distance due to a weaving target will be evaluated by determining the magnitude of the steady-state component when the input, target acceleration, is a unit harmonic signal of frequency ω (e.g., $n_T = 1g\ sin\omega t$; g is acceleration of gravity), i.e., from the expression (5.26).

The case of the "ideal" neoclassical guidance we discuss considering an example of a tail-controlled missile with

$$G_2(s) = \frac{1 - \dfrac{s^2}{\omega_z^2}}{(\tau_1 s + 1)\left(\dfrac{s^2}{\omega_M^2} + \dfrac{2\zeta}{\omega_M}s + 1\right)} \tag{6.4}$$

and the compensator

$$G_1(s) = -\tau_1 s + 1 \tag{6.5}$$

The negative sign of derivative $-\tau_1 s$ is stipulated by the condition $b_1 > 0$. According to the theorem, this type of correction gives the zero-miss-distance.

Now, instead of (6.5) we consider a physically realizable unit:

$$G_1(s) = \frac{-\tau_1 s + 1}{\varepsilon s + 1} \tag{6.6}$$

where ε is a small parameter.

The expression for the amplitude characteristic [see equations (5.17)–(5.19), (5.26)] has the form:

$$|P(t_F, \omega)| = g\omega^{N-2}(\omega^2 + \varepsilon^{-2})^{\frac{B_2 N}{2\varepsilon}}(\omega^2 + 1/\tau_1^2)^{B_1 N/2\tau_1}((\omega_M^2 - \omega^2)^2 \tag{6.7}$$
$$+ 4\omega_M^2\zeta^2)^{CN\omega_M^2/4}\exp(.)$$

where exp(.) is given by equation (5.27).

For $N = 3$, $\omega_M = 20\ rad/s$, $\omega_z = 5\ rad/s$, $\tau_1 = 0.5\ s$, and $\zeta = 0.7$ the maximum miss distance (peak miss) of 234 m corresponds to a target maneuver

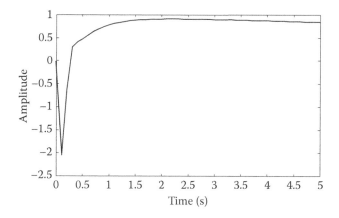

FIGURE 6.2 Step response for $\tau_1 = 0.2$ s and $\varepsilon = 0.01$ s.

with the frequency of 1.4 rad/s, i.e., with the period of 4.48 s. For $\tau_1 = 0.2$ s and $\varepsilon = 0.01$ s we obtain the miss distances of order $O(10^{-6})$, i.e., a very good accuracy. However, the step response of the flight control system with the transfer function $W(s) = G_1(s)G_2(s)$ (see Figure 6.2) shows that the dynamic characteristics of the flight control system does not correspond to the design requirements. The signal generated by the negative derivative amplifies the "wrong way tail effect," so that dynamics of the modified system becomes inadmissible.

REMARK: In [2] the nonlinear planar missile guidance system model was considered taking into account saturation due to aerodynamic or structural constraints. The positive realness (PR) condition was imposed for no saturation to occur. Since the PR condition is widely used in the nonlinear control theory as a stability condition, its relationship with finite time stability was established [1], so that its combination with the accuracy condition [1] enables us to obtain appropriate dynamic characteristics of the flight control system. However, the PR condition restricts significantly the class of missile systems where, theoretically and under zero initial conditions, zero-miss-distance can be achieved. The important class of tail-controlled missiles does not satisfy the PR condition.

6.3 PSEUDOCLASSICAL MISSILE GUIDANCE

The proportional navigation guidance law is so popular that it is considered as classical. Below we describe its modification by using the results of classical control theory. The approach offered is based on utilizing feedforward/feedback control signals to make the real missile acceleration close to the commanded acceleration generated by the PN law. The performance of the modified guidance law is equivalent to the performance of

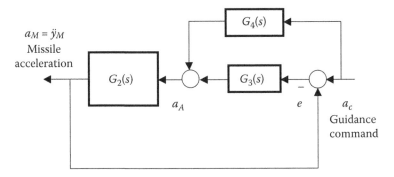

FIGURE 6.3   Modified guidance law.

the PN law applied to a fictitious flight control system with better dynamic characteristics.

We will consider the modified PN guidance law based on analysis of the following structure widely used in control theory (see Figure 6.3). Figure 6.3 presents in details the part of the structure in Figure 6.1 that contains the symbol "?." Here the new commanded acceleration a_A (a new guidance law) is formed as a sum of the feedforward signal $G_4(D)a_c$ and the feedback signal $G_3(D)(a_c - a_M)$, i.e.,

$$a_A = G_4(D)a_c + G_3(D)(a_c - a_M) \qquad (6.8)$$

(D is the differential operator; transfer functions $G_3(s)$ and $G_4(s)$ characterize the feedback and feedforward channels, respectively.)

The transfer functions $W_\Sigma(s)$ characterizing the input-output relations between a_c and a_M can be presented as:

$$W_\Sigma(s) = \frac{G_2(s)(G_3(s) + G_4(s))}{1 + G_2(s)G_3(s)} \qquad (6.9)$$

where $G_4(0) = 1$ (it follows from the condition $W_\Sigma(0) = 1$, which is similar to $W(0) = 1$).

We will consider the fictitious flight control system $W_\Sigma(s)$ with better dynamic characteristics than the original flight control system $W(s)$ with respect to the commanded acceleration (the PN law).

The analysis of the guidance system with the new guidance law (6.8) is equivalent to the analysis of the guidance system in Figure 6.1 with the commanded acceleration a_c and the fictitious flight control system with the transfer function $W_\Sigma(s)$.

The problem of designing the new guidance law, which performance is better than the performance of the PN guidance law, reduces to determining $W_\Sigma(s)$ (the transfer functions of the feedback and feedforward channels

$G_3(s)$ and $G_4(s)$) that gives the smaller miss distance $y(t_F)$ than in the case of the initial $W(s)$ and has the transient response satisfying the design specifications.

According to equation (6.1), the miss distance due to a weaving target is evaluated by determining its steady-state component when the input is a unit harmonic signal of target acceleration with frequency ω, i.e., from the expression

$$P\,(t_F, i\omega) = \exp\left(N\int_{\infty}^{i\omega} H(\sigma)d\sigma \right) \frac{g}{(i\omega)^2} \qquad (6.10)$$

where $P(t_F, i\omega)$ is the frequency response relating the miss distance at moment t_F to the target acceleration a_T.

We will evaluate the steady-state component of equation (6.1), when the input is a sinusoidal signal; the peak miss distance characterizes its amplitude.

The integral in equation (6.10) can be presented in the following form:

$$\int_{i\infty}^{i\omega} H(\sigma)d\sigma = \int_{\infty}^{\omega} i(\operatorname{Re} H(i\omega) + i\operatorname{Im} H(i\omega))d\omega$$

$$= \int_{\omega}^{\infty} (-i\operatorname{Re} H(i\omega) + \operatorname{Im} H(i\omega))d\omega \qquad (6.11)$$

Since the absolute value of $\exp(N\int_\infty^{i\omega} H(\sigma)d\sigma)$ equals $\exp(N\int_\omega^\infty \operatorname{Im} H(i\omega)d\omega)$, we will analyze the expression for $\int_\omega^\infty \operatorname{Im} H(i\omega)d\omega$. Taking into account equation (6.2), we have

$$H(i\omega) = \frac{\operatorname{Re} W(i\omega)}{i\omega} + i\frac{\operatorname{Im} W(i\omega)}{i\omega} = \frac{\operatorname{Im} W(i\omega)}{\omega} - i\frac{\operatorname{Re} W(i\omega)}{\omega} \qquad (6.12)$$

so that

$$\exp\left(N\int_\omega^\infty \operatorname{Im} H(i\omega)d\omega \right) = \exp\left(-N\int_\omega^\infty \frac{\operatorname{Re} W(i\omega)}{\omega}d\omega \right) \qquad (6.13)$$

THEOREM: The peak miss distance under the new guidance law a_A of equation (6.8) is less than under the PN guidance law if $W_\Sigma(s)$ of equation (6.9) has no poles in the right half-plane of the complex variable s and

$$\exp\left(\int_\omega^\infty \frac{\operatorname{Re} W_\Sigma(i\omega)}{\omega}d\omega \right) > \exp\left(\int_\omega^\infty \frac{\operatorname{Re} W(i\omega)}{\omega}d\omega \right) \qquad (6.14)$$

PROOF: The first condition is required for the existence of the integral in equation (6.10). The inequality (6.14) follows immediately from equations (6.10) and (6.13).

COROLLARY 1 OF THEOREM: If $W_\Sigma(s)$ is not a strictly proper rational function and its nominator and denominator are polynomials of the same order, then the peak miss distance equals zero.

PROOF: In the case where the nominator and denominator have the same order, $W_\Sigma(s)$ and, hence, $\text{Re}W_\Sigma(i\omega)$ contain a positive constant term, so that the integral in the left part of the inequality (6.14) equals infinity (i.e., the exponential term of equations (6.10) and (6.13) equals zero). Therefore, the peak miss equals zero.

Zero-miss-distance for a nonstrictly proper rational transfer function was discussed in the previous section. Here the conditions of ZMD were proved in a different way. Moreover, the above statement relates only to the steady-state component of the miss distance. As mentioned earlier, the class of nonstrictly proper rational transfer functions is very sensitive to noise and its realization requires "pure" differential units that cannot be realized in practice. That is why we will consider only strictly proper rational functions $W_\Sigma(s)$.

COROLLARY 2 OF THEOREM: The peak miss distance under the new guidance law a_A [see equation (6.8)] is less than under the PN guidance law for target maneuver frequencies ω_T, if $W_\Sigma(s)$ has no poles in the right half-plane of the complex variable s and

$$\text{Re}\,W_\Sigma(i\omega) > \text{Re}\,W(i\omega), \quad \omega \geq \omega_T \qquad (6.15)$$

PROOF: Because the denominator of the integrand of (6.14) is positive, the condition (6.14) is satisfied if the condition (6.15) holds.

In practice, it is easier to use the condition (6.15) than the condition (6.14). However, it is difficult (in the general case simply impossible) to find physically realizable units $G_3(s)$ and $G_4(s)$ to satisfy the condition (6.15) for all $\omega \geq \omega_T$. That is why it is reasonable, first, to try to choose $G_3(s)$ and $G_4(s)$ from the condition (6.15) for $\omega \in [0, \omega_c]$, where ω_c characterizes the guidance system $W(i\omega)$ bandwidth, and then check whether the condition (6.14) is satisfied. If the condition (6.14) is not satisfied, $G_3(i\omega)$ and $G_4(i\omega)$ should be chosen from the condition (6.15) for a higher range of ω.

The conditions (6.14) and (6.15) were obtained for the steady-state mode. Because of the correlation between the transient and frequency responses, it is plausible to assume that the new guidance laws satisfying these conditions will decrease the miss distance for small times of flight as well.

Below the described approach is demonstrated by using simple enough control structures, so that the new guidance laws can be easily realized in practice. Both transient and frequency responses are considered.

6.4   EXAMPLE SYSTEMS

The bounded input – bounded output (BIBO) stability conditions for the considering class of the PN guidance structures were discussed in Chapter 5. The transfer functions $G_3(s)$ and $G_4(s)$ should be chosen so that $W_\Sigma(s)$ is asymptotically stable.

6.4.1   PLANAR MODEL OF ENGAGEMENT

For a missile guidance model in Figure 6.1, $G_1(s) = 1$ and $G_2(s)$ is described by equation (4.12). The feedforward and feedback units in Figure 6.2 are chosen as

$$G_3(s) = \frac{k_1(\tau_{10}s + \mu)}{\tau_2 s + 1} \tag{6.16}$$

and

$$G_4(s) = k_2 \tag{6.17}$$

where τ_{10}, τ_2 and k_1 are constant parameters, $\mu = 1$ or 0, $k_2 = 1$ or 0.
Based on equation (6.9) the transfer function $W_\Sigma(s)$ equals

$$W_\Sigma(s) = \frac{((k_1\tau_{10} + k_2\tau_2)s + k_1\mu + k_2)a(s)}{(\tau_2 s + 1)(\tau_1 s + 1)\left(\dfrac{s^2}{\omega_M^2} + \dfrac{2\zeta}{\omega_M}s + 1\right) + k_1(\tau_{10}s + \mu)a(s)} \tag{6.18}$$

The procedure of finding $G_3(s)$ and $G_4(s)$ is demonstrated on examples of tail-controlled missiles where the right half-plane airframe zero can significantly influence dynamics of the flight control system. The two cases $\omega_z = 30$ *rad/s* and $\omega_z = 5$ *rad/s* will be considered. As mentioned earlier, the last case corresponds to high altitudes of missile flight. The other parameters of $G_2(s)$ are chosen as $\zeta = 0.7$, $\omega_M = 20$ *rad/s*, and $\tau_1 = 0.5$ *s*. The effective navigation ratio $N = 3$; it can be changed as shown below.

Formally, it is possible to present the condition (6.15), accompanied by the conditions for the poles of $W_\Sigma(i\omega)$ to guarantee a certain transient response, as a part of a mathematical programming problem to determine the unknown parameters of $W_\Sigma(i\omega)$. However, we will employ a standard engineering approach utilizing elements of control theory and Matlab software.

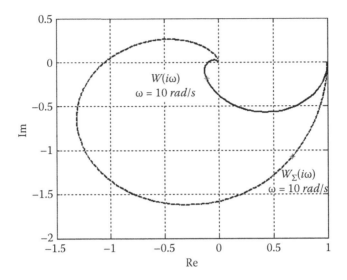

FIGURE 6.4   Frequency response $W(i\omega)$ and $W_\Sigma(i\omega)$ for $\omega_z = 30$ *rad/s*.

The frequency response of $W(i\omega)$ for $\omega_z = 30$ *rad/s* is given in Figure 6.4 (solid line). Dynamic properties of a system with such response are satisfactory and the system has a sufficient margin of stability. According to control theory, the increase of gain k_1 (see Figure 6.3) decreases the steady-state error $e = a_c - a_M$. As seen from Figure 6.4, for $\omega \geq 10$ *rad/s* $W(i\omega) < < 1$ and Re $W(i\omega) < 0$, so that the condition (6.14) should be checked for the range of frequencies $\omega \in [0, 10]$.

First, we consider the case of $G_3(s) = k_1$. Analysis of equation (6.9) shows that by choosing, for example, $k_1 = 5$ we significantly decrease $e = n_c - n_l$ (i.e., make a_M closer to a_c). There exist two realizations of $W_\Sigma(i\omega)$ for $k_1 = 5$: $k_2 = 0$ and $k_2 = 1$. In the first case, the gain of $W_\Sigma(i\omega)$ equals 5/6, so that to satisfy the condition $W_\Sigma(0) = 1$ the effective navigation ratio should be increased by factor 6/5. In the future, we assume the corresponding increase (if necessary) of N, so that the gain of $W_\Sigma(s)$ equals 1. The frequency response $W_\Sigma(i\omega)$ in Figure 6.4 (dashed line) shows that $\text{Re}W_\Sigma(i\omega) > \text{Re}W(i\omega)$ for $\omega \in [0, 10]$ (see also Figure 6.5). The real frequency response in Figure 6.5, as well as the step response in Figure 6.6, shows that the modified system $W_\Sigma(i\omega)$ has better dynamic characteristics than the original one.

Comparing the peak miss characteristics for the missile system with the PN guidance law and the modified guidance law given in Figure 6.7, we can conclude that the guidance laws [see equation (6.8)]:

$$a_A = 5(a_c - a_M), \quad N = 3 * 6/5, \quad a_c = Nv_{cl}\dot{\lambda} \tag{6.19}$$

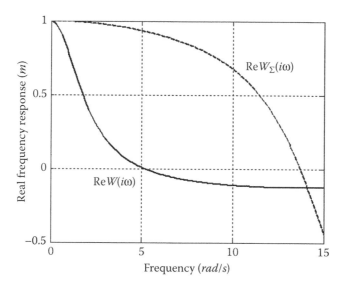

FIGURE 6.5 Real frequency response $W(i\omega)$ and $W_\Sigma(i\omega)$ for $\omega_z = 30$ *rad/s*.

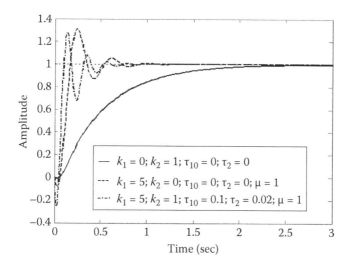

FIGURE 6.6 Step response of $W(s)$ and $W_\Sigma(s)$ for $\omega_z = 30$ *rad/s*.

or

$$a_A = a_c + 5(a_c - a_M), \quad N = 3, \quad a_c = Nv_{cl}\dot{\lambda} \tag{6.20}$$

significantly decrease the miss distance.

By using a phase-lead network with parameters τ_{10} and τ_2 that increases $\mathrm{Re}W_\Sigma(i\omega)$ more and, as a result, by using a more complicated guidance

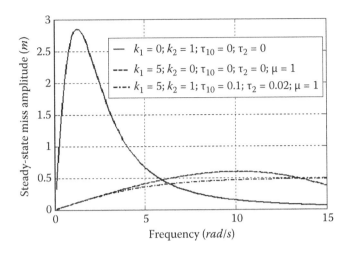

FIGURE 6.7 Comparison of the peak miss for the PN and modified guidance law for $\omega_z = 30$ *rad/s*.

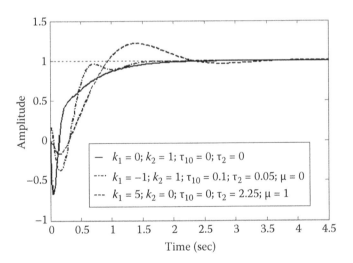

FIGURE 6.8 Step response of $W(s)$ and $W_\Sigma(s)$ for $\omega_z = 5$ *rad/s*.

law (see Figure 6.6 and Figure 6.7; the case $k_2 = 1$; $\tau_{10} = 0.1$ s; $\tau_2 = 0.02$ s; $\mu = 1$) we get an additional decrease of the miss distance. However, the additional change is not significant.

The step and frequency responses of the flight control system for $\omega_z = 5$ *rad/s* are given in Figures 6.8 and 6.9. At high altitudes the "wrong way tail effect" of tail-controlled endoatmospheric interceptors is substantial, so that their dynamic characteristics at high altitudes are significantly worse than at lower altitudes. Given in Figure 6.10, the

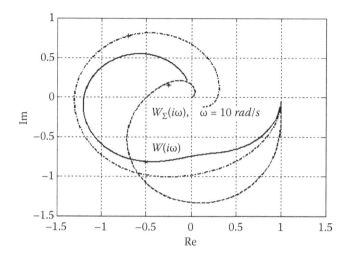

FIGURE 6.9   Frequency response $W(i\omega)$ and $W_\Sigma(i\omega)$ for $\omega_z = 5$ *rad/s*.

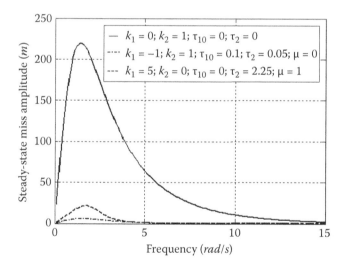

FIGURE 6.10   Comparison of the peak miss for the PN and modified guidance law for $\omega_z = 5$ *rad/s*.

amplitude characteristic of the frequency response for the missile with the PN guidance law shows that the peak miss for $\omega_z = 5$ *rad/s* is significantly higher than for the case $\omega_z = 30$ *rad/s* (see Figure 6.6). The analysis of the frequency response $W(i\omega)$ in Figure 6.9 (solid line) shows that gains $k_1 > 0.8$ would make the flight control system $W_\Sigma(s)$ unstable. Moreover, compared to the case $\omega_z = 30$ *rad/s*, here $W(i\omega)$ has a wider bandwidth and the domain, where $\mathrm{Re}W(i\omega) < 0$.

Creating a feedback system with a gain k_1, similar to the case $\omega_z = 30$ *rad/s*, would require $G_3(s)$ to narrow the bandwidth significantly (i.e., $\tau_1 = 0$ and the time constant τ_2 should be big enough). The analysis similar to the case $\omega_z = 30$ *rad/s* shows that $\tau_2 = 2.25$ *s* enables us to decrease the miss distance significantly (see Figures 6.8–6.10; dashed lines). The guidance law has the following form:

$$2.25\dot{a}_A + a_A = 5(a_c - a_M), \quad N = 3 \cdot 6/5, \quad a_c = Nv_{cl}\dot{\lambda} \qquad (6.21)$$

As seen from the step response (Figure 6.8), the wrong way tail effect acts as a "negative force" that can be compensated by an opposite directed force, which can be realized by a positive feedback unit with a transfer function $\tau_{10}s/(\tau_2 s + 1)$.

The frequency and step responses and the amplitude characteristic of the frequency response for the parameters $k_2 = -1$; $k_2 = 1$ $\tau_{10} = 0.1$ *s*; $\tau_2 = 0.05$ *s*; $\mu = 0$, chosen from the condition (6.15), are given in Figures 6.8–6.10 (dash-dot lines). It follows that the guidance law

$$0.05\dot{a}_A + a_A = -0.05\dot{a}_c + a_c + 0.1\dot{a}_M, \quad a_c = Nv_{cl}\dot{\lambda} \qquad (6.22)$$

enables us to obtain a lower peak miss than under the guidance law (6.21).

6.4.2 MULTIDIMENSIONAL MODEL OF ENGAGEMENT

The above results were obtained and tested for the considered linear planar model assuming zero initial conditions of the flight control system coordinates and a constant closing velocity. The new guidance laws are tested also on a more precise multidimensional nonlinear model of engagement, similar to that considered in Chapter 3, with the following parameters:

- A target initial condition $R_{T1} = 4500$ *m*, $R_{T2} = 2500$ *m*, $R_{T3} = 0$

$$V_{T1} = -350 \ m/s, \ V_{T2} = 30 \ m/s; \ V_{T3} = 0$$

- A missile initial condition $R_{M1} = R_{M2} = R_{T3} = 0$;

$$V_{M1} = 165 \ m/s, \ V_{M2} = 475 \ m/s; \ V_{M3} = 0$$

- A target acceleration $a_{T1} = 0$, $a_{T2} = 3g \sin 1.31t$, $a_{T3} = 0$
- A missile acceleration limit $\|a_c\| \le 10g$
  (R_i, V_i, $i = 1 - 3$, are distance and velocity coordinates)

The flight control system has the same parameters as in the linear planar model.

TABLE 6.1

Comparative Analysis of Guidance Laws

Case #	Parameters	Time of Intercept (s)	Miss (m)
1	$\omega_z = 30 rad/s$; $N = 3$; $k_1 = 0$; $k_2 = 1$; $\tau_{10} = 0$; $\tau_2 = 0$	8.0	2.01
2	$\omega_z = 30 rad/s$; $N = 3*6/5$; $k_1 = 5$; $k_2 = 0$; $\tau_{10} = 0$; $\tau_2 = 0$; $\mu = 1$	8.0	0.52
3	$\omega_z = 5 rad/s$; $N = 3$; $k_1 = 0$; $k_2 = 1$; $\tau_{10} = 0$; $\tau_2 = 0$	8.0	5.58
4	$\omega_z = 5 rad/s$; $N = 3*6/5$; $k_1 = 5$; $k_2 = 0$; $\tau_{10} = 0$; $\tau_2 = 2.25s$; $\mu = 1$	8.0	0.81
5	$\omega_z = 5 rad/s$; $N = 3$; $k_1 = -1$; $k_2 = 1$; $\tau_{10} = 0.1s$; $\tau_2 = 0.05s$; $\mu = 0$	8.0	0.29

The simulation results are presented in Table 6.1. The miss distance and the time of intercept correspond to the moment of time, when the closing velocity became positive.

As seen from the table, the considered new guidance laws decrease the miss distance significantly (i.e., increase the missile performance).

In Chapter 3, the class of the guidance laws developed based on the Lyapunov method was discussed. Below we consider as the commanded acceleration the guidance law containing the "cubic" power of the LOS rate components λ_i ($i = 1 - 3$), i.e.,

$$a_{ci} = N v_{cl} \dot{\lambda}_i + N_{1i} \dot{\lambda}_i^3, \quad N > 2, \quad N_{1i} > 0 \quad (i = 1, 2, 3) \quad (6.23)$$

where N_{1i} are chosen as in the example of Chapter 3.

As mentioned earlier, even from a purely physical consideration we can assume that the missile guidance system with a variable gain, which is bigger, when LOS rate is big, and smaller, when LOS rate is small, would act better than the traditional PN system. The component (6.23) with a properly chosen N_{1i} serves this purpose.

Because of the different "ideology" of the guidance laws considered in this and the previous chapters, we will test the effectiveness of the "cubic" term by considering the commanded acceleration (6.23) and the guidance laws (6.19)–(6.22), i.e., instead of the PN law with $N = 3$ the commanded acceleration has the form of equation (6.23). Formally, the nonlinear term in equation (6.23) does not allow us to rely on the analytical expressions (6.14) and (6.15), because they were obtained for the linear model of the missile guidance system. However, the basic idea to make the actual acceleration closer to the commanded acceleration enables us to assume that the guidance laws (6.19)–(6.22) can be used with the "cubic" term in equation (6.23). The simulation results are presented in Table 6.2.

TABLE 6.2

Analysis of Influence of Nonlinear "Cubic" Term

Case #	Parameters	Time of Intercept (s)	Miss (m)
1	$\omega_z = 30 rad/s$; $N = 3$; $k_1 = 0$; $k_2 = 1$; $\tau_{10} = 0$; $\tau_2 = 0$	7.2	2.01
2	$\omega_z = 30 rad/s$; $N = 3*6/5$; $k_1 = 5$; $k_2 = 0$; $\tau_{10} = 0$; $\tau_2 = 0$; $\mu = 1$	6.1	0.5
3	$\omega_z = 5 rad/s$; $N = 3$; $k_1 = 0$; $k_2 = 1$; $\tau_{10} = 0$; $\tau_2 = 0$	7.0	0.59
4	$\omega_z = 5 rad/s$; $N = 3*6/5$; $k_1 = 5$; $k_2 = 0$; $\tau_{10} = 0$; $\tau_2 = 2.25s$; $\mu = 1$	6.0	0.29
5	$\omega_z = 5 rad/s$; $N = 3$; $k_1 = -1$; $k_2 = 1$; $\tau_{10} = 0.1s$; $\tau_2 0.05s$; $\mu = 0$	6.0	0.28

The simulation results show that the guidance laws (6.8), (6.19)–(6.22) can work successfully with other guidance laws, which contain the PN law as their component.

In contrast to the acceleration feedback used in autopilot systems (its influence is reflected in $G_2(s)$), the additional acceleration feedback described above is used to generate the new guidance law (6.8). It is assumed that the new guidance would work with existing autopilots. However, the expression of (6.8) and, especially, the acceleration operator $G_4(s)$ can also be used for the integrated design of guidance and autopilot systems.

The traditional approach to missile guidance and control system design has been to neglect interactions between these systems. Usually, individual missile systems are designed separately. Then they are assembled together. If the whole system performance is unsatisfactory, the individual subsystems are redesigned to improve the system performance. Because of its iterative nature, the design process can be highly time-consuming and expensive. The approach discussed in this chapter can be considered as an important component of design methods for integrated guidance-autopilot systems.

REFERENCES

1. Gurfil, P., Jodorkovsky, M., and Guelman, M. Neoclassical Guidance for Homing Missiles, *Journal of Guidance, Control, and Dynamics* 24, no. 3 (2001): 452–59.
2. Gurfil, P., Jodorkovsky, M., and Guelman, M. Design of Nonsaturating Guidance Systems, *Journal of Guidance, Control, and Dynamics* 23, no. 4 (2000): 693–700.
3. Dorf, R. C. *Modern Control Systems*. New York, NY: Addison-Wesley, Inc., 1989.
4. Meerov, M. V. *Structural Synthesis of High Accuracy Automatic Control Systems*. Moscow, Russia: Nauka, 1965.

5. Smith, O. *Feedback Control Systems.* New York, NY: McGraw-Hill, Inc., 1958.
6. Shneydor, N. A. *Missile Guidance and Pursuit,* Chichester, UK: Horwood Publishing, 1998.
7. Yanushevsky, R. Analysis of Optimal Weaving Frequency of Maneuvering Targets, *Journal of Spacecraft and Rockets,* 41, no. 3 (2004): 477–79.
8. Yanushevsky, R. Frequency Domain Approach to Guidance System Design, *IEEE Transactions on Aerospace and Electronic Systems,* 43 (2007).
9. Yanushevsky, R. *Theory of Linear Optimal Multivariable Control Systems.* Moscow, Russia: Nauka, 1973.
10. Yanushevsky, R. On the Robustness of the Solution of the Problem of the Analytical Construction of Controls, *Automation and Remote Control* 3 (1966): 356–63.
11. Yanushevsky, R. Robust Stabilizability of Linear Differential-Difference Systems with Unstable D-Operator, *IEEE Transactions on Automation Control* 37, no. 6 (1992): 652–53.
12. Zarchan, P. *Tactical and Strategic Missile Guidance, Progress in Astronautics and Aeronautics.* Vol. 176. Washington, DC: American Institute of Astronautics and Aeronautics, Inc., 1997.
13. Zarchan, P. Proportional Navigation and Weaving Targets, *Journal of Guidance, Control, and Dynamics* 18, no. 5 (1995): 969–74.
14. Zarchan, P., Greenberg, E., and Alpert, J. Improving the High Altitude Performance of Tail-Controlled Endoatmospheric Missiles, *AIAA Guidance, Navigation and Control Conference,* AIAA Paper 2002-4770, August 2002.

7 Guidance Law Performance Analysis under Stochastic Inputs

7.1 INTRODUCTION

The guidance law analysis in the previous chapters was strictly deterministic. It was assumed that the information about their parameters contains no errors and the components of the missile guidance system have fixed parameters, so that no uncertainties exist. This approach is very useful at the initial stage of design.

The realization of the discussed guidance laws requires information about the line-of-sight (LOS) rate, closing velocity, and target acceleration. This information is received from the sensors that measure the variables that are present in the guidance laws. As with any measurements, these measurements are accompanied with noises that can significantly increase the miss distance, if the necessary means are not taken to decrease their influence on vehicle performance.

Here we consider noises that distort the guidance laws and their influence on the miss distance. As to the evaluation of the effect of uncertainties induced by the autopilot and airframe parameters, such types of problem are considered in control theory [1,5,10]. The appropriate results can be used when designing autopilot systems.

Because of the random nature of noises, the analysis of their influence on the miss distance requires the utilization of the mathematical apparatus of random functions, analysis of random processes driven by noise.

Random processes related to the guidance problems can be also stipulated by the random character of target maneuvers.

Below we present basic facts from the theory of stochastic processes, which are necessary for understanding the following material. The characterization of the main sources of noise that influence the proportional navigation (PN) guidance law performance is given. The effect of random disturbances (measurement noise and random maneuvers) on the guided system performance is evaluated by the root-mean-square miss distance. The analytical expressions of the miss distance under the above-mentioned stochastic inputs will be obtained. Analytical expressions for the root-mean-square miss are examined in details for a simple first-order model of the guidance system. For the higher order models, the computational procedure and corresponding algorithms based on the obtained analytical expressions are discussed. The advantage of the approach considered is that it excludes the necessity of simulating the adjoint system to analyze the miss distance under the stochastic inputs (see, e.g., [3,11]) and, as a result, significantly simplifies the computational process.

Analytical expressions and related computational algorithms are given for the PN guided vehicles. However, the same expressions can be used for analysis of the more sophisticated guidance laws considered in Chapter 6 if, instead of the real flight control system, we will operate with a fictitious one [see equation (6.9)].

In Chapters 3–6 we considered the laws applied to missile guidance. In the next chapter we will show that they can be applied directly or with some modification to a wide class of unmanned aerial vehicles. Moreover, since the PN guidance law is the main component of these laws, the analysis of its influence on the miss distance is useful at the initial stage of design of all guided unmanned aerial vehicles.

7.2   BRIEF DISCUSSION OF STOCHASTIC PROCESSES

The theory of stochastic signals in its most general form is extremely abstract, and a rigorous presentation requires a degree of mathematical sophistication beyond the scope of this book. Our main objective here is to present a specific set of results related to random signals that will be used later. Although the material related to random signals cannot be considered rigorous, we will summarize the important results and the mathematical assumptions accompanying them. We assume that the reader is familiar with fundamental concepts of the theory of probability such as random variables, probability distributions, and averages.

The general concept of a stochastic process can be stated in the following way. Let U be a set of elementary events and t is a continuous parameter. A stochastic process $\eta(t)$ is defined as the function of two arguments:

$$\eta(t) = f(e,t), \quad e \in U, \quad t \in T \tag{7.1}$$

For every moment of time t, the function $f(e, t)$ is a function of e only and, consequently, is a random variable. For every fixed value of the argument e (i.e., for every elementary event), $f(e, t)$ depends only on t (i.e., is a function of time). Every such function is called a realization, or sample function, of the stochastic process $\eta(t)$. A stochastic process can be regarded either as a collection of random variables $\eta(t)$ that depend on the parameter t, or as a collection of the realizations of the process $\eta(t)$. To define a process it is necessary to specify a probability measure (e.g., a set of probability distribution functions) in the functional space of its realizations.

The probability law of a random variable η can always be specified by stating its distribution function or the probability density function $p(\eta)$.

The average or mean of a random variable η is defined by:

$$m_\eta = E[\eta] = \int_{-\infty}^{\infty} \eta p(\eta) d\eta \tag{7.2}$$

The variance of η is defined by:

$$\text{Var}[\eta] = E[(\eta - E[\eta])^2] = \sigma^2[\eta] = E[\eta^2] - E^2[\eta] \tag{7.3}$$

The standard deviation of η is defined by:

$$\sigma[\eta] = \sqrt{\text{Var}[\eta]} \tag{7.4}$$

It can be shown that, if random variables η_i are independent, the mean m_0 and the variance σ_0^2 of the sum are the sum of the means and variances, respectively, i.e.,

$$m_0 = E\left[\sum \eta_i\right] = \sum E[\eta_i] = \sum m_{\eta_i} \tag{7.5}$$

and

$$\sigma_0^2 = E\left[\left(\sum \eta_i - m_0\right)^2\right] = \sum E\left[(\eta_i - m_{\eta_i})^2\right] = \sum \sigma_{\eta_i}^2 \tag{7.6}$$

Below we present two important probability density functions: the uniform distribution $p_{\text{uniform}}(\eta)$ and the Gaussian or normal distribution $p_{\text{normal}}(\eta)$.

For the uniform distribution we have:

$$p_{\text{uniform}}(\eta) = \frac{1}{b-a}, \quad m_{\text{uniform}} = \frac{1}{b-a}\int_a^b \eta d\eta = \frac{b+a}{2} \tag{7.7}$$

and

$$\sigma_{\text{uniform}}^2 = E[\eta^2] - m_{\text{uniform}}^2 = \frac{b^3 - a^3}{3(b-a)} - \left(\frac{b+a}{2}\right)^2 = \frac{(b-a)^2}{12} \tag{7.8}$$

where the variable $\eta \in [a, b]$.

For the normal distribution we have:

$$p_{\text{normal}}(\eta) = \frac{1}{\sqrt{2\pi}\sigma_{\text{normal}}} \exp\left[-\frac{(\eta - m_{\text{normal}})^2}{2\sigma_{\text{normal}}}\right] \qquad (7.9)$$

where m_{normal} and σ_{normal}^2 are the mean and variance, respectively.

Analogous to the averages that characterize a random variable, the ensemble averages such as the average or mean of a process:

$$m(t) = E[\eta(t)] = \int_{-\infty}^{\infty} \eta(t)p(\eta,t)d\eta \qquad (7.10)$$

the mean-square value:

$$E[\eta^2(t)] = \int_{-\infty}^{\infty} \eta^2(t)p(\eta,t)d\eta \qquad (7.11)$$

the rms, the root-mean-square (rms) value:

$$rms = \sqrt{E[\eta^2(t)]} \qquad (7.12)$$

and the variance:

$$\sigma_\eta^2(t) = E[(\eta(t) - E[\eta(t)])^2] = E[\eta^2(t)] - E^2[\eta(t)] \qquad (7.13)$$

are introduced for stochastic processes, where $p(\eta, t)$ is the probability density function of a stochastic process introduced formally analogous to the probability density function of a random variable.

The square root of the variance $\sigma_\eta^2(t)$ is also known as the standard deviation. For random processes with zero mean the rms value coincides with the standard deviation.

The role played for a random variable η by its mean and variance is played for a stochastic process by its mean value function and its covariance kernel:

$$Cov[\eta(\tau),\eta(t)] = E[(\eta(\tau) - E[\eta(\tau)])(\eta(t) - E[\eta(t)])] \qquad (7.14)$$

A stochastic process $\eta(t)$ is called stationary if the probability distribution functions for two finite groups of variables $\eta(t_1)$, $\eta(t_2),\ldots,\eta(t_n)$ and $\eta(t_1 - k)$, $\eta(t_2 - k),\ldots,\eta(t_n - k)$ coincide and, hence, are independent of k.

Clearly, any numerical characteristic of a stationary process $\eta(t)$ is independent of the time t, i.e., for the expectation and variance we have:

$$E[\eta(t_i)] = m_\eta, \quad \text{Var}[\eta(t_i)] = \sigma_\eta^2 \quad (-\infty < t_i < \infty) \tag{7.15}$$

The covariance kernel $Cov[\eta(\tau), \eta(t)]$ of a stationary process is a function of the absolute difference $|t - \tau|$, i.e.,

$$Cov[\eta(t), \eta(t + \tau)] = R_\eta(\tau) \tag{7.16}$$

where $R_\eta(\tau)$ is called the covariance function.

The stochastic process $\eta(t)$ was defined as the function of two arguments. It can be shown that for a class of stochastic processes, called the ergodic processes, which includes the stationary processes (more rigorous formulation can be found, e.g., in [2]), averages computed from a sample of a stochastic process can be identified with corresponding ensemble averages.

Given a finite sample $\{\eta(t),\ 0 \le t \le T\}$ of the process, we define the sample covariance function for stationary stochastic processes with zero mean as:

$$R_{\eta T}(\tau) = \frac{1}{T} \int_{t_0}^{t_0 + T} \eta(t)\eta(t + \tau)dt \tag{7.17}$$

Based on the ergodic property:

$$\lim_{T \to \infty} R_{\eta T}(\tau) = R_\eta(\tau) \tag{7.18}$$

where $R(\tau)$ is called the autocorrelation function.

Frequency-domain methods, based on the Fourier transform, are widely used for analysis of deterministic signals. The Fourier transform of the autocorrelation function plays the significant role in analysis of the stationary random signals also.

The Fourier transform of the function $(1/2\pi) R_\eta(\tau)$:

$$\Phi_\eta(\omega) = \frac{1}{2\pi} \int_{-\infty}^{\infty} R_\eta(\tau)e^{-i\omega\tau}d\tau \tag{7.19}$$

of the stationary random function $\eta(t)$ is called the power spectral density of $\eta(t)$ or simply the spectral density.

Using the expression of the inverse Fourier transform and equations (7.13), (7.17), and (7.18), we can obtain

$$\sigma_\eta^2 = \int_{-\infty}^{\infty} |\eta(t)|^2 dt = R_\eta(0) = \int_{-\infty}^{\infty} \Phi_\eta(\omega)d\omega \qquad (7.20)$$

This expression is widely used to determine the mean-square value of a stationary random function.

The spectral density of a signal exists, if and only if the signal is a wide-sense stationary [2]. If the signal is not stationary, then the same methods used to calculate the spectral density can still be used, but the result cannot be called the spectral density. Based on the above expression, it can be shown that the power spectral density $\Phi_\eta(\omega)$ of the signal $\eta(t)$ is the square of the magnitude of the Fourier transform $\Pi(\omega)$ of the signal (Parseval's theorem), i.e.,

$$\Phi_\eta(\omega) = \frac{\Pi(\omega)\Pi^*(\omega)}{2\pi} = \left| \frac{1}{2\pi} \int_{-\infty}^{\infty} \eta(t)e^{-i\omega t}dt \right|^2 \qquad (7.21)$$

where the symbol "*" indicates the complex conjugate operation.

The random process is said to be a white noise process, if it possesses a constant power spectral density Φ_η. White noise is so called as an analogy with white light, which contains all frequencies. An infinite-bandwidth white noise signal is purely a theoretical construct. By having power at all frequencies, the total power of such a signal is infinite. White noise is the abstract, physically unrealizable, random process with autocorrelation function equal to a delta-function (as mentioned, it is equivalent to a constant spectral density). Despite its abstract nature, white noise is widely used for analysis of real systems, when the real noise bandwidth significantly exceeds the system bandwidth (i.e., in practice, a signal can be "white" with a flat spectrum over a defined frequency band).

The above definition of white noise states only that it has equal energy at all frequencies and refers to correlations at two distinct times, which are independent of the noise amplitude distribution. While the frequency distribution may be the same, the amplitude distribution can be different. Noise with a Gaussian amplitude distribution (normal distribution) is called Gaussian noise. This says nothing of the correlation of the noise in time or of the spectral density of the noise. It is often incorrectly assumed that Gaussian noise is necessarily white noise. Gaussianity refers to the way signal values are distributed, while the term "white" refers to correlations at two distinct times, which are independent of the noise amplitude distribution. Gaussian white noise (white noise with a Gaussian amplitude distribution also called the pseudowhite noise) is a good approximation of many real-world situations and generates mathematically tractable models. A useful relationship between the desired white noise spectral density Φ_η and the standard deviation σ of the pseudowhite noise (white noise has infinite standard deviation), i.e., the Gaussian random numbers generated every interval Δ, is given by [11]:

$$\Phi_\eta = \sigma^2 \Delta \tag{7.22}$$

The output $y(t)$ of a linear system with the impulse response $P(t, \tau)$, when the input signal $\eta(t)$ is white noise with the spectral density Φ_η, equals

$$y(t) = \int_{-\infty}^{t} P(t, \tau) \eta(\tau) d\tau \tag{7.23}$$

The mean-square value of $y(t)$ can be presented as:

$$E[y^2(t)] = \int_{-\infty}^{t} \int_{-\infty}^{t} P(t, \tau_1) P(t, \tau_2) E[\eta(\tau_1) \eta(\tau_2)] d\tau_1 d\tau_2 \tag{7.24}$$

Taking into account that the autocorrelation function (7.17) of white noise equals

$$E[\eta(\tau_1) \eta(\tau_2)] = \Phi_\eta \delta(\tau_1 - \tau_2) \tag{7.25}$$

the previous expression can be simplified as:

$$E[y^2(t)] = \Phi_\eta \int_{-\infty}^{t} P^2(t, \tau) d\tau \tag{7.26}$$

It follows from equation (7.26) that the mean-square response of a linear system influenced by white noise with the spectral density Φ_η is proportional to the integral of the square of the impulse response.

7.3   RANDOM TARGET MANEUVERS

In Chapter 4 we obtained the analytical expression for the miss distance for a constant maneuver a_T. This type of maneuver is convenient for the analytical analysis but is far from reality. In Chapter 5 we considered sinusoidal target maneuvers and determined the miss peak for the various times of flight t_F. The shape of this maneuver policy is deterministic and quite realistic. It corresponds to the so-called barrel-roll strategy, which in contrast to the sinusoidal deterministic maneuver is random. The miss peak enables us to evaluate, at a certain degree, the worst case. A more realistic scenario corresponds to sinusoidal target maneuvers with random starting times of the maneuver (i.e., with a random phase).

The result of [3], showing how to use the concept of a shaping filter to the statistical representation of signals with known form but random starting time, is used here to write the expressions for the random step and sinusoidal signals, if their starting time is uniformly distributed over the flight time.

The ability of using shaping filters exited by white noise to generate random signals follows from equations (7.23)–(7.26) and (7.21) and based on the fact that random processes that have the same mean and autocorrelation functions are mathematically equivalent for problems dealing with the mean-square values of random signals.

A signal $x(t)$ of a known form $a(t)$ with random starting time T can be presented as:

$$x(t) = a(t - T)s(t - T) \qquad (7.27)$$

where $s(t)$ is the unit step function, i.e., $s(t) = 0$, $t < 0$ and $s(t) = 1$, $t \geq 0$.

In the case of uniformly distributed starting time over the flight time t_F, the probability density function $p_T(t)$ of T is given by:

$$p_T(t) = \begin{cases} 1/t_F, & 0 \leq t \leq t_F \\ 0, & t > t_F \end{cases} \qquad (7.28)$$

Hence, the autocorrelation function of the signal (7.27) with random starting time is:

$$R_x(t_1, t_2) = E[x(t_1)x(t_2)] = \int_{-\infty}^{\infty} x(t_1)x(t_2)p_T(T)dT \qquad (7.29)$$

For a random step signal a_T we have:

$$R_x(t_1, t_2) = \int_0^{t_F} a_T^2 s(t_1 - T)s(t_2 - T)dT/t_F \qquad (7.30)$$

Assuming $0 \leq t_1 \leq t_2 \leq t_F$, the above expression can be simplified to:

$$R_x(t_1, t_2) = \frac{a_T^2}{t_F} \int_0^{t_1} s(t_1 - T)s(t_2 - T)dT \qquad (7.31)$$

The autocorrelation function of the output $y(t)$ of the linear time-invariant system with the impulse response $P(t)$, when the input signal $\eta(t)$, $0 \le t \le t_F$, is white noise with the spectral density Φ_η, equals [see equations (7.17), (7.18), (7.23), and (7.25)]:

$$R_y(t_1,t_2) = \int_0^{t_1} \int_0^{t_2} P(t_1 - \tau_1) P(t_2 - \tau_2) \Phi_\eta \delta(\tau_1 - \tau_2) d\tau_1 d\tau_2 \quad (7.32)$$

Assuming $0 \le t_1 \le t_2 \le t_F$, the above expression becomes:

$$R_y(t_1,t_2) = \Phi_\eta \int_0^{t_1} P(t_1 - \tau_1) P(t_2 - \tau_1) d\tau_1 \quad (7.33)$$

Equations (7.31) and (7.33) are equivalent, if

$$\Phi_\eta = a_T^2 / t_F \quad \text{and} \quad P(t) = 1 \quad (7.34)$$

For the sinusoidal target maneuver that starts at time T, we have:

$$x(t) = a_T \sin(\omega_T t - T) s(t - T) \quad (7.35)$$

so that, instead of equation (7.30), the autocorrelation function of this signal with random starting time is:

$$R_x(t_1,t_2) = \int_0^{t_F} a_T^2 \sin(\omega_T t_1 - T) \sin(\omega_T t_2 - T) s(t_1 - T) s(t_2 - T) dT / t_F$$

$$(7.36)$$

or, assuming $0 \le t_1 \le t_2 \le t_F$:

$$R_x(t_1,t_2) = \frac{a_T^2}{t_F} \int_0^{t_1} \sin(\omega_T t_1 - T) \sin(\omega_T t_2 - T) dT \quad (7.37)$$

Equations (7.33) and (7.37) are equivalent, if

$$\Phi_\eta = a_T^2 / t_F \quad \text{and} \quad P(t) = \sin \omega_T \quad (7.38)$$

The above consideration shows that step and sinusoidal maneuvers of amplitude a_T, whose starting time is uniformly distributed over the flight time t_F, i.e., the probability density function is given by equation (7.28), have the same autocorrelation function as a linear network, driven by white noise with spectral density $\Phi_\eta = a_T^2/t_F$, with transfer functions:

$$W_{filter}(s) = \frac{1}{s} \qquad (7.39)$$

and

$$W_{filter}(s) = \frac{1/\omega_T}{s^2/\omega_T^2 + 1} \qquad (7.40)$$

respectively.

7.4 ANALYSIS OF INFLUENCE OF NOISES ON MISS DISTANCE

As mentioned earlier, the ability to use the adjoint system to obtain the analytical expression (4.13) for the miss distance is stipulated by specifics of the considered model of the missile guidance system. Its state matrix [see, e.g., equation (3.8)] is a function of $t_F - t$. Its impulse response as a function of t_F can be analyzed as a function of σ ($0 \leq \sigma \leq t_F$), where σ is the impulse application time, by varying $t_F - \sigma$. The relationship between the impulse responses of the adjoint (more precisely, modified adjoint system) and original systems $P_{ma}(t_F - \sigma, t_F - t_0) = P(t_0, \sigma)$, where t_0 is the impulse observation time, enables us to examine $P(t_F, \sigma)$ as a function of t_F by considering $P_{ma}(t_F - \sigma, t_F - t_F) = P_{ma}(t_F - \sigma, 0)$ as a function of σ ($0 \leq \sigma \leq t_F$).

Here the adjoint system impulse response is used for statistical analysis of the original system in the presence of stochastic inputs.

The root-mean-square (rms) response $y(t_F)$ of a linear time-varying system with the impulse function $P_0(t, \sigma)$ at the finite time t_F, stipulated by the white noise input with the spectral density Φ_n, is presented [see equation (7.26)] by:

$$\text{rms} = \{E[y^2(t_F)]\}^{1/2} = \sqrt{\Phi_n \int_0^{t_F} P_0^2(t_F, \sigma) d\sigma} \qquad (7.41)$$

The analytical expressions for $P_0(t_F, \sigma)$ will be obtained by the method of adjoints. Earlier we used the method of adjoints to obtain analytical expressions of the miss distance assuming that the target acceleration is a deterministic function of time, here we will evaluate the effect of random disturbances (measurement noise and random target maneuvers) on the guided missile system performance, choosing the root-mean-square miss criterion. Assuming the linearized engagement planar model, the PN law, and the linear guidance system dynamics (see Figure 5.3) we will use the

expression (5.10) and (5.34) to analyze the influence of the random disturbances on the miss distance.

Figure 7.1 is similar to Figure 5.6. It contains only additional stochastic
inputs, which will be discussed in details.

Target tracking represents an estimation of position, velocity, and acceleration of a target. The estimation must handle different perturbations.

One of the perturbations is glint noise. Glint noise occurs, when radar is
used in target tracking, because of interference between the reflected radar
waves. In real radar target tracking systems, changes in the target aspect
with respect to the radar can cause the apparent center of radar reflections
(direction "seen" by the antenna) to wander significantly. The random wan-
dering of the apparent radar reflecting center gives rise to noisy or jittered
angle tracking. This form of measurement noise is called angle fluctuations or target glint. Glint noise mainly affects the performance of radar-
guided missiles and to a smaller extent the performance of missiles with
electro-optical seekers, where infrared radiance fluctuations are smaller
than radar reflection fluctuations [6].

Glint affects the measurement components (mostly the angles) by producing heavy-tailed, non-Gaussian disturbances, which may severely affect the
tracking accuracy. Glint noise is non-Gaussian, which makes the estimation
more difficult. One of the most common target tracking methods today is the
Kalman filter. This method assumes the perturbations to be Gaussian, but this
will not be true for glint noise. In target tracking, the measurement noise is
usually assumed to be Gaussian. However, as mentioned above, the distribution of glint noise is non-Gaussian long-tailed. Moreover, it may be highly
correlated, so that, to be rigorous, it cannot be modeled as white noise.

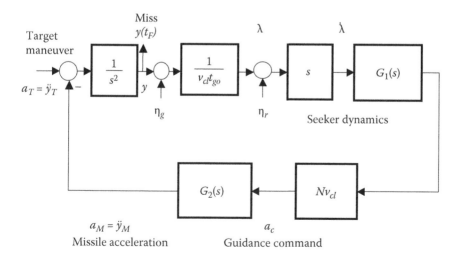

FIGURE 7.1 Missile guidance model.

Nevertheless, to get the analytical expression that enables us to evaluate approximately the rms due to glint noise η_g (see Figure 7.1), we assume that η_g is white noise with the power density Φ_{gn}.

The model of the system adjoint to the system in Figure 7.1 is presented in Figure 7.2. The output $P_g(t_F, t)$ corresponds to the impulse response of the PN missile guidance system to the white noise input $\eta_g(t)$. The analytical expression for $P_g(t_F, s)$ can be written based on the earlier obtained expression of

$$P_T(t_F, s) = \exp\left(\int_{\infty}^{s} NH(\sigma)d\sigma\right)$$

(see (5.10)). For the structure in Figure 7.2 with the input $\delta(t)$ and the output $P_g(t_F, t)$ we have:

$$P_g(t_F, t) = \delta(t) - P_T(t_F, t)$$

or

$$P_g(t_F, s) = 1 - P_T(t_F, s) \tag{7.42}$$

Hence, based on equation (7.41), the rms miss distance due to glint noise can be obtained from the equation:

$$\frac{rms^2}{\Phi_{gn}} = \frac{E[y^2(t_F)]_{glint}}{\Phi_{gn}} = \int_0^{t_F} \left\{L^{-1}\left[1 - P_T(t_F, s)\right]\right\}^2 dt \tag{7.43}$$

where L^{-1} is the symbol of the inverse Laplace transform.

Noises that accompany range measurements are divided usually on range independent and range dependent. Within the structure in Figure 7.1, we will link the range independent and dependent noise with the LOS measurements. For the midcourse guidance, when the missile is guided by radar, the information concerning the LOS is transmitted based on measurements of the range components [see equation (1.8)]. During the terminal stage for semiactive systems in which the target is illuminated by a transmitter (illuminator) situated not on the missile, range dependent

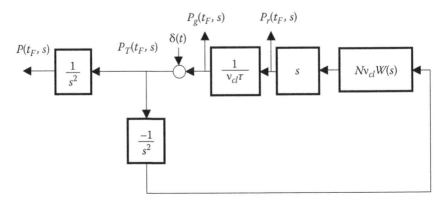

FIGURE 7.2 Modified block diagram of adjoint system.

noise, sometimes called fading noise, is the terminal noise produced in the missile receiver and also caused by various factors (e.g., by signal processing effects or a source independent of the radar, such as a jammer). For active systems, in which the missile has its own radar, range dependent noise, also called receiver noise, is also the terminal noise produced in the missile receiver. In both cases the noise dependence on range follows from the radar range equation, which shows that the signal noise ratio (SNR = $\sigma^2_{signal}/\sigma^2_{noise}$) is inversely proportional to the fourth power of range r [4]. The factor r^{-4} in the radar equation characterizes the divergence of the electromagnetic radiation with range r (on the outward and return pass). In contrast to active systems, semiactive systems operate only with the return pass. In this case, a received signal power is inversely proportional to the second power of range. That is why the active-receiver noise for active systems is proportional to the square of the distance between the missile and the target, and the passive-receiver noise for semiactive systems is proportional to the distance between the missile and the target.

The output $P_r(t_F, t)$ in Figure 7.1 corresponds to the impulse response of the PN missile guidance system to the white noise input $\eta_r(t)$, which models the range independent LOS angular noise. The analytical expression for $P_r(t_F, s)$ is given based on the earlier obtained expression of $P_T(t_F, s)$ and following the rules of operations with transfer functions in control theory [1,5]. We have

$$P_r(t_F,s) = P_T(t_F,s)\left(-\frac{1}{s^2}\right)Nv_{cl}W(s)s \qquad (7.44)$$

It is easy to check that the absolute value of the above expression equals the absolute value of the product of the derivative of $P_T(t_F, s)$ and the closing velocity [see also equation (4.14) describing the relationship between $H(s)$ and $W(s)$]:

$$\frac{dP_T(t_F,s)}{ds} = \exp\left(\int_\infty^s NH(\sigma)d\sigma\right)NH(s)$$

$$= \exp\left(\int_\infty^s NH(\sigma)d\sigma\right)N\frac{W(s)}{s} = P_T(t_F,s)N\frac{W(s)}{s} \qquad (7.45)$$

By comparing equations (7.44) and (7.45) we can obtain the rms miss distance due to range independent noise from:

$$\frac{\text{rms}^2}{\Phi_{fn}} = \frac{E[y^2(t_F)]_{independent\ noise}}{\Phi_{fn}} = \int_0^{t_F}\left\{L^{-1}\left[v_{cl}\frac{dP_T(t_F,s)}{ds}\right]\right\}^2 dt \qquad (7.46)$$

where Φ_{fn} is the power spectral density of range independent noise.

As seen from equation (7.46), the rms value is proportional to the closing velocity, i.e., higher closing velocity yields more miss distance due to range independent noise.

The block-diagram of the missile guidance system for the case of range dependent noise is given in Figure 7.3. Usually spectral density of range dependent noise is given for a certain reference range r_0, so that the noise level is estimated with respect to the chosen reference level. The white noise signal $\eta_r(t)$ inputs the unit with the gain equal the range $(r/r_0)^i$ of power i, where $i = 1$ for semiactive systems and $i = 2$ for active systems.

Using the known expression $r = v_{cl}(t_F - t)$, we present the system, adjoint to the given one in Figure 7.3, with the output $P_{ri}(t_F, s)$ in the form shown in Figure 7.4.

Taking into account that:

$$P_r(t_F, s) = -v_{cl} \frac{dP_T(t_F, s)}{ds} \tag{7.47}$$

and that for a function $f(t)$ the Laplace transform

$$L\{t^i f(t)\} = (-1)^i \frac{d^i L\{f(t)\}}{ds^i}$$

using equation (7.46) we can write the expression for the rms miss distance due to passive-receiver noise:

$$\frac{rms^2}{\Phi_{pn}} = \frac{E\left[y^2(t_F)\right]_{passive}}{\Phi_{pn}} = \int_0^{t_F} \left\{ L^{-1}\left[\frac{v_{cl}^2}{r_0} \frac{d^2 P_T(t_F, s)}{ds^2} \right] \right\}^2 dt \tag{7.48}$$

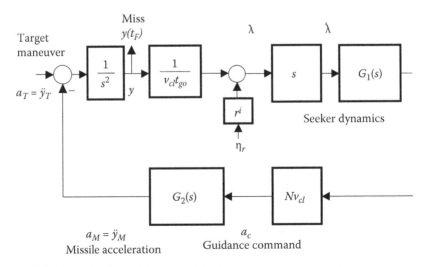

FIGURE 7.3 Missile guidance model with range dependent noise.

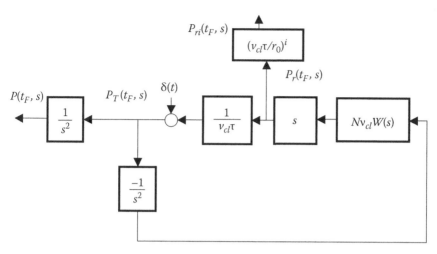

FIGURE 7.4 Modified block diagram of adjoint system with range dependent noise.

and for the rms miss distance due to active-receiver noise equals:

$$\frac{rms^2}{\Phi_{an}} = \frac{E[y^2(t_F)]_{active}}{\Phi_{an}} = \int_0^{t_F} \left\{ L^{-1} \left[\frac{v_{cl}^3}{r_0^2} \frac{d^3 P_T(t_F,s)}{ds^3} \right] \right\}^2 dt \qquad (7.49)$$

where Φ_{pn} and Φ_{an} are the power spectral density of the passive-receiver noise and the active-receiver noise, respectively.

As seen from the above expressions (7.47)–(7.49), for the passive- and active-receiver noise the rms value is proportional to the square and cube of the closing velocity, respectively, i.e., higher closing velocity yields more miss distance due to range dependent noise.

The neoclassical guidance discussed in the previous chapter is based on the structure of the missile guidance system that corresponds to $P_T(t_F, s) = 0$ [see equations (5.10), (6.1), (6.4), and (6.5)]. As it follows from equations (7.46), (7.48), and (7.49), the neoclassical guidance suppresses range independent and dependent noises, i.e., eliminates their influence on the miss distance. The case of glint noise is different. For long flight times the right part of equation (7.43) is close to 1 (for $t_F \to \infty$ it tends to 1), so that the neoclassical guidance does not reduce the influence of glint on the miss distance. However, even the "theoretical ability" of the neoclassical guidance to nullify the influence of range independent and dependent noises is difficult to realize in practice, especially for fin-controlled missiles. As indicated earlier, the neoclassical guidance structure requires proportional-derivative controllers, which are sensitive to noise, i.e., they create additional noise sources whose effect on the miss distance can be significant.

7.5 EFFECT OF RANDOM TARGET MANEUVERS ON MISS DISTANCE

It was shown above that step maneuvers of amplitude a_T, whose starting time is uniformly distributed over the flight time t_F, and the random-phase sinusoidal maneuvers $a_T(t) = a_T \sin_T(\omega_T t + \varphi_T)$, where φ_T is a uniformly distributed random variable, can be represented as a white noise process with the power spectral density $\Phi_\varphi = a_T^2/t_F$ passing through the shaping filter with the transfer functions $W_{filter}(s) = 1/s$ and $W_{filter}(s) = (s^2/\omega_T + \omega_T)^{-1}$, respectively [see equations (7.39) and (7.40)]. The missile guidance model that reflects random target maneuvers is shown in Figure 7.5. The adjoint model is given in Figure 7.6.

It follows immediately from Figure 7.6 and the expressions (7.34) and (7.38)–(7.40) that for the step maneuvers with starting time uniformly distributed over the flight time t_F the rms miss distance is given by:

$$\frac{rms^2}{\Phi_\varphi} = \frac{E\left[y^2(t_F)\right]_\varphi}{\Phi_\varphi} = \int_0^{t_F} \left\{ L^{-1}\left[\frac{P_T(t_F,s)}{s^3}\right]\right\}^2 dt \qquad (7.50)$$

and for the random-phase sinusoidal maneuvers the rms miss distance is:

$$\frac{rms^2}{\Phi_\varphi} = \frac{E\left[y^2(t_F)\right]_\varphi}{\Phi_\varphi} = \int_0^{t_F} \left\{ L^{-1}\left[(s^2/\omega_T + \omega_T)^{-1}\frac{P_T(t_F,s)}{s^2}\right]\right\}^2 dt \qquad (7.51)$$

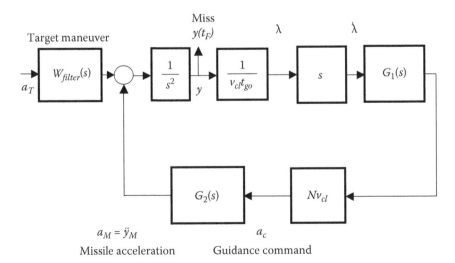

FIGURE 7.5 Missile guidance model with random target maneuvers.

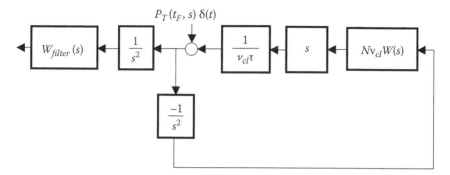

FIGURE 7.6   Block diagram of adjoint system for random maneuvers.

Since for the neoclassical guidance $P_T(t_F, s) = 0$, based on equations (7.50) and (7.51) we can conclude that such guidance is the best remedy against maneuvering targets. However, as mentioned in the previous chapter and in the previous section, the seeming simplicity of the neoclassical guidance is deceptive.

The rms miss distance in equations (7.50) and (7.51) corresponds to the steady-state miss (i.e., when the transient response of the missile guidance system has disappeared) [7–9]. The expressions (7.43), (7.46), and (7.48)–(7.51) enable us to analyze the influence of random disturbances on the miss distance and to design filters that would decrease this influence.

7.6 COMPUTATIONAL ASPECTS

The analytical expressions obtained for the rms miss distance can be easily transformed into computational algorithms. It is important to mention that the computational programs based on these algorithms are more compact and the time of computing is less (currently computers are very fast, so that the time factor does not play a dominant role) than by using the existing modeling procedure of the method of adjoints in the time domain. Moreover, the computational programs can be flexible enough to analysis the influence of various parameters of the missile guidance system's performance.

In Chapter 5 we showed that the impulse response $P(t_F, t)$ can be calculated based on the real part of the frequency response $\mathrm{Re}[P(t_F, i\omega)]$ [see equation (5.60)]. Analogous to equation (5.60), the expression of the impulse response $P_T(t_F, t)$ can be obtained, i.e.,

$$P_T(t_F, t) = \frac{2}{\pi} \int_0^\infty \mathrm{Re}[P_T(t_F, i\omega)] \cos \omega t \, d\omega \qquad (7.52)$$

Multiplying

$$\text{Re}[P_T(t_F,i\omega)], \quad \text{Re}\left[\frac{dP_T(t_F,i\omega)}{ds}\right],$$

$$\text{Re}\left[\frac{d^2P_T(t_F,i\omega)}{ds^2}\right], \quad \text{and} \quad \text{Re}\left[\frac{d^3P_T(t_F,i\omega)}{ds^3}\right]$$

by $\cos\omega t$ and integrating similarly to equation (7.52), we can obtain the expressions of the impulse responses corresponding to the expressions in parentheses of equations (7.43), (7.46), (7.48), and (7.49), respectively. It is impossible to obtain analogously the expressions of the inverse Laplace transform of equations (7.50) and (7.51), because the Fourier transform corresponding to the Laplace transform of equations (7.50) and (7.51) does not exist. However, taking into account that [see equations (5.10) and (5.11)]:

$$\frac{P_T(t_F,s)}{s^3} = \frac{P(t_F,s)}{s} \tag{7.53}$$

the inverse Laplace transform of $P_T(t_F, s)/s^3$ can be obtained analogous to the step miss using equation (5.62). The approximation of equation (7.51) will be discussed later.

The numerical integration of the modified equations (7.43), (7.46), and (7.48)–(7.51), does not present any difficulties.

The above-indicated computational procedure can be simplified, if in all operations we would operate with the real and imaginary parts of $P_T(t_F, i\omega)$. Taking into account that:

$$\left.\frac{dP_T(t_F,s)}{ds}\right|_{s=i\omega} = -i\frac{dP_T(t_F,i\omega)}{d\omega}$$

$$\left.\frac{d^2P_T(t_F,s)}{ds^2}\right|_{s=i\omega} = (-i)^2\frac{d^2P_T(t_F,i\omega)}{d\omega^2}$$

and

$$\left.\frac{d^3P_T(t_F,s)}{ds^3}\right|_{s=i\omega} = (-i)^3\frac{d^3P_T(t_F,i\omega)}{d\omega^3}$$

and presenting $P_T(t_F, i\omega) = \text{Re}[P_T(t_F, i\omega)\} + i\text{Im}[P_T(t_F, i\omega)]$, after simple operations we can write:

$$\text{Re}\left(\frac{d^kP_T(t_F,i\omega)}{ds^k}\right) = (-i)^{k-1}\frac{d^k\,\text{Im}[P_T(t_F,i\omega)]}{d\omega^k}, \quad k=1,3$$

and

$$\text{Re}\left(\frac{d^2 P_T(t_F, i\omega)}{ds^2}\right) = -i\frac{d^2 \text{Re}[P_T(t_F, i\omega)]}{d\omega^2}, \quad k = 2 \qquad (7.54)$$

Since the expressions (7.43), (7.46), and (7.48)–(7.51) include the squares of the inverse Laplace transform values, we can ignore signs in the above expressions.

The expressions (7.43), (7.46), and (7.48)–(7.49) can be reduced to:

$$\frac{E[y^2(t_F)]_{glint}}{\Phi_{gn}} = \int_0^{t_F}\left\{\frac{2}{\pi}\int_0^\infty \text{Re}[1 - P_T(t_F, i\omega)]\cos\omega t d\omega\right\}^2 dt \qquad (7.55)$$

for the rms miss distance due to glint noise;

$$\frac{E[y^2(t_F)]_{independent\ noise}}{\Phi_{fn}} = \int_0^{t_F}\left\{\frac{2}{\pi}v_{cl}\int_0^\infty\frac{d\,\text{Im}[P_T(t_F, i\omega)]}{d\omega}\cos\omega t d\omega\right\}^2 dt$$

$$(7.56)$$

for the rms miss distance due to range independent noise;

$$\frac{E[y^2(t_F)]_{passive}}{\Phi_{pn}} = \int_0^{t_F}\left\{\frac{2}{\pi}\frac{v_{cl}^2}{r_0}\int_0^\infty\frac{d^2\,\text{Re}[P_T(t_F, i\omega)]}{d\omega^2}\cos\omega t d\omega\right\}^2 dt \qquad (7.57)$$

for the rms miss distance due to passive-receiver noise;

$$\frac{E[y^2(t_F)]_{active}}{\Phi_{pn}} = \int_0^{t_F}\left\{\frac{2}{\pi}\frac{v_{cl}^3}{r_0^2}\int_0^\infty\frac{d^3\,\text{Im}[P_T(t_F, i\omega)]}{d\omega^3}\cos\omega t d\omega\right\}^2 dt \qquad (7.58)$$

for the rms miss distance due to active-receiver noise.

It is important to underline that the above expressions are valid, because the Fourier transform exists (see Chapter 5 and Appendix B). As mentioned earlier, the expressions for the rms miss distance due to random maneuvers (7.50) and (7.51) cannot be written analogously because the Fourier transform of functions having the Laplace transform

$$\frac{P_T(t_F, s)}{s^3} \quad \text{and} \quad \left(\frac{s^2}{\omega_T} + \omega_T\right)^{-1}\frac{P_T(t_F, s)}{s^2}$$

does not exist. Based on equation (7.53), we can present equation (7.50) analogous to the expression (4.62) for the step miss, i.e., for the step maneuvers with starting time uniformly distributed over the flight time t_F the rms miss distance is:

$$\frac{E[y^2(t_F)]_\varphi}{\Phi_\varphi} = \int_0^{t_F} \left\{ \frac{2}{\pi} \int_0^\infty \frac{\mathrm{Re}[P(t_F,i\omega)]}{\omega} \sin\omega t d\omega \right\}^2 dt \qquad (7.59)$$

For the random-phase sinusoidal maneuvers, the simplified form of equation (7.51) can be written only under assumption that we can neglect the transient of $P(t_F, s)$, so that only the stationary term of the inverse Laplace transform is considered. The approximate value of the rms miss distance (in some cases the mistake can be significant) is given by:

$$\frac{E[y^2(t_F)]_\varphi}{\Phi_\varphi} = \int_0^{t_F} \{|P(t_F,i\omega_T)| \sin(\omega_T t + \varphi(t_F,i\omega_T))\}^2 dt \qquad (7.60)$$

where the amplitude $|P(t_F,i\omega_T)|$ and phase $\varphi(t_F, i\omega_T)$ are determined from equations (5.26) and (5.28), respectively.

The described approach is used below to analyze the rms miss distance for the missile guidance systems with parameters similar to the considered in the previous chapters.

7.7   EXAMPLES

For the first-order model of the missile guidance system, the expressions (7.43), (7.46), and (7.48)–(7.51) can be obtained in the analytical form. We will write them and compare with the results obtained based on the algorithmic procedure described below. This procedure will be applied to more realistic models of the missile guidance system to demonstrate how misleading the results obtained from the analysis of a simple first-order model can be. For this model we have [see equations (3.19) and (4.21)]:

$$P_T(t_F,s) = \left(\frac{s}{s+1/\tau} \right)^N$$

so that

$$1 - P_T(t_F,s) = 1 - \left(\frac{s}{s+1/\tau} \right)^N \qquad (7.61)$$

$$\frac{dP_T(t_F,s)}{ds} = \frac{N}{\tau} \frac{s^{N-1}}{(s+1/\tau)^{N+1}} \qquad (7.62)$$

$$\frac{d^2 P_T(t_F,s)}{ds^2} = \frac{N}{\tau} \frac{s^{N-2}(-2s+(N-1)/\tau)}{(s+1/\tau)^{N+2}} \tag{7.63}$$

$$\frac{d^3 P_T(t_F,s)}{ds^3} = \frac{N}{\tau} \frac{s^{N-3}(4s^2 - 6s(N-1)/\tau + (N-1)(N-2)/\tau^2)}{(s+1/\tau)^{N+3}} \tag{7.64}$$

$$\frac{P_T(t_F,s)}{s^3} = \frac{s^{N-3}}{(s+1/\tau)^N} \tag{7.65}$$

and

$$(s^2/\omega_T + \omega_T)^{-1} \frac{P_T(t_F,s)}{s^2} = (s^2/\omega_T + \omega_T)^{-1} \frac{s^{N-2}}{(s+1/\tau)^N} \tag{7.66}$$

The below calculations are made for an effective navigation ratio $N = 3$ and a guidance time constant $\tau = 0.5$ s; random quantities are assumed to have zero mean.

The inverse Laplace transform of equation (7.61) equals

$$L_{glint}^{-1} = (4t^2 - 12t + 6)e^{-2t}$$

Integrating the square of the above expression, from equation (7.43) we obtain the rms miss distance due to glint noise with spectral density 0.4 m^2/Hz shown in Figure 7.7 (solid line).

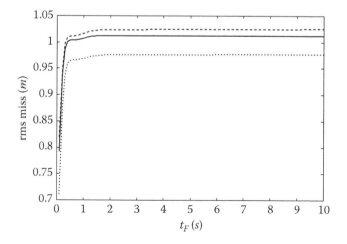

FIGURE 7.7   The rms miss distance due to glint noise.

The inverse Laplace transform of equation (7.62) equals

$$L^{-1}_{independent} = (6t - 12t^2 + 4t^3)e^{-2t}$$

Integrating the square of the above expression and assuming a closing velocity of 1500 m/s, from equation (7.46) we obtain the rms miss distance due to the independent range noise with spectral density $6.5 \cdot 10^{-8}$ rad^2/Hz, as shown in Figure 7.8 (solid line).

The inverse Laplace transform of equation (7.63) equals

$$L^{-1}_{passive} = (-4t^4 + 12t^3 - 6t^2)e^{-2t}$$

Integrating the square of the above expression and assuming a closing velocity of 1500 m/s, from equation (7.48) we obtain the rms miss distance due to passive-receiver noise with spectral density $6.5 \cdot 10^{-4}$ rad^2/Hz at a reference range 10,000 m, as shown in Figure 7.9 (solid line).

The inverse Laplace transform of equation (7.64) equals

$$L^{-1}_{active} = (6t^3 - 12t^4 + 4t^5)e^{-2t}$$

Integrating the square of the above expression and assuming a closing velocity of 1500 m/s, from equation (7.49) we obtain the rms miss distance due to active-receiver noise with spectral density $6.5 \cdot 10^{-4}$ rad^2/Hz at a reference range 10,000 m, as shown in Figure 7.10 (solid line).

The inverse Laplace transform of equation (7.65) equals

$$L^{-1}_{\varphi1} = 0.5t^2e^{-2t}$$

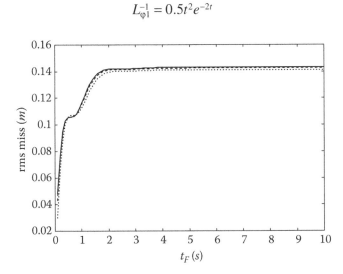

FIGURE 7.8   The rms miss distance due to independent angle noise.

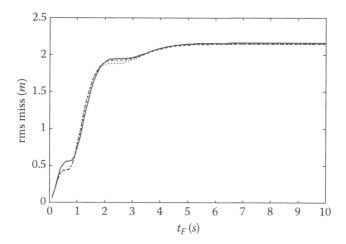

FIGURE 7.9 The rms miss distance due to passive-receiver noise.

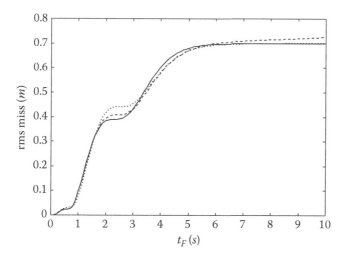

FIGURE 7.10 The rms miss distance due to active-receiver noise.

Integrating the square of the above expression and assuming a 3-g step maneuver with uniformly distributed starting time, from equation (7.50) we obtain the rms miss distance as shown in Figure 7.11 (solid line).

The inverse Laplace transform of equation (7.66) for $\omega_T = 1.4$ *rad/s* equals

$$L_{\varphi2}^{-1} = -0.025\cos 1.4t + 0.093\sin 1.4t + (-0.235t^2 - 0.08t + 0.025)e^{-2t}$$

Integrating the square of the above expression and assuming a 3-g amplitude of a target acceleration, from equation (7.51) we obtain the rms miss

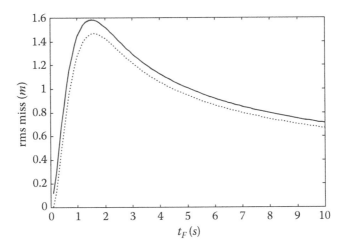

FIGURE 7.11   The rms miss for uniformly distributed 3-g step maneuver.

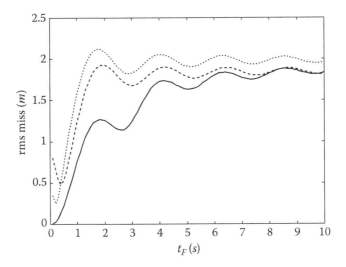

FIGURE 7.12   The rms miss for 3-g random-phase sinusoidal maneuver.

distance due to the uniformly distributed random-phase sinusoidal maneuver with the frequency $\omega_T = 1.4$ *rad/s*, as shown in Figure 7.12 (solid line).

   The above analytical expressions and the corresponding relationships between the rms miss and the time of flight t_F can be obtained by using the Laplace transform tables or Maple software. However, even the mentioned sophisticated software cannot help in a case of more realistic models of the missile guidance system, which contain complex-conjugated poles stipulated by the flight-control system damping and natural frequency [see equation (5.21)].

The precise solution (solid lines in Figures 7.7–7.12) for the first-order model was compared with the results of simulation based on equations (7.55)–(7.60) and the analytical expressions (5.35), (5.37), (5.42), and (5.43) for $P_T(t_F, i\omega)$ and $P(t_F, i\omega)$ (dashed lines in Figures 7.7–7.12). In most cases the error does not exceed 1%–3%. For the uniformly distributed step maneuver (Figure 7.11), the simulation result is very close to the precise solution so that the solid and dashed lines coincide. The largest error is obtained for the random-phase distributed maneuver (Figure 7.12). As expected, it becomes smaller after the transient (i.e., $t_f > 3$ s).

Below we present the results of a MATLAB rms miss simulation based on the expressions (5.35), (5.37), (5.42), (5.43), and (7.55)–(7.60). Parallel with the considered above simple model with a time constant 0.5 s, more realistic models will be considered:

i. $\omega_z = 30$ *rad/s*, $\zeta = 0.7$, $\omega_M = 20$ *rad/s*, $\tau = 0.5$ s
ii. $\omega_z = 5$ *rad/s*, $\zeta = 0.7$, $\omega_M = 20$ *rad/s*, $\tau = 0.5$ s
iii. $\omega_z = 30$ *rad/s*, $\zeta = 0.7$, $\omega_M = 20$ *rad/s*, $\tau = 0.1$ s
iv. $\omega_z = 5$ *rad/s*, $\zeta = 0.7$, $\omega_M = 20$ *rad/s*, $\tau = 0.5$ s

Comparative analysis of the rms distance for the missile guidance systems due to glint noise is given in Figure 7.13. The solid line, which coincides with the solid line in Figure 7.7, corresponds to the case of a first-order system. The dotted and dashed lines correspond to the case of a tail-controlled missile system with the airframe zero frequency $\omega_z = 30$ *rad/s*.

In both cases, the rms miss distance is more than for the simple first-order model. The larger is flight control system time constant, the smaller is the rms miss distance.

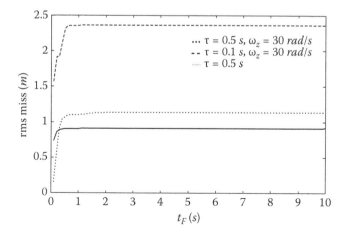

FIGURE 7.13   Comparative analysis of the rms miss due to glint noise.

The significant increase of the rms miss distance for the airframe zero frequency $\omega_z = 5$ *rad/s* (see Figure 7.14), which corresponds to high-altitude endoatmospheric interceptors, is stipulated by the "wrong way tail effect."

As mentioned in the previous chapters, because of the wrong way tail effect the dynamic characteristics of tail-controlled missiles operating at high altitude are significantly worse than at low frequencies. However, if in the deterministic case the decrease of the flight control system time constant decreases the miss distance at low altitudes (high values of ω_z) and increases it at high altitudes (low values of ω_z), the rms miss distance due to glint noise increases with the decrease of the flight control system time constant for all considered cases.

The above-mentioned effect of glint noise may look strange, especially for the first-order model, because the decrease of the time constant makes it closer to the ideal inertialess case. However, the described effect can be predicted by analyzing equation (7.55) and the above-given expression of the first-order system $P(t_F, s)$ [see also equation (3.19)] as a function of τ.

Although the results shown in Figure 7.14 are far from reality and because glint noise may be highly correlated and should not be modeled as white noise, they are very informative. As seen from Figure 7.13 and Figure 7.14, the rms miss distance increases substantially at high altitude and the decrease of the flight control system time constant brings an additional drastic increase. As seen from equations (5.13) and (7.43), the rms miss due to glint noise increases with the increase of the effective navigation ratio.

Analysis of the influence of the independent range noise presented in Figures 7.15 and 7.16. The solid line in Figure 7.15 coincides with the solid

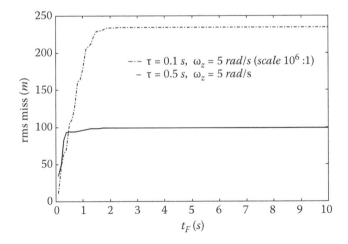

FIGURE 7.14   Comparative analysis of the rms miss due to glint noise.

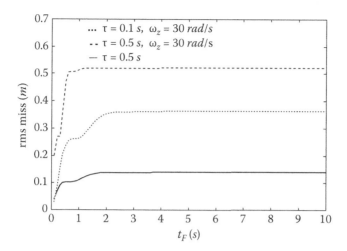

FIGURE 7.15   Comparative analysis of the rms miss due to independent noise.

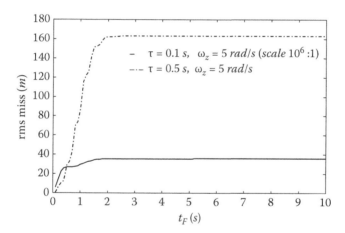

FIGURE 7.16   Comparative analysis of the rms miss due to independent noise.

line in Figure 7.8 built based on the analytical expression for the first-order guidance system model. The dotted and dashed lines correspond to the case of a tail-controlled missile system with the airframe zero frequency $\omega_z = 30$ *rad/s*. In contrast to the case of glint noise, the decrease of the flight control system time constant decreases the rms miss distance due to independent noise at low altitudes (high values of ω_z) and increases it at high altitudes (low values of ω_z), i.e., analogous to the deterministic case.

   Correlation between the above-considered cases for range independent noise is not changed qualitatively for passive- and active-range dependent noises. As follows from Figures 7.17 and 7.18, the rms miss distance for active-receiver noise is less than the passive-receiver noise miss. In the

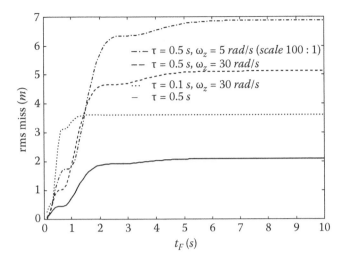

FIGURE 7.17   Comparative analysis of the rms miss due to passive-receiver noise.

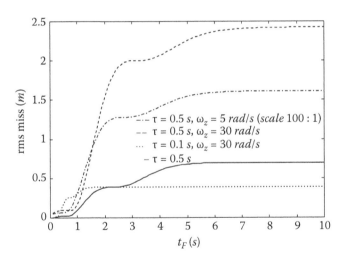

FIGURE 7.18   Comparative analysis of the rms miss due to active-receiver noise.

case of range dependent noise, the decrease of the flight control system time constant at high altitude increases the rms miss significantly less than due to range independent noise.

Since the range independent and dependent noise miss distances depend on a closing velocity, they can be significant for ballistic missiles.

Comparative analysis of the rms distance for the step maneuvers with starting time uniformly distributed over the flight time according to equation (7.59), and for the random-phase sinusoidal maneuvers based on equation (7.60) is presented in Figure 7.19 and Figure 7.20.

As seen from Figures. 7.19 and 7.20, the rms miss distance is very sensitive to the flight-control system time constant. The solid lines in Figures 7.19 and 7.20 correspond to the precise solution for the first-order model.

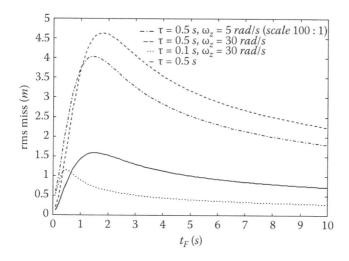

FIGURE 7.19 Comparative analysis for uniformly distributed 3-g step maneuver.

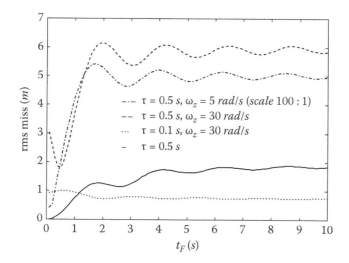

FIGURE 7.20 Comparative analysis for 3-g random-phase sinusoidal maneuver.

By choosing an appropriate quantization in the time h and frequency h_0 domains, the operations of integration and differentiation are approximated by the summation and difference operations, respectively. In practice, the upper infinite limit of integration in equations (7.55)–(7.60) is changed to ω_{c0}, the frequency being chosen so that for $\omega \geq \omega_{c0}$ the values of the integrands of equations (7.55)–(7.60) are less than 5%–10% of their maximum values, respectively. In the above examples, $\omega_c = 19 \ rad/s$, $h_0 = 0.1 \ rad/s$ and $h = 0.1 \ s$.

The methods of computational mathematics enable us to increase accuracy of the operations of integration and differentiation. It is known that the first, second, and third derivatives of a continuous function $y(t_0)$ at t_0 can be approximated by:

$$h\dot{y}(t_0) = Diff(y_0, 1) - \frac{1}{2} Diff(y_0, 2) + \frac{1}{3} Diff(y_0, 3) - \cdots$$

$$h^2\ddot{y}(t_0) = Diff(y_0, 2) - Diff(y_0, 3) + \frac{11}{12} Diff(y_0, 4) - \cdots$$

$$h^3\dddot{y}(t_0) = Diff(y_0, 3) - \frac{3}{2} Diff(y_0, 4) + \frac{7}{4} Diff(y_0, 5) - \cdots$$

where h is a quantization step size and $Diff(y_0, i)$ denotes the i-th order difference of $y(t_0)$ at t_0.

The approximation of the third derivative by the third-order difference, i.e., by using only the first term of the $h^3\dddot{y}(t_0)$ approximation, can give a tangible error. Instead of using the high order differences, the expression (7.58) can be transformed into:

$$\frac{E[y^2(t_F)]_{active}}{\Phi_{pn}} = \int_0^{t_F} \left\{ \frac{2}{\pi} \frac{v_{cl}^3}{r_0^2} t^2 \int_0^\infty \frac{d \, Im[P_T(t_F, i\omega)]}{d\omega} \cos\omega t \, d\omega \right\}^2 dt \qquad (7.67)$$

This expression is easily obtained by using integration by parts of equation (7.58) and taking into account that $Im^{(i)}[P_T(t_F, 0)] = Im^{(i)}[P_T(t_F, \infty)] = 0] = 0$, $i = 0, 1, 2$, where i is the order of derivative. The above equation, as well as equations (7.56) and (7.57), can be simplified analogously, taking into account $Re^{(1)}[P_T(t_F, 0)] = 0$.

Instead of equation (7.56), we can use

$$\frac{E[y^2(t_F)]_{independent \ noise}}{\Phi_{fn}} = \int_0^{t_F} \left\{ \frac{2}{\pi} v_{cl} t \int_0^\infty Im[P_T(t_F, i\omega)] \sin\omega t \, d\omega \right\}^2 dt \qquad (7.68)$$

Instead of equation (7.57), we can use

$$\frac{E\left[y^2(t_F)\right]_{passive}}{\Phi_{pn}} = \int_0^{t_F} \left\{ \frac{2}{\pi} \frac{v_{cl}^2}{r_0} t \int_0^\infty \frac{d\,\mathrm{Re}[P_T(t_F,i\omega)]}{d\omega} \sin\omega t d\omega \right\}^2 dt \qquad (7.69)$$

However, the accuracy of the computational procedure based on equations (7.67)–(7.69) is very sensitive to ω_{c0}, especially for big t_F because of t factors in equations (7.67)–(7.69).

The expressions of $\mathrm{Re}[P_T(t_F, i\omega)]$ and $\mathrm{Im}[P_T(t_F,i\omega)]$ can be obtained directly from equations (5.42) and (5.43) or from equation (5.10) [see also equations (6.10)–(6.13)]:

$$P_T(t_F,i\omega) = \exp\left(-N\int_\omega^\infty \frac{W(i\omega)}{\omega} d\omega\right)$$

$$= \exp\left(-N\int_\omega^\infty \frac{\mathrm{Re}[W(i\omega)]}{\omega} d\omega\right) \exp\left(-iN\int_\omega^\infty \frac{\mathrm{Im}[W(i\omega)]}{\omega} d\omega\right)$$

so that the amplitude $|P_T(t_F,i\omega)|$ and phase $\varphi_T(t_F, i\omega)$ of $P_T(t_F, i\omega)$ equal

$$|P_T(t_F,i\omega)| = \exp\left(-N\int_\omega^\infty \frac{\mathrm{Re}[W(i\omega)]}{\omega} d\omega\right) \qquad (7.70)$$

and

$$\varphi_T(t_F,i\omega) = -N\int_\omega^\infty \frac{\mathrm{Im}[W(i\omega)]}{\omega} d\omega \qquad (7.71)$$

The computational procedure based on equations (7.70) and (7.71) is simpler than the above-described procedure using the analytical expressions for the frequency response (5.35), (5.37), (5.42), and (5.43). It requires only the knowledge of the frequency response $W(i\omega)$ of the missile guidance system. Moreover, $W(i\omega)$ can be presented approximately based on experimental data. The results of simulation for the first-order model based on equations (7.70) and (7.71), instead of equations (5.42) and (5.43) (see dotted lines in Figures 7.7–7.12), show that the accuracy of this procedure is a little bit lower than the above-described procedure employing the analytical expressions $P(t_F, i\omega)$ and $P_T(t_F, i\omega)$. The accuracy can be improved by increasing the upper limit ω_{c0} of integration in equations (7.70) and (7.71).

In all considered cases, the first-order system analysis gives a significantly less miss estimate than a more realistic model with a high airframe frequency ω_z. The first-order model gives absolutely unacceptable results in a case of small values of the airframe frequency.

7.8 FILTERING

Above-described noises corrupt seeker measurements. To get reliable information for the line-of-sight rate required by the PN guidance law or its modifications considered in the previous chapters, it is necessary to use a filter that would decrease the rms miss distance contributed by all noise sources. Assuming independence of the random variable considered, the total rms miss distance can be determined as the square root of the sum of the variances of the miss distances from glint, independent, and dependent noises.

Considering $G_1(s)$ as the filter transfer function (see Figures 7.1–7.3), the filtering problem can be formulated as the problem of finding $G_1(s)$ that would decrease the total rms miss distance. Such formulation has a certain drawback, because it ignores the miss distance stipulated by the target maneuver. Dynamic characteristics of the filter influence this component of the total miss distance.

For highly maneuvering targets, the assumption that the phase angle of target weave, which is associated with initial conditions at the start of the missile's terminal guidance, can be treated as a random variable uniformly distributed between 0 and 2π over a set of engagement is quite realistic, and the corresponding rms miss distance can be considered as the measure of effectiveness in analyzing missile performance against weaving targets. It means that the transfer function $G_1(s)$ of a filter should be chosen to minimize the total rms miss including the rms for the random phase sinusoidal maneuvers.

Below we describe an engineering approach for choosing a filter with the transfer function $G_1(s)$ to improve the performance of a tail-controlled missile with $\omega_z = 30$ *rad/s*, $\zeta = 0.7$, $\omega_M = 20$ *rad/s* and $\tau = 0.5$ *s*. From the comparative analysis of the rms miss due to range independent and dependent noises and the rms for the random phase sinusoidal maneuvers (see Figures 7.15, 7.17, 7.18, and 7.20) for the given missile guidance system and the system with parameters $\omega_z = 30$ *rad/s*, $\zeta = 0.7$, $\omega_M = 20$ *rad/s* and $\tau = 0.1$ *s* we can conclude that the decrease of τ significantly decreases all rms miss components, but the component due to glint and the total rms miss is significantly less for the missile guidance system with $\tau = 0.1$ *s*. Based on this analysis, it is easy to conclude that by applying the filter with the transfer function

$$G(s) = \frac{0.5s + 1}{0.1s + 1},$$

we would improve significantly the performance of the given missile guidance system at low altitudes. As indicated, at high altitudes its dynamic parameters are changed so that the filter parameters should be changed as well.

We will not consider here any optimal filtering problems. Simple constant gain and optimal digital filters widely used in practice will be discussed in the next chapter.

REFERENCES

1. Chen, C. *Linear System Theory and Design*. New York, NY: CBS College Publishing, 1984.
2. Gihman, I. I., and Skorohod, A. V. *The Theory of Stochastic Processes I*. New York, NY: Springer, 1974.
3. Fitzgerald, R. J. Shaping Filters for Disturbances with Random Starting Times, *Journal of Guidance and Control* 2 (1979): 152–54.
4. Skolnik, M. I. *Radar Handbook*. New York, NY: McGraw-Hill, Inc., 1990.
5. Smith, O. *Feedback Control Systems*. New York, NY: McGraw-Hill, Inc., 1958.
6. Shneydor, N. A. *Missile Guidance and Pursuit*. Chichester, UK: Horwood Publishing, 1998.
7. Yanushevsky, R. Analysis of Optimal Weaving Frequency of Maneuvering Targets, *Journal of Spacecraft and Rockets* 41, no. 3 (2004): 477–79.
8. Yanushevsky, R. Analysis and Design of Missile Guidance Systems in Frequency Domain, *44th AIAA Space Sciences Meeting*, Paper AIAA 2006-825, Reno, Nevada, 2006.
9. Yanushevsky, R. Frequency Domain Approach to Guidance System Design, *IEEE Transactions on Aerospace and Electronic Systems*, 43 (2007).
10. Yanushevsky, R. *Theory of Linear Optimal Multivariable Control Systems*. Moscow, Russia: Nauka, 1973.
11. Zachran, P. Complete Statistical Analysis of Nonlinear Missile Guidance Systems, *Journal of Guidance and Control*, 12, no. 1 (1979): 71–78.

8 Guidance of UAVs

8.1 INTRODUCTION

As mentioned in Chapter 1, the United States currently possesses five major UAVs: the Air Force's Predator and Global Hawk, the Navy and Marine Corps's Pioneer, and the Army's Hunter and Shadow.

Predator is a medium-altitude (7.6 km), long-endurance (24 hours), 8.23 m long UAV that typically operates at a 3–4.5 km altitude to get the best imagery from its video cameras, although it has the ability to reach a maximum altitude of 7.6 km. It launches and lands like a regular aircraft, but is controlled by a pilot on the ground using a joystick. The vehicle's operational radius is about 740 km and airspeed up to 400 km/h. The Predator's primary function is airborne reconnaissance and target acquisition. To accomplish this mission, the Predator is outfitted with a 200 kg surveillance payload, which includes two electro-optical (E-O) cameras and one infrared (IR) camera for use at night. It also includes synthetic aperture radar (SAR), which allows the UAV to operate in severe weather conditions. The Predator's satellite communications provide for beyond line-of-sight (LOS) operations. The Predator's primary satellite link consists of a 6.1 meter satellite dish and associated support equipment. The satellite link provides communications between the ground station and the aircraft when it is beyond the line-of-sight and is a link to networks that disseminate secondary intelligence. In 2002, the Predator's military designation was changed from RQ-1B (reconnaissance unmanned) to the MQ-1 (multimission unmanned) due to its added capabilities of laser designation and missile firing. New types of Predators have a multispectral targeting system that will add a laser designator to the E-O/IR payload. The so-called B variant, also referred to as the "hunter-killer," is a larger version of the Predator (length 10.97 m) with improvements in altitude, payload, speed, and range. The MQ-9 can fly at altitudes of 13.5–16 km, and carry eight Hellfire missiles (compared to two for the MQ-1).

The Global Hawk is the largest (13.54 m long), high altitude (20 km), long endurance (35 hours) remotely piloted UAV. The vehicle's operational radius is 24,985 km and airspeed 650 km/h. It provides near real-time imagery of large geographic areas. Besides the obvious size difference between the Predator and Global Hawk, another significant difference between them is that the Global Hawk flies autonomously from takeoff to landing. Its high quality imagery is delivered by more sophisticated integrated suite of sensors compared to the Predator, which significantly increased its pay-load (880 *kg*). A ground station for controlling the Global Hawk, monitoring its status and making operational changes if necessary consists of two segments: a Mission Control Element (MCE) and Launch and Recovery Element (LRE). The MCE is used for mission planning and execution, command and control, and image processing and dissemination; an LRE-for controlling launch and recovery and associated ground support equipment. The LRE provides precision differential global positioning system corrections for navigational accuracy during takeoff and landings, while the GPS and INS are used during mission execution. Both ground segments are equipped with external antennas for line-of-sight and satellite communications with the air vehicles.

At 4.25 *m* long, the Pioneer is roughly half the size of the Predator. It can reach maximum altitudes of 4.5 *km* but flies an optimal altitude of 1–1.5 *km* above its target with a speed 200 *km/h*. The Pioneer can stay aloft for five hours during the daytime, and has a range of 185 km. Its mission is to provide real-time intelligence and a reconnaissance capability to the field commander. Its 30–45 *kg* payload includes an electro-optical and IR camera and may also include a meteorological sensor, a mine detection sensor, and a chemical detection sensor. The vehicle is launched by rocket assist (shipboard), by catapult, or from a runway; it recovers into a net (shipboard) or with arresting gear.

Possessing the standard UAV mission of reconnaissance and surveillance, the 7 *m* long Hunter is a medium-altitude (4.5 *km*) UAV with an operating range of 265 *km*, a 111–148 *km/h* speed, and about 10 hours endurance. It is equipped with an E-O/IR sensor payload (90 *kg*) for day/night operations. The vehicle is controlled by two operators—one controlling flight and the other controlling the payload functions. It can be launched from a paved or semipaved runway or it can use a rocket assisted (RATO) system, where it is launched from a zero-length launcher using a rocket booster. The RATO launch is useful on board small ships and in areas where space is limited. The vehicle can land on a regular runway, grassy strip, or highway using arresting cables.

The Shadow is a 3.4 *m* long UAV with an average flight duration of six hours, a 111–148 *km/h* speed, and the operational range is 78 *km*. Although it can reach a maximum altitude of 4.5 *km*, its optimum level is about 2.5 *km*. The Shadow provides real-time reconnaissance, surveillance, and target acquisition information. Weighing 127.3 *kg*, the vehicle is catapulted from a rail on a launcher and has an automatic takeoff and landing capability; it recovers on a runway via a tail hook. The Shadow's 12.25 *kg* payload consists of an E-O/IR sensor turret that produces day or night video, and can relay its data to a ground station in real-time via a LOS data link.

If the considered UAVs require additional support equipment and staff for launch and landing operations, a simplified launch and recovery within a confined area, operations at low speeds for precise targeting and special operations, or an ability to loiter within a large speed range are reached in some VTOL UAVs by incorporating certain features of helicopters.

The Eagle Eye takes off like a helicopter, but then tilts up its rotor to fly like a plane. This remote-controlled aircraft is 5.46 m in length and weighs around 900 kg (depending on payload, which is usually 98 kg). It can fly at up to 400 km/h for up to six hours at altitudes of 0.5–6 km and has an operational radius of roughly 500 km. The vehicle's potential applications are surveillance and reconnaissance of both land and sea, border patrol, NBC (nuclear, biological, and chemical) detection and contamination area operations, delivery of critical supplies, fire detection, fisheries protection, and so on.

Similar to the Eagle Eye, the Dragon Warrior is a vertical takeoff and landing *UAV* resembling a small helicopter. With a weight of 120 kg, a payload capacity of 11–16 kg, and at 2.7 m long it has a range of 90 km and an endurance of 3 hours. It flies at an altitude of 1.3–1.7 km with a speed of 185 km/h. Its payload includes an E-O/IR (electro-optical/infrared) sensor and laser designator. The Dragon Warrior is envisioned to play a major role in urban reconnaissance.

The Dragon Warrior and Eagle Eye realize all useful features of helicopters were being used. Although helicopters are used mostly for transportation, military transport helicopters can be modified or converted to conduct specific missions such as combat search and rescue and can be armed with weapons for attacking ground targets. Weapons used on attack helicopters can include auto-cannons, machine guns, rockets, and guided Hellfire missiles. The helicopter's ability to take off and land almost anywhere, fly very slow, and hover in one place for extended periods of time make them very attractive in accomplishing tasks that fixed-wing aircraft cannot perform. The above-mentioned features motivated the development of unmanned helicopters.

The helicopter-derived Fire Scout is about 7 m long UAV with an altitude ceiling of 6 km, and an endurance of more than six hours that flies at a speed of 200 km/h. It carries an E-O/IR sensor payload that incorporates a laser designator. The Fire Scout is capable of autonomous takeoff, landing, and flight; but of course it can be controlled remotely as well.

The A-160 Hummingbird unmanned helicopter (weight 1,815 kg) operates autonomously at a range of 3,700 km and flies at 200–260 km/h at a maximum altitude of 9 km, with a hover capability up to 4,570 m for up to 30 hours while carrying a 135 kg payload. Potential missions include surveillance, reconnaissance, target acquisition, communications and data relay, lethal and nonlethal weapons delivery, and special operations missions.

Various operational tactical requirements, technological progress, and the desire to decrease the cost of UAVs brought to life the so-called mini-UAVs that usually fly below 300 *m* and because of their small size have much less chance of a collision threat than larger UAVs. The mini-UAV can be carried, launched, and recovered by soldiers. They generally have ranges up to 20 *km*, and an endurance of not more than three hours. These UAVs are not designed to do depth reconnaissance, but to look over the next hill, watch a neighborhood in a city before troops enter it, patrol a base's outer perimeter, and so on. Even smaller micro-UAVs are in development that focus more tightly on local objects.

Among mini-UAVs, the Organic Air Vehicle (OAV) is designed to provide small combat teams and individual soldiers with the capability to detect the enemy forces concealed in forests or hills, around buildings in urban areas, or in places where the shooter does not have a direct line-of-sight. The OAV weighs about 35 *kg* and has a mission endurance of 25 minutes, a range of up to 2 *km* and speeds of 80–100 *km/h*. Its 3–3.5 *kg* payload comprises of E-O sensors and can be upgraded to include infrared and acoustic sensors and mine detectors. Smaller portable versions of the OAV are under development. These vehicles should weigh up to 10 *kg*, including 0.5 *kg* payload. Powered by a diesel fuel engine, it will be required to perform relatively short missions, of up to 15 minutes with a range of 1 *km*. The operational ceiling will be 2.4 *km* above sea level.

The 1.3 *m* and 4.5 *kg* hand launchable Pointer can fly at an altitude of 3–10 m and can stay airborne for about 90 minutes at a range of 4.8 *km* from its ground station.

The 0.9 *m* and 2.27 *kg* Dragon Eye provides over-the-hill reconnaissance, surveillance, and target acquisition at the tactical level. The UAV can be stored in a backpack and launched by either hand or bungee cord. The reconnaissance UAV can fly 65 *km/h*, cover a combat radius of 10 *km*, and stay aloft for an hour. Its operating altitude is 90–150 *m*. Interchangeable payloads for the Dragon Eye include a daytime camera, an infrared camera, and a low-light black and white camera.

The Desert Hawk UAV is a battery powered 3 *kg* mini-UAV. It has a length of 1 *m*, flies at altitudes of less than 330 *m*, can carry a payload of 0.5 *kg*, and has an operational radius of 10 *km*, with an endurance of 60–90 minutes. Launched into the air by two people using a bungee cord as a slingshot, the mini-UAV flies its mission fully autonomously, at speeds of 40 to 80 *km/h*, following a flight path that has been plotted out beforehand on a laptop using GPS coordinates. The plane can be directed to circle over an area of interest, or the operator can alter its flight path while the plane is in the air. Its payloads comprise of interchangeable systems, including an infrared thermal imaging system for night use or a set of three color cameras for daylight. The ground control system features a touch-screen laptop allowing operators to easily place waypoints. Mission information is immediately updated when the unmanned vehicle is re-tasked to support a dynamically changing tactical environment.

The rapidly increasing fleet of UAVs, along with the widening sphere of their applications, puts new problems before designers of the unmanned aerial vehicles. The more complicated the problems the more sophisticated the guidance required. Advanced guidance algorithms development is essential and necessary for meeting new requirements with the increasing area of UAV applications and for defining future UAV concepts and associated critical technologies.

The more autonomous ability a UAV has, the more complex its guidance and control system is and, as a result, the higher is its weight, size, and cost, the less—its endurance, a combat radius, and/or speed. The desire to reduce the operator's load leads to an increase of a UAV's payload, so that it becomes more difficult to reach its best potential performance characteristics. A well justified tradeoff between the operator and autonomous UAV's functions remains the most important problem that would allow minimizing the size and cost of future UAVs.

Many new models of UAVs will be flight-tested within 5 to 10 years. There is a tendency to focus on building small, even microscopic UAVs with smaller weapons that can hunt in swarms, engage targets in the close quarters of urban battlefields and as soon as possible. Nevertheless, recently Israel built a new powerful 4.5 ton mega-UAV, the Eitan, with 24–36 hour endurance and the ability to operate above 12 *km*. The United States is developing the UAV with close characteristics of the Eitan to be used as a part of boost-phase intercept systems, which are considered in Chapter 11.

8.2   BASIC GUIDANCE LAWS AND VISION-BASED NAVIGATION

The UAV path planning is considered in terms of global (mission) planning and local (trajectory) planning. The global planning determines the main most important requirements to the UAV flight path, its length, time of flight, and evaluates possible environment uncertainties. The local planning deals with the trajectory algorithm development based on the available information. The UAV's trajectory can be divided in three parts: the initial uncontrolled part immediately after its launch, the controlled part in accordance with the predetermined flight path, and the terminal part corresponding to its landing that can be completely or partially controllable. Depending upon the mission requirements and especially in the case of surveillance and reconnaissance flights, the trajectory can contain the low speed parts. Moreover, in special cases, hovering UAVs should stay in a particular position/area for a certain period of time.

The increasing areas of UAV applications and, as a result, a variety of UAV missions make the flight path preparation a nonstandard procedure and the most important factor determining the success of the mission. Usually, the flight path planner tries to take into account all details available concerning the area of interest. For UAVs that perform patrolling or surveillance functions along a definite narrow ground area of a simple configuration, it is not difficult to design a detailed flight path with necessary UAV velocities at its separate parts. However, the trajectory planning in the case of a complex environment presents a difficult problem.

Realistic mission scenarios may include obstacles and other no-fly zones. Additionally, the UAV must be able to overcome environmental uncertainties such as modeling errors, external disturbances, and an incomplete situational awareness. Path planning problems have been actively studied in robotics. The problem of planning a path is formulated such as finding a collision-free path in an environment with static or dynamic obstacles. As in robotics, the UAV path planning in adversarial environments (adversaries are considered obstacles) has the objective to complete the given mission efficiently while maximizing the safety of the UAV. The goal of the trajectory planner is to compute, within an appropriate time window, an optimal or suboptimal path for surviving penetration through the environment while satisfying mission objectives. The planner considers terrain data, threat information, fuel constraints, time constraints, and other constraints specified based on available information.

Typically, a trajectory planner returns waypoint locations and estimated time of arrivals, headings, resource utilization (e.g., fuel consumption, sensor constraints), and metrics indicating the characteristics of the route such as risk and effectiveness. Trajectory planning can be considered as a draft of the future guidance law. The optimal approach is used to make the trajectory, for example, energy efficient to increase the UAV endurance or to realize the minimum time trajectory, and so on [1,2,14]. The operations research methods enable us to formulate and solve such types of problems. If one has perfect information of the threats that will be encountered, a safe path can always be constructed by solving an optimization problem. If there are uncertainties in the information, usually several scenarios are considered and a tradeoff solution is offered. In all cases the flight path can be presented as a sequence of waypoints with additional requirements (constraints) to separate parts of the trajectory.

The hypothetical trajectory of a UAV is given in Figure 8.1. The flight plan comprises a series of waypoints (for simplicity only four waypoints are indicated) assumed to be joined by straight line trajectory segments, originating at the climb phase just after takeoff and terminating at the landing phase. The main difference between industrial robot path planning and fixed-wing UAV path planning is that the UAV must maintain

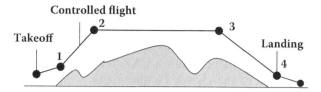

FIGURE 8.1   Standard flight profile.

its velocity above a minimum velocity, which implies that it cannot follow a path with sharp turns or vertices. Given the waypoints, a possible flyable route can be generated. But is it necessary? Usually, it is impossible to predict all possible flight situations and the "common sense" approach reinforced by training—the way a pilot acts—is the best solution.

The guidance algorithms considered in the previous chapters [see, e.g., equations (3.16), (3.74), and (3.92)] can be applied to navigate the UAV sequentially from the initial waypoint to the next one, and so on. In this case, each waypoint is considered as a dummy target and the LOS rate can be calculated based on equation (1.11). Because of relatively small distances between the waypoints, usually the UAV trajectory is close to a straight line and significant LOS changes can take place (excluding the case of unexpected obstacles) only in the vicinity of the waypoints. The trajectory planner should transfer to the UAV onboard computer only the sequence of the waypoints with the speed constraints for the trajectory segments and the acceleration limits, usually only in the vicinity of the waypoints. (Avoidance of obstacles and the corresponding algorithms are considered in Section 8.5.)

The guidance system implementing the considered guidance laws will stabilize and control the air vehicle flight between the chosen waypoints. The cubic term in equation (3.16) amplifies the effect of the proportional navigation (PN) term and is efficient in the case of maneuvering targets. For many UAV missions, there is no need to use this term and the proportional navigation algorithms is sufficient to guide the UAV between the waypoints. Specifics for applying the above-considered guidance laws and corresponding algorithms to UAVs are in the necessity of imposing limits on the UAV speed either during the whole flight or on its separate parts. During the autonomous flight, the airborne computer system collects the data in real-time from the sensing units such as the GPS/IMU to generate the commands to drive servo systems controlling the UAV flight. (IMU, an inertial measurement unit, is an electronic device that measures velocity, orientation, and gravitational forces, using a combination of accelerometers and gyroscopes.)

Usually, the terminal landing is the most difficult part of the UAV trajectory. The guidance laws considered in the previous chapters do not work in this case. They are to straightforward and do not take into account many specific details that accompany the landing process. The most promising

approach is to develop a control system replicating the forced landing function as performed by a human pilot. A vision-based navigation approach
and related computational algorithms should be a part of this control system. First of all, video images allow a UAV to identify a safe landing site,
and then keep track of the chosen landing point as the vehicle descends
toward the chosen landing site. The vision-based navigation ability to
detect and avoid obstacles is an important feature that can also be used
during the landing process. However, the main advantage of using vision-
based navigation during the landing phase is that the landing environment
makes the UAV positioning information from GPS or any other similar
positioning devices unreliable. Reasons for this may vary from signal jam-
ming to obstacles that occlude positioning signals.

Many researchers have suggested computer-vision systems to provide a
more accurate estimate of the vehicle's changing position. The difficulty
that arises in applying vision technologies is that a human's experience and
intuition cannot be adequately transferred to a computer program. To compensate for a computer's "slow-wittedness," it must be given extra data,
such as a second camera image, to replace human intuition with explicit
calculations. The primary goal of stereo vision analysis is to determine the
3-D form of an environment that is represented by 2-D images, since one
image is insufficient for this type of deduction. However, if another pic-
ture is taken of the same scene from a slightly different position or angle,
then the two images can be compared and the distance can be estimated
between the camera and the objects represented in the images.

The analysis of the vision-based algorithms and their application to
perform the tasks of landing-site selection and position estimation can be
found in other material [4,15,19–23].

The possibility of using the guidance laws discussed in the previous
chapters to guide UAVs we demonstrate on an example of a hypothetical
surveillance problem assuming that a UAV should travel along a circular
path of a 50 km radius. Approximating the circle by a regular octagon we
obtain nine waypoints $PIPi(j)$ $(i = 1 - 3; j = 1 - 9)$ with the following coordinates (distances are given in meters):

$PIP1(1) = 500;$	$PIP2(1) = 0;$	$PIP3(1) = 1000$
$PIP1(2) = 15,150;$	$PIP2(2) = 35,550;$	$PIP3(2) = 1000$
$PIP1(3) = 50,500;$	$PIP2(3) = 50,000;$	$PIP3(3) = 1000$
$PIP1(4) = 85,850;$	$PIP2(4) = 35,550;$	$PIP3(4) = 1000$
$PIP1(5) = 100,500;$	$PIP2(5) = 0;$	$PIP3(5) = 1000$
$PIP1(6) = 85,850;$	$PIP2(6) = -35,550;$	$PIP3(6) = 1000$
$PIP1(7) = 50,500;$	$PIP2(7) = -50,000;$	$PIP3(7) = 1000$
$PIP1(8) = 15,150;$	$PIP2(8) = -35,550;$	$PIP3(8) = 1000$
$PIP1(9) = 500;$	$PIP2(9) = 0;$	$PIP3(9) = 1000$

We assume that the controlled part of the UAV trajectory (R_{M1}, R_{M2}, R_{M3}) starts at $R_{M1} = 250$ m; $R_{M2} = 0$; $R_{M3} = 500$ m and the velocity components $V_{M1} = 20$ m/s; $V_{M2} = 40$ m/s, and $V_{M3} = 0$. The chosen acceleration limit equals 2-g. The fight control dynamics are assumed to be presented by a third-order transfer function with damping $\zeta = 0.7$, natural frequency $\omega_M = 100 rad/s$ and the flight control system time constant $\tau = 0.1$ s.

The PN law generated trajectory in the horizontal plane for the navigation ratio $N = 3$ and the corresponding UAV speed and acceleration are given in Figures 8.2–8.4. Formally, $\lambda(t)$ does not exist at the waypoint, since we deal with a discontinuous function. However, a discrete model enables us to evade this "inconvenience." The round trip takes about 10 hours. Since the PN law, considered here as the only source of motion, produces the lateral motion, the initial energy dissipates at the waypoints and the UAV speed decreases. Correspondingly, the spikes of the commanded acceleration become smaller. The above-mentioned could be easily predicted since the simulated flight is without any permanently acting thrust force. That is why for more realistic initial conditions ($V_{M1} = 20$ m/s; $V_{M2} = 0$ and $V_{M3} = 40$ m/s) the circular flight cannot be finished; simulation results show that it fails after the third waypoint).

In contrast to many missiles that have uncontrollable thrust, UAVs control their propulsion force, which is a dominant factor of their longitudinal motion. Moreover, as we indicated earlier, different parts of the trajectory

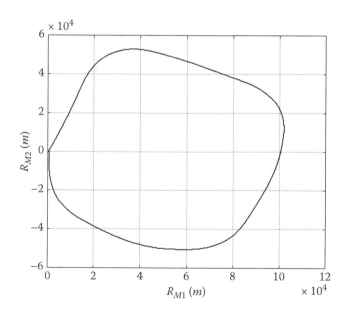

FIGURE 8.2 UAV trajectory for the PN law.

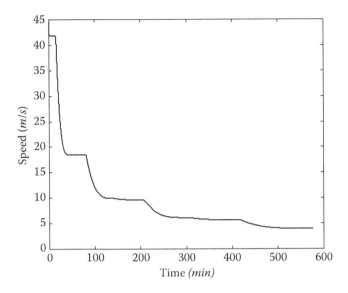

FIGURE 8.3   UAV speed for the PN law.

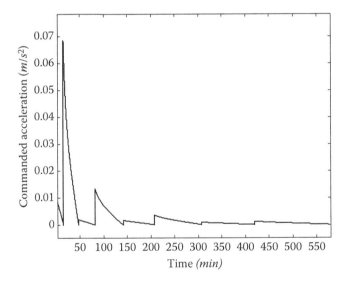

FIGURE 8.4   UAV commanded acceleration for the PN law.

may require a different UAV speed. While for many missiles the change of their axial acceleration depends on time (it is determined by specifics of rocket engines) and in some types it depends also upon the target motion, in the case of UAVs the longitudinal motion and its velocity are planned in advance and it is supposed that the UAV has a certain speed in each part of the flight. This determines the required longitudinal acceleration and the corresponding propulsion forces.

This requirement should be reflected in the guidance law. Figures 8.5–8.7 correspond to the modified guidance law:

$$a_{Ms}(t) = 3v_{cl}\dot{\lambda}_s(t) + k(v_{M0}(t)\lambda_s(t) - v_{Ms}(t)) \qquad (s = 1, 2, 3) \qquad (8.1)$$

where $v_{M0}(t)\lambda_s(t)$ $(s = 1, 2, 3)$ are the desired components of the UAV velocity, $v_{M0}(t)$ is a preferable speed, and k is a constant coefficient.

The simulation results correspond to the case $v_{M0}(t) = 50$ *m/s* and $k = 0.005$. The flight time is less than two hours and the UAV speed is close to the desired value during the most part of the flight. It is clear that better initial conditions, which depend on the quality of the launch operation, more waypoints and the tuned properly gain k will allow to get a more precise trajectory.

The PN law in the considered guidance model, based on a sequence of waypoints the UAV should follow, acts as a corrector in the vicinity of the waypoints. In the modified guidance law we combined the lateral and longitudinal motions as we did in Chapter 3.

Usually, these motions are controlled by different systems, so that methods of analysis and design of the PN guided systems considered in Chapters 4–7 applied traditionally to missiles can be also employed for UAVs.

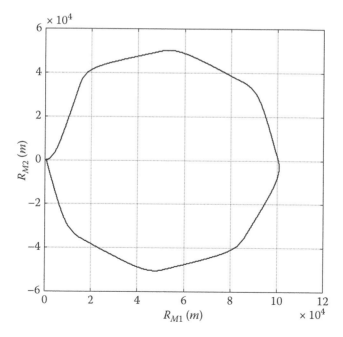

FIGURE 8.5 UAV trajectory for the modified law.

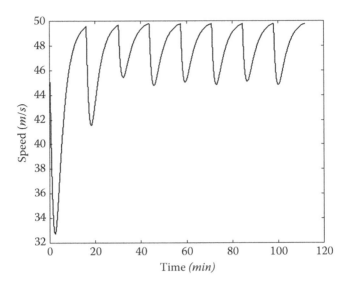

FIGURE 8.6 UAV speed for the modified law.

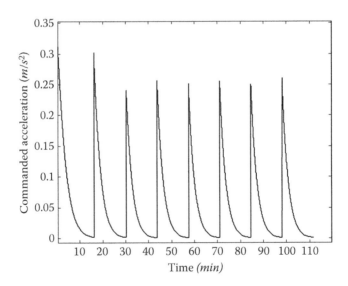

FIGURE 8.7 UAV acceleration for the modified law.

In Chapter 3 the guidance laws were developed based on a rigorous mathematical approach; here the modified guidance was obtained based on physical rather than rigorous mathematical consideration. The mathematical justification of this guidance law follows from the material of the next section.

8.3 GENERALIZED GUIDANCE LAWS FOR UAVS

In Chapter 3 the motion of an unmanned aerial vehicle was presented consisting of two components—lateral and longitudinal. The lateral and longitudinal motions are analysed from the following equations [see equations (3.87) and (3.88)]:

$$\ddot{\lambda}_s(t)r(t) + 2\dot{r}(t)\dot{\lambda}_s(t) + r(t)\sum_{s=1}^{3}\dot{\lambda}_s^2(t)\lambda_s(t) = a_{Tts}(t) - a_{Mts}(t) \qquad (s=1,2,3)$$
(8.2)

and

$$\ddot{r}(t) - r(t)\sum_{s=1}^{3}\dot{\lambda}_s^2(t) = a_{Tr}(t) - a_{Mr}(t)$$
(8.3)

where, applied to UAVs, $a_{Tr}(t)$ and $a_{Mr}(t)$ are the target and UAV radial accelerations, respectively; the target can be a waypoint or an unmanned aerial vehicle.

8.3.1 WAYPOINT GUIDANCE PROBLEM

Returning to the waypoint guidance considered in the previous section, we should modify only the longitudinal motion acceleration $a_{Mr}(t)$ (3.91) to meet the requirement for the axial speed $v_M(t) = v_{M0}$.

Similar to the approach in Chapter 3 by introducing a pseudoacceleration:

$$a_{Mr1}(t) = a_{Mr}(t) - r(t)\sum_{s=1}^{3}\dot{\lambda}_s^2(t)$$

we decouple the dynamics of longitudinal and lateral motions reducing equations (8.2) and (8.3) to:

$$\ddot{\lambda}_s(t)r(t) + 2\dot{r}(t)\dot{\lambda}_s(t) + r(t)\sum_{s=1}^{3}\dot{\lambda}_s^2(t)\lambda_s(t) = a_{Tts}(t) - a_{Mts}(t) \qquad (s=1,2,3)$$
(8.4)
$$\ddot{r}(t) = a_{Tr}(t) - a_{Mr1}(t)$$

In the case of the UAV waypoint guidance, the system (8.4) has the form:

$$\ddot{\lambda}_s(t)r(t) + 2\dot{r}(t)\dot{\lambda}_s(t) + r(t)\sum_{s=1}^{3}\dot{\lambda}_s^2(t)\lambda_s(t) = -a_{Mts}(t) \qquad (s=1,2,3) \quad (8.5)$$

and

$$\ddot{r}(t) = -a_{Mr1}(t) \tag{8.6}$$

and the longitudinal motion should satisfy the indicated speed condition $v_{Mi}(t) = v_{M0i}$ for each linear segment i of the trajectory between two consecutive waypoints (v_{M0i} and $v_{Mi}(t)$ are the UAV required and actual axial speeds).

This condition can be achieved by choosing for each segment i:

$$a_{Mr1}(t) = k_{2i}(t)(v_{M0i} - v_{Mi}(t)) = k_{2i}(t)(v_{M0i} + \dot{r}(t)) \tag{8.7}$$

where $k_{2i}(t)$ has constant positive values at each segment of the trajectory.

In the case of the missile guidance problem, the condition (3.91) guarantees $\lim_{t \to t_F} r(t) \to 0$; here the system (8.6) and (8.7) is only partially asymptotically stable with respect to $\dot{r}(t)$, similarly to the partial asymptotically stability of the system (8.5) with respect to $\lambda_s(t)$ [see the discussion related to the system (2.36) and (2.37)]. The analysis of equation (8.5) shows that the considered earlier guidance laws realize the parallel navigation if $\dot{r}(t) < 0$, i.e., there exists t_F, which corresponds to $r(t_F) = 0$. The time t satisfying $r(t) = 0$ can be found from the solution of equations (8.5) and (8.6). It means that there exists such $k_{2i}(t)$ that the guidance law [see also equation (3.96)]:

$$a_{Ms}(t) = N v_{cl} \dot{\lambda}_s(t) + N_1 \dot{\lambda}_s^3(t) + (1 - N_{2s})r(t)\sum_{s=1}^{3} \dot{\lambda}_s^2(t)\lambda_s(t) \tag{8.8}$$

$$+ k_{2i}(t)(v_{M0i} - v_{Mi}(t))\lambda_s(t) \qquad (s = 1, 2, 3; \ i = 1, 2...)$$

at least approximately (with a small error) will satisfy the formulated requirements to the UAV trajectory. The considered earlier guidance law follows from equation (8.8) if $N_1 = 0$ and $N_{2s} = 1$.

8.3.2 RENDEZVOUS PROBLEM

As indicated in Chapter 1, rendezvous is the guidance when, in addition to the coincidence of an object's and a target's position, the object's velocity equals the target velocity. The lateral motion (8.4) enables the object to reach the target if $\dot{r}(t) < 0$. To satisfy the equality of velocities, the tangential component of the guidance law should contain the term that equals the tangential target acceleration $u_{s3}(t) = a_{Tts}(t)$ ($s = 1, 2, 3$) (i.e., in equations (3.92) and (3.95) $N_{3s} = 1$). The rendezvous problem is reduced to the choice of the longitudinal acceleration [see equation (8.4)] that would guarantee the mentioned additional condition and $\dot{r}(t) < 0$.

We will consider the longitudinal component of the guidance law in the form:

$$a_{Mr1}(t) = k_2(v_{Tr}(t) - v_{Mr}(t)) + a_{Tr}(t) + k_3 r(t) = k_2\dot{r}(t) + a_{Tr}(t) + k_3 r(t) \quad (8.9)$$

where $v_{Mr}(t)$ and $v_{Tr}(t)$ are axial components of the UAV and target velocity vectors, respectively; k_2 and k_3 are positive coefficients.

Substituting $a_{Mr1}(t)$ in equation (8.4) we have

$$\ddot{r}(t) = -k_2\dot{r}(t) - k_3 r(t) \qquad (8.10)$$

This equation is asymptotically stable, i.e., the distance between the UAV and target tends to zero, as well as $\dot{r}(t) \to 0$. Moreover, for $\dot{r}(0) < 0$ it is possible to choose such k_2 and k_3 so that $\dot{r}(t) < 0$. It is clear from the expression (1.12) that for $V_{Ms}(t) = V_{Ts}(t)$ ($s = 1, 2, 3$) $\dot{r}(t) = 0$.

Based on equations (3.92), (3.96), and (8.9), the guidance law for the rendezvous problem can be presented in the form:

$$a_{Ms}(t) = Nv_{cl}\dot{\lambda}_s(t) + N_1\dot{\lambda}_s^3(t) + (1 - N_{2s})r(t)\sum_{s=1}^{3}\dot{\lambda}_s^2(t)\lambda_s(t)$$

$$+ (k_2\dot{r}(t) + k_3 r(t))\lambda_s(t) + a_{Ts}(t) \qquad (s = 1, 2, 3)$$

(8.11)

where the choice of N, N_1, N_{2s} and the corresponding expressions were discussed earlier.

One possible application of the considered rendezvous problem is the so-called aerial refueling, the process of transferring fuel during the flight from one aerial vehicle (commonly called the tanker) to another (the receiver). Aerial refueling capability is a critical component of the U.S. military's ability to operate military aerial vehicles (e.g., bombers, fighters, or surveillance aircraft) in the theater with maximum effectiveness, to deploy to distant theaters of operation quickly, and to remain in the air longer while operating in those theaters.

Usually, to complete an aerial refueling, the pilot of the receiver vehicle begins by flying formation in a position directly below and approximately 15 m behind the boom nozzle. The boom is flown in the trail position at 30° below horizontal, on the tanker's centerline with the nozzle extended about 1 m.

UAVs can be used both as receivers and tankers. The UAV receiver should follow the above-described procedure (i.e., it should fly below

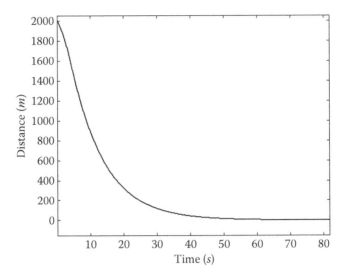

FIGURE 8.8   Distance between objects.

the tanker). The UAV-tanker should position itself above the receiver approximately 15 m ahead. Since the mentioned approximate distance between the vehicles is small enough, we will apply the guidance law (8.11) for the solution of the refueling problem considering it as the rendezvous problem. A more precise solution can be obtained assuming a nonzero distance between the moving objects; it will be presented in the next section.

Figures 8.8 and 8.9 correspond to the simulation results of the rendezvous problem. For simplicity we consider the planar case, i.e., it is assumed that the UAV operates remaining in the vertical plane with the target aircraft and using the guidance law:

$$a_{Ms}(t) = 3v_{cl}(t)\dot{\lambda}_s(t) - k_2 v_{cl}(t)\lambda_s(t) + k_3 r(t)\lambda_s(t) + a_{Ts}(t) \qquad (s = 1, 2)$$

which is the particular case of equation (8.11), tries to rendezvous with the target. The initial position and velocity vectors are: $R_M = (110, 700)$; $R_T = (2000, 0)$; $V_M = (15, 45)$; $V_T = (0, 150)$. It is also assumed that the target accelerates (its acceleration components equal 0.1-g and 0, respectively) and the UAV acceleration limit equals 3-g. As in the previous example, the UAV fight control dynamics are presented by a third-order transfer function with damping $\zeta = 0.7$, natural frequency $\omega_M = 100$ *rad/s* and the flight control system time constant $\tau = 0.1$ *s*. For $k_2 = 10$ and $k_3 = 1$ (the problem of finding their optimal values was not considered) rendezvous took a little

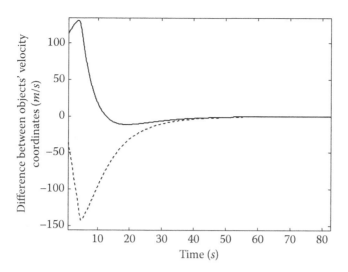

FIGURE 8.9 Difference between objects' velocity components.

bit more than one minute. Solid and dashed lines in Figure 8.9 show the planar components $V_{T1} - V_{M1}$ and $V_{T2} - V_{M2}$, respectively.

8.3.3 CONDITIONAL RENDEZVOUS PROBLEM

In contrast to the above-considered rendezvous problem, conditional rendezvous assumes a specified distance between an object and a target, i.e., after a certain period of time the object and target start moving synchronously (i.e., with the same velocity) and keeping their relative position unchanged.

The relative position between two objects can be described by indicating the distance r_0 between them and their line-of-sight vector $\lambda_{0s}(t)$ ($s = 1, 2, 3$). By specifying this vector, the problem of conditional rendezvous can be reduced to the above-considered rendezvous problem. It means that the UAV trying to implement conditional rendezvous with the target $r_T(t) = (R_{T1}, R_{T2}, R_{T3})$ can be considered as a moving object that tries to rendezvous a fictitious target with coordinates:

$$R_{Ts}^f(t) = R_{Ts}(t) - r_0\lambda_{0s}(t) \qquad (s = 1, 2, 3) \tag{8.12}$$

For the lateral motion the guidance law (8.11) can be written as:

$$a_{Mts}(t) = Nv_{cl}^f\dot{\lambda}_s^f(t) + N_1\dot{\lambda}_s^{f3}(t) + (1 - N_{2s})r^f(t)\sum_{s=1}^{3}\dot{\lambda}_s^{f2}(t)\lambda_s^f(t) + a_{Tts}(t) \tag{8.13}$$

where the LOS $\lambda_s^f(t)$ and the closing velocity $v_{cl}^f(t)$ should be determined with respect to the fictitious target, i.e.,

$$r^f(t) = \sqrt{\sum_{s=1}^{3}(R_{Ts}^{f2}(t) - R_{Ms}^2(t))}$$

$$\lambda_s^f(t) = (R_{Ts}^f(t) - R_{Ms}(t))/r^f(t)$$

$$\dot{\lambda}_s^f(t) = (V_{Ts}(t) - V_{Ms}(t) + \lambda_s^f(t)v_{cl}^f(t))/r^f(t) \qquad (8.14)$$

$$v_{cl}^f(t) = -\frac{\sum_{s=1}^{3}(R_{Ts}^f(t) - R_{Ms}(t))(V_{Ts}(t) - V_{Ms}(t))}{r^f(t)}$$

(Notations in the above equations are obvious [see equations (1.8)–(1.12)]; the upper index "*f*" indicates the fictitious target.)

Similar to equation (8.9), the longitudinal component of the guidance law is presented in the form:

$$a_{Mr1}(t) = k_2\dot{r}^f(t) + a_{Tr}(t) + k_3r^f(t) \qquad (8.15)$$

and substituting $a_{Mr1}(t)$ in equation (8.5) we have:

$$\ddot{r}^f(t) = -k_2\dot{r}^f(t) - k_3r^f(t) \qquad (8.16)$$

This equation is asymptotically stable. It is clear from the $v_{cl}^f(t)$ expression (8.14) that $V_{Ms}(t) = V_{Ts}(t)$ ($s = 1, 2, 3$) for $\dot{r}^f(t) = 0$. It follows also from equation (8.14) that $r^f(t) = 0$ corresponds to $R_{Ts}^f(t) = R_{Ms}(t)$ or $R_{Ts}(t) - R_{Ms}(t) = r_0\lambda_{0s}(t)$ ($s = 1, 2, 3$) [see equation (8.12)], i.e., the distance between the UAV and target tends to r_0 and the UAV positions itself properly according to the required LOS vector $\lambda_0(t)$.

Based on equations (8.13) and (8.15), the guidance law for the conditional rendezvous problem can be presented in the form:

$$a_{Ms}(t) = Nv_{cl}^f\dot{\lambda}_s^f(t) + N_1\dot{\lambda}_s^{f3}(t) + (1 - N_{2s})r^f(t)\sum_{s=1}^{3}\dot{\lambda}_s^{f2}(t)\lambda_s^f(t)$$

$$+ (k_2\dot{r}^f(t) + k_3r^f(t))\lambda_s^f(t) + a_{Ts}(t) \qquad (s = 1, 2, 3) \qquad (8.17)$$

where the choice of N, N_1, N_{2s} and the corresponding expressions were discussed earlier.

Usually in practice, it is more convenient to indicate the desired angles between the target velocity vector and the LOS vector rather than the required LOS vector $\lambda_0(t)$, since $\lambda_0(t)$ changes if the target accelerates or decelerates. However, $\lambda_0(t)$ can be easily determined if the target velocity vector is known. Various computational procedures can be developed. It is possible to make small changes in the guidance law (8.17) based on the material of Chapter 3 related to the modified generalized guidance laws.

8.4   GUIDANCE OF A SWARM OF UAVs

The cooperative control of groups of small inexpensive UAVs is of special interest in military and civilian applications for their abilities to coordinate simultaneous coverage of large areas or cooperate to achieve goals such as mapping, patrolling, search and rescue, surveillance, and communications relaying. It has been shown that having multiple UAVs flying in formation can be used for applications such as interferometric imaging that could not be performed by a single UAV [11]. The mentioned tasks may be repetitive or dangerous, making them ideal for autonomous aerial vehicles.

When UAVs perform a cooperative task by flying as a group, they can be considered flying in formation. A formation may be precisely defined by desired relative position vectors or more globally defined such as through artificial potential field methods [3,5,7–12,17,18,20]. Formations must safely reconfigure in response to changing missions or environments. A formation flight guidance law can guarantee the UAVs safe flight by specifying relative position requirements (a fixed formation) or establishing general rules governing UAV interaction [9,10,18]. The transition process must reassign UAVs to positions in the new formation and provide trajectories from the initial formation positions to the new ones. These trajectories must guarantee UAVs safety and be compatible with their dynamics.

Encouraged by studies of insect colony dynamics, the motion of fish schools or bird flocks, researchers have produced numerous papers and simulations investigating how the examined swarming behavior can be applied to the control of UAVs. In [6,21], the concept of a behavioral control architecture taking inspiration from natural behaviors was introduced. In [7] it was applied to control the behavior of a swarm of UAVs. A set of rules was established and applied to all the vehicles in the group. As stated in [5], by having this form of control, a system controlled through relatively simple laws can achieve a desired behavior and have the advantages of being scalable, robust, and flexible. Artificial potential fields are an example of behavioral control architecture [10,11,13,18] and were introduced in [13] in the area of obstacle avoidance for manipulators. More recently they have been applied successfully in the area of autonomous robot motion planning in [9] and in space applications in [3,12]. The basic idea behind artificial potential fields is to create a workspace where each UAV is attracted toward a goal state with a repulsive potential ensuring collide avoidance [9]. Graph theory was used in [17] as a basis for the analysis and control of large groups of cooperating agents, the local interactions, and spatial distribution of a swarm of UAVs. The virtual structure approach that treats each UAV as a particle that attempts to maintain a fixed geometric relationship was offered in [11].

Trying to be down-to-earth we consider a more practical approach to guidance of a swarm of UAVs. To the above-mentioned advantages of using a group of UAVs working together we add one more important factor. Swarms can be more self-sufficient, since one human would look after more UAVs. Cooperation and coordination in UAV groups will allow increasingly large numbers of UAVs to be operated by a single user.

Unfortunately, now the Predator ground control station (GCS) only controls one Predator at a time, and it can only control and process information from Predator vehicles. The RQ-4 Global Hawk GCS controls and processes information only from Global Hawks. Other UAVs use their own proprietary GCS systems. But commanders in the theater need access to information gathered by all types of UAVs that are flying missions. The current trend is to develop a common open GCS architecture supporting everything from the MQ-8 Fire Scout unmanned helicopter to the long-range Global Hawk for controlling multiple types of UAVs and not be blocked by proprietary walls.

The UAVs require human guidance to varying degrees and often through several operators, which is what essentially defines a unmanned aerial system (UAS). For example, the Predator and Shadow each require a crew of two to be fully operational. There has been an increasing effort to design systems that decrease an operator's workload and, as a result, the operators to vehicles ratio. Supervisory control of the UAV performs two functions: (i) to keep the UAV in stable flight; and (ii) to meet mission constraints such as routes to waypoints, time on targets, and avoidance of threat areas and no-fly zones.

The more autonomous ability a UAV has, the more complex its guidance and control systems is and, as a result, the higher is its size and weight, the less—its endurance, a combat radius, and/or speed. Payload capacity and endurance (fuel capacity) are inversely related. The desire to reduce the operator's load leads to an increase of a UAV's payload that worsens its dynamic properties, so that it becomes more difficult to reach its best potential performance characteristics.

A hierarchical control system, where a local operator controls a group of UAVs and an operator-manager controls the actions of local operators, can provide the best solution of the coordinated control of the fleet of UAVs. Moreover, the use of a manned aerial vehicle with an operator as

the leader of a local UAV group, instead of using a ground-based operator, can significantly reduce the UAV's payload and increase their performance quality.

Based on the above mentioned, we will consider the guidance of a swarm
of UAVs as guidance implemented by a local operator who determines the
trajectory of the leading UAV, coordinated with an operator-manager, and
relative positions of separate UAVs within this group with respect to the
leading UAV. In this case, the operator's main workload is focused on the
leading vehicle, which should be able to communicate with other members
of its group (it should not be a big data exchange) and the direct operator's
commands to separate UAVs can be used only on special occasions.

By setting up the desired relative positions of the separate UAVs, members of the group with respect to the leading UAV, the guidance of the
swarm of UAVs reduces to the conditional rendezvous problem discussed
in the previous section.

The efficiency of the guidance law (8.17) is demonstrated in two examples of three UAVs with the UAV leader, which using the previous terminology will be considered as the target, and two other UAVs that should
follow the leader in a certain formation. For simplicity we consider the
planar motion, which can be realistic for many practical applications.

The initial position of the vehicles is (distances are given in m, velocities
in m/s, accelerations in m/s^2):

$$R_{T1} = 1000, \quad R_{T2} = 0; \quad R_{M11} = 110, \quad R_{M12} = 700;$$

$$R_{M21} = 110, \quad R_{M22} = -700;$$

$$V_{T1} = 0, \quad V_{T2} = 60; \quad V_{M11} = 15, \quad V_{M12} = 45; \quad V_{M21} = 15, \quad V_{M22} = 45;$$

$$a_{T1} = 0.1g, \quad a_{T2} = 0.$$

It is assumed that the UAVs have a 4-g acceleration limit and dynamics
similar to that in other examples of this chapter; the guidance law parameters $N = 3$, $N_1 = 0$, $N_{2s} = 1$, $k_2 = 10$, and $k_3 = 1$, the same as in the previous
example.

The first formation (see Figure 8.10) corresponds to the case when all
three UAVS should move synchronously being on the same vertical (in the
horizontal plane) line and the distance between the leader (solid line) and
two others UAVs $r_0 = 500$ m. For this case $\lambda_{01} = 0$, $\lambda_{02} = \pm 1$. In Figure 8.10
the UAVs positions are indicated with a 10 s interval. The synchronous
motion was reached in about 60 s.

The second formation corresponds to the case when two UAVs should follow the leader (solid line) being a 500 m behind and forming a 45° (0.707 rad)

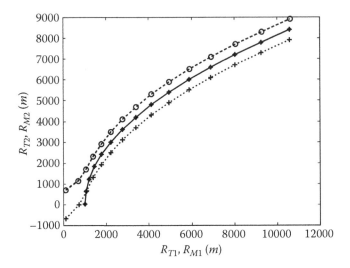

FIGURE 8.10 Conditional rendezvous of three UAVs ($r_0 = 500$ *m*; $\lambda_{01} = 0$, $\lambda_{02} = \pm 1$).

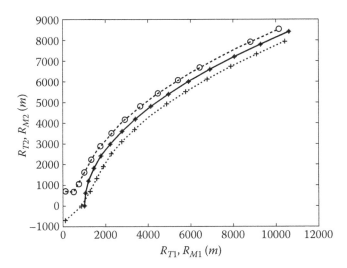

FIGURE 8.11 Conditional rendezvous of three UAVs ($r_0 = 500$ *m*; a 45° horizontal flight path angle).

angle with his flight path in the horizontal plane. In this case, λ_{01} and λ_{02} are functions of time and they were determined using the values of the UAV leader velocity vector (to determine the unit velocity vector, i.e., its direction) and the required 45° angle. In Figure 8.11 the UAV's positions are indicated with a 10 *s* interval. The synchronous motion was reached in less than 40 *s*.

8.5   OBSTACLE AVOIDANCE ALGORITHMS

Compared to the guidance problem, the avoidance obstacles problem—at least on a theoretical level—looks significantly simpler, since the admissible avoidance area is usually incomparably larger than the admissible area dictated by the guidance accuracy requirements. The avoidance problem becomes more complicated if additional limitations are imposed on the admissible area (e.g., the requirement to minimize the obstacle avoidance path, the energy spent on the obstacle avoidance operation, etc.).

An obstacle may take any form and may be either inert or hostile (including seeking a collision). Moving obstacle avoidance is more difficult than stationary ones, since the avoidance operation success depends on how precise the future position of the moving obstacle is predicted. The most difficult moving objects are missiles fired to destroy the UAV. In this case, the UAV should possess high maneuverability to escape a possible intercept.

The ability to see and avoid obstacles is a necessary condition for flight in civil airspace and effective path planning should guarantee that the UAV conforms to the federal aviation regulations (FAR) governing flight within the National Airspace System (NAS).

Once an obstacle has been detected, the flight path must be altered in order to ensure the UAV safety while minimizing deviation from the predetermined path and continuing to ensure collision avoidance. Planning around stationary obstacles is rather trivial and as we indicated in this chapter the operation research methods can be used and are used to plan optimal paths around stationary objects. Path planning around moving obstacles, however, is not a trivial task.

The UAV's ability to avoid obstacles depends largely on sensors used to provide necessary information about obstacles. Preliminary knowledge of stationary obstacles allows path planners to design the flight path properly. Vision navigation, discussed earlier, presents a reliable way to avoid close obstacles. Optical flow sensors can work altogether with other sensors to obtain more precise obstacle information.

Sensors used in existing collision avoidance systems can be broken into two types: active and passive. Examples of passive sensors include optical flow sensors (charge coupled device CCD and complementary metal oxide semiconductor CMOS, image sensors that capture images digitally with embedded optical flow algorithms), monocular or stereo vision systems using object detection and/or extraction and infrared cameras, which can be used to identify obstacles by the thermal wavelengths they emit. Active sensors include ultrasonic devices, SONAR (sound navigation and ranging), active infrared devices, radar/laser, and Doppler radar-based devices. In selecting a suitable sensor for obstacle detection it is imperative to take into account the size, speed, onboard power resources, and payload capabilities of the UAV. Environmental factors such as the available light and obstacle density also impact upon the choice of a suitable sensor system. The UAV's onboard system should use GPS/IMU data, terrain data to identify areas of probable obstacles complemented by using real-time higher resolution visual (E-O/IR, radar/laser, etc.), and nonvisual (RF, acoustic, etc.) data to detect stationary as well as moving obstacles in its path. The vision information, as another source of sensor information, helps to refine (or correct) the motion measurements from other sensors.

A possible solution of the obstacle avoidance problem consists of determining the direction of the obstacle with respect to the UAV's path, its size and speed, if it is approaching the UAV, and directing the UAV in the area, where the intersection with the obstacle becomes impossible. Slightly modified guidance laws considered in Chapter 3 can be used to reach this area. Properly installed and tuned collision avoidance sensors should indicate when the UAV should resume its course to reach the previously targeted waypoint.

In light of the guidance laws discussed in this book, the simplest approach to avoid obstacles follows from the analysis of the guidance law related to the lateral motion [see equations (3.92)–(3.95)]. As indicated earlier, this type of motion, under the condition (2.5), implements parallel navigation, i.e., an object quite effectively moves toward the target. It is quite clear that the acceleration (force) $a_{MOs}(t)$ acting in an opposite direction, i.e.,

$$a_{MOs}(t) = -Nv_{cl}\dot{\lambda}_s(t) - N_{1s}\dot{\lambda}_s^3(t) + N_2 r(t) \sum_{s=1}^{3} \dot{\lambda}_s^2(t)\lambda_s(t)$$

$$- N_{3s}a_{Tts}(t) \qquad (s = 1, 2, 3) \tag{8.18}$$

will direct the UAV from the obstacle. Usually, there is no need to use the target acceleration term in equation (8.18) and the desired goal can be achieved only by using the linear term; the additional cubic term can provide more rapid response.

In equation (8.18), the LOS $\lambda_s(t)$ ($s = 1, 2, 3$) characterizes the line between the UAV, which should avoid an obstacle by lateral maneuver, and the obstacle. The presented collision avoidance law doesn't work if the vehicle runs into the obstacle, i.e., $\dot{\lambda}_s(t) = 0$ ($s = 1, 2, 3$). Such situation should be determined by an appropriate sensor and the UAV should be removed immediately from this LOS. An additional lateral acceleration term, shifting the vehicle from a "dead course," should be added. In a case of moving obstacles (e.g., missiles attacking the UAV) the required lateral acceleration should satisfy the condition $\dot{r} \geq 0$, where r is the distance between the UAV and the obstacle.

When the UAV leaves this "dead" zone, the law (8.18) starts working and the UAV moves away from the obstacle until the installed sensors would signal that the direction from the current UAV's position to the way-point is safe.

The computational algorithm for the whole class of the considered guidance and obstacle avoidance problems can be presented in the following form:

$$a_{Ms}(t) = K_1\dot{\lambda}_s(t) + K_2\dot{\lambda}_s^3(t) + K_3 r(t) \sum_{s=1}^{3} \dot{\lambda}_s^2(t)\lambda_s(t) + (K_4 V_{M0} + K_5\dot{r}(t)$$

$$+ K_6 r(t))\,\lambda_s(t) + K_7 a_{Trs}(t) + K_8 a_{Tts}(t) \qquad (s = 1, 2, 3)$$

(8.19)

where the coefficients K_i $(i = 1\text{--}8)$ are determined by the given earlier expressions.

The ability to use a kind of generic algorithms (8.19) makes the discussed approach very attractive. The launch and landing operations are specific for various UAVs and these operations are not discussed in this book.

REFERENCES

1. Anderson, E., Beard, R., and McLain, T. Real-Time Dynamic Trajectory Smoothing for Unmanned Air Vehicles, *IEEE Transactions on Control Systems Technology* 13, no. 3 (2005): 471–77.
2. Babaei, A., and Mortazavi, M. Fast Trajectory Planning Based On In-Flight Waypoints for Unmanned Aerial Vehicles, *Aircraft Engineering and Aerospace Technology* 82, no. 2 (2010): 107–15.
3. Badawy, A., and McInnes, C. On-Orbit Assembly Using Super Quadric Potential Fields, *Journal of Guidance, Control, and Dynamics* 31, no. 1 (2008): 30–43.
4. Bahnu, B., Das, S., Roberts, B., and Duncan, D. A System for Obstacle Detection During Rotorcraft Low Altitude Flight, *IEEE Transactions on Aerospace and Electronic Systems* 32, no. 3 (1996): 875–97.
5. Balch, T., and Arkin, R. Behavior-Based Formation Control for Multi Robot Teams, *IEEE Transactions on Robotics and Automation* 14, no. 6 (1998): 926–39.
6. Brooks, R. A Robust Layered Control System for a Mobile Robot, *IEEE Journal of Robotics and Automation* 2, no. 1 (1986): 14–23.
7. Crowther, W. Rule-Based Guidance for Flight Vehicle Flocking, *Proceedings of the Institution of Mechanical Engineers—Part G: Journal of Aerospace Engineering* 218, no. 2 (2004): 111–24.
8. D'Orsogna, M., Chuang, Y., Bertozzi, A., and Chayes, S. The Road to Catastrophe: Stability, and Collapse in 2D Driven Particle Systems, *Physical Review Letters* 96, no. 10 (2006): 104302.

9. Ge, S., and Cui, Y. Dynamic Motion Planning for Mobile Robots Using Potential Field Method, *Autonomous Robots* 13 (2002): 207–22.

10. Han, K., Lee, J., and Kim, Y. Unmanned Aerial Vehicle Swarming Control Using Potential Functions and Sliding Mode Control, *Proceedings of the Institution of Mechanical Engineers—Part G: Journal of Aerospace Engineering* 222 (2008): 721–30.

11. Ilaya, O., Bil, C., and Evans, M. Control Design for Unmanned Aerial Vehicle Swarming, *Proceedings of the Institution of Mechanical Engineers—Part G: Journal of Aerospace Engineering* 222, no. 4 (2008): 549–67.

12. Izzo, D., and Pettazi, L. Autonomous and Distributed Motion Planning for Satellite Swarm, *Journal of Guidance, Control, and Dynamics* 30, no. 2 (2007): 449–59.

13. Khatib, O. Real-Time Obstacle Avoidance for Manipulators and Mobile Robots, *The International Journal of Robotics Research* 5, no. 1 (1986): 90–98.

14. Liu, C., Li, W., and Wang, H. Path Planning for UAVs Based on Ant Colony, *Journal of the Air Force Engineering University* 2, no. 5 (2004): 9–12.

15. Merino, L., Wiklund, J., Caballero, T., et al. Vision-Based Multi-UAV Position Estimation, *IEEE Robotics & Automation Magazine* 13, no. 3 (2006): 53–62.

16. Mettler, B., Tischler, M., and Kanade, K. System Identification Modeling of a Small-Scale Unmanned Rotorcraft for Fight Control Design, *American Helicopter Society Journal* 47, no. 1 (2002): 50–63.

17. Olfati-Saber R. Flocking of Multi-Agent Dynamic Systems: Algorithms and Theory, *IEEE Transactions on Automation Control* 51, no. 3 (2006): 401–20.

18. Reif, J., and Wang, H. Social Potential Fields: A Distributed Behavioral Control for Autonomous Robots, *Robots and Autonomous Systems* 27, no. 3 (1999): 171–94.

19. Reissell, L., and Pai, D. Multi Resolution Rough Terrain Motion Planning, *IEEE Transactions on Robotics and Automation* 14, no. 1 (1998): 19–33.

20. Reynolds, C. Flocks, Herds and Schools: A Distributed Behavioral Model, *Computer Graphics* 21, no. 4 (1987): 25–34.

21. Strelow, D., and Singh, S. Motion Estimation from Image and Inertial Measurements, *International Journal of Robotics Research* 23, no. 12 (2004): 1157–95.

22. Trucco, E., and Verri, A. *Introductory Techniques for 3-D Computer Vision.* Upper Saddle River, NJ: Prentice Hall, 1998.

23. Yakimovsky, Y., and Cunningham, R. A System for Extracting Three Dimensional Measurements From a Stereo Pair of TV Cameras, *Computer Graphics and Image Processing* 7 (1978): 195–210.

9 Testing Guidance Laws Performance

9.1 INTRODUCTION

Any simulation focuses on achieving certain goals—to check the validity of some ideas, to make a preliminary estimate of design efficiency, and so on. Each simulation model deals with a certain scenario. For missiles, the presence of a target, its parameters and trajectory, is a necessary part of the model, and the intercept accuracy is one of the most important parameters determined by simulation. For UAVs, the model should test how precise the vehicle follows the prescribed flight path presented by a sequence of waypoints.

There are generally three phases to the engagement and interception of a target. The first launch phase is usually uncontrolled. During this stage, the rocket motor is initiated and the missile is boosted up to its operating velocity at the direction of a target. This is followed by the midcourse phase, if the missile is not locked onto the target. Usually, during this phase the missile is guided by radar into an area that allows it to lock onto the target with its own sensor. During the terminal (homing) guidance phase, the missile is guided onto the target using its local sensor measurements. Depending on the interceptor and mission, the terminal phase can begin anywhere from tens of seconds down to a few seconds before intercept. The purpose of the terminal phase is to remove the residual errors accumulated during the prior phases and ultimately to reduce the final distance between the interceptor and threat below some specified level. For systems that use a fuze and fragmentation warhead, this final miss distance must be less than the warhead lethal radius. A direct-hit missile can tolerate only very small "misses" relative to a selected aimpoint. In either case, during the terminal phase of flight the interceptor must have a high degree of accuracy and a quick reaction capability. Moreover, near the very end of the terminal phase (often referred to as the endgame), the interceptor may be required to maneuver to maximum capability in order to converge on and hit a fast moving evasive target.

Threat missile systems continue a steady evolution in technical sophistication leading to increased capability and, consequently, the ability to perform a wider range of missions. For example, tactical ballistic missiles can have high velocity and, upon reentry, can exhibit complex coning motion and slowdown as they move through the atmosphere. Likewise, high performance cruise missiles can fly at supersonic speeds, have high lateral acceleration capability, and can execute maneuvers that are difficult to anticipate. The diversity of these threat missile systems and missions poses a significant challenge to missile interceptor design.

Modern missiles operate over a wide range of flight conditions, which vary with altitude, speed, and engine thrust. Aerodynamic missiles use aerodynamic forces to maintain their flight path. Ballistic missiles contain a part of their trajectory that is not influenced by propulsion or control. They are categorized according to their range—the maximum distance measured along Earth's surface from the point of launch to the point of impact of the last element of the payload.

The United States divides missiles into five range classes [24]. Battlefield short-range ballistic missiles (BSRBM) have a range up to 150 km. Short-range ballistic missiles (SRBM) have a range up to 1000 km. Medium-range ballistic missiles (MRBM) have a range of 1000–2400 km. The intermediate-range ballistic missile (IRBM) operational range is 2400–5500 km. Intercontinental ballistic missiles (ICBM) operate at distances over 5500 km. Cruise missiles present a special class of missiles targeting mostly surface objects and possessing a very high accuracy. The computer that guides a cruise missile is programmed prior to launch with information about the ground terrain between the point where the missile is launched and its intended destination. Using sensors on the missile, it uses various terrain references to find the target. A cruise missile has sophisticated tracking equipment, including various sensors such as cameras and satellite data receivers that allow it to determine its position. Thus a cruise missile can sense its environment, process that information, decide what to do next, and execute that decision. Unlike a ballistic missile, which is usually guided for only a small initial part of its flight after which it follows a trajectory governed only by the gravitational field, a cruise missile requires continuous guidance since both the velocity and the direction of its flight can be unpredictably influenced by a variety of factors.

Various types of missiles have their specific features that should be reflected in simulation models. Usually the six degree simulation models (6-DOF) are used to imitate the engagement scenarios and to test the effectiveness of guidance laws. When nonlinear guidance laws and non-linearities in a missile guidance system (e.g., acceleration saturation) as well as the uncertainties in missile dynamics are considered, it becomes necessary to resort to Monte Carlo techniques (repeated simulation trials) to arrive at the rms miss distance.

Proportional navigation (PN) is known to be an optimal solution, which under the assumption of a constant closing velocity, absence of autopilot lags and absence of target maneuvers minimizes the linear quadratic cost functional of the miss distance and missile acceleration. More advanced guidance laws depend on more detailed models of the target and missile and assumptions concerning the intercept scenario. The more realistic optimal problems have been investigated and optimal guidance laws were developed. However, their performance is dependent on the estimation of the time-to-go, which is assumed to be known and commonly approximated as the range between the target and missile divided by the closing velocity. Typically, the estimates of the range and closing velocity are obtained from radar or other ranging devices. In reality, this data are contaminated by noise from radar-jamming devices or the processing electronics. This affects accuracy of the estimation of the time-to-go, which causes errors in the terminal miss distance.

The linear approach, based on the assumption that the deviations from a nominal collision course are small, fails when the interception kinematics are highly nonlinear. Guidance system saturation occurs when the system demands (e.g., a commanded lateral acceleration 40-g) exceed the missile capability (e.g., the missile is only capable of 30-g). This situation arises in short-range engagements where the missile is far from the nominal collision course and in the case of highly maneuvering targets.

We considered the case of 2-DOF sinusoidal maneuvers, also known as weave maneuvers. This type is a useful starting point for analysis of intercept scenarios that involve ballistic missiles, although the ballistic target dynamics may involve an arbitrary periodic motion in three dimensions when re-entering the atmosphere. Instead of considering the 3-DOF PN problem, in many cases we assumed that lateral and longitudinal maneuver planes were decoupled by the means of roll-control, so that the consideration of the 2-DOF problem was justified. It was assumed also that the gravitational component of the total missile lateral acceleration is negligible. Such simplifications are possible only on the initial stage of analysis and design. Moreover, just after booster burnout, axial acceleration, center of gravity, and mass moment of inertia characteristics are changed. These variations should be incorporated in aerodynamic models.

UAVs capability varies significantly and can be categorized based on their type (fixed-wing or rotating-wing), payload or mission profile (altitude, range, duration). The fixed-wing aircraft have been favored as the platform for UAV because they are simple in structure, efficient, and easy to build and maintain. The autopilot design is easier for fixed-wing aircrafts than for rotary-wing aircraft because the fixed-wing aircrafts dynamics are simpler. The rotorcraft-based UAVs are desirable for certain applications where the unique flight capability of the rotorcraft (to takeoff and land within limited space; to hover and cruise at very low speed) is required. The trend in design of modern helicopters is toward more lightweight and smaller vehicles while retaining payload capability and performance parameters. This often requires that the rotor be closer to the airframe, and the higher rotor speeds required often result in higher disk loading. However, aerodynamics of such vehicles are very complicated and their simulation models do not describe precisely their dynamic features. In many cases, simulation of special flight modes of UAVs does not require using complicated 6-DOF models. Simplified 3-DOF or 2-DOF models give satisfactory results and can be used on the initial stage of design.

Although the described earlier guidance laws exclude the interference of an operator in the UAV flight, the real UAV systems should have two operation modes: automatic mode and manual mode. The mode switch should be controlled by the manual operation system. If necessary, an operator can switch from the automatic mode to manual mode to avoid disorder of the flight of the UAV.

9.2 FORCES ACTING ON UNMANNED AERIAL VEHICLES

Thrust is the main forward force acting on a missile or an aircraft and generated by a propulsion system. The thrust equation for missiles and aircraft can be derived from the general form of Newton's second law (i.e., force equals the rate of change of momentum with time):

$$T = \frac{d(mV)}{dt} \tag{9.1}$$

where m is the mass and V is the velocity of an object.

For a moving fluid (gas), the important parameter is the mass flow rate $\dot{m}_p$ that is defined as:

$$\dot{m} = \rho V S \tag{9.2}$$

where ρ, V, and S are the density, velocity, and area, respectively.

Taking into account that a net change of pressure in the flow produces an additional change in momentum, thrust T for missiles and aircraft with jet engines equals the sum of two components—the momentum thrust and the pressure thrust:

$$T = m_p v_e - m_a v_a + (p_e - p_a)A_e \tag{9.3}$$

where m_p is the mass expelled in unit time, m_a is the input mass flow; v_e is the exhaust velocity (the average actual velocity of the exhaust gases), v_a is an input gas velocity; p_e is the exhaust pressure, p_a the ambient pressure, and A_e the area of the exit of the motor nozzle.

The nozzle of turbine engines is usually designed to make the exit pressure equal to the ambient pressure, so that their pressure thrust equals zero, and the thrust for a turbojet engine is created due to a change in the momentum of air.

Propellers, which are used to drive many lightweight aircraft, generate thrust by internal combustion engines; small UAVs use electric engines. They act as rotating wings creating a lift force (aerodynamic forces and corresponding equations will be considered later).

The forward thrust of fixed-wing aircraft is proportional to the mass of the airflow multiplied by the velocity of the airflow [see equation (9.3)]. Rotary wing aircraft use engine thrust to support the weight of the aircraft and direct some of this thrust to control forward speed. In helicopters, the main rotor thrust T_M is the source for vertical lift and horizontal force.

For rockets, thrust is produced by the expulsion of a reaction mass, such as the hot gas products of a chemical reaction; in equation (9.3) m_p is the propellant mass flow rate and $m_a = 0$, so that for missiles the second term of equation (9.3) drops out.

Even when the propellant flow rate and exhaust velocity are constant so that the thrust force is constant, a rocket will accelerate at an increasing rate because the missile's overall mass decreases as propellant is used up. The change in velocity depends on the missile initial total weight, glide weight (its final weight after the propellant is expended), thrust magnitude, and the rate at which the propellant is burning.

If the missile is launched from the air, it already possesses a large initial speed. In contrast to such types of missiles, those launched from the ground (ground-based missiles), i.e., having zero initial speed need more propellant to reach the same speed. Ground-based strategic interceptors with a large operational range consist of one or two boost stages boosters.

The thrust acceleration profile of a single boost stage missile is given in Figure 9.1. Here the boost phase lasts about 7.5 s. Next, the so-called sustain phase continues until 25 s. The last glide phase corresponds to $T = 0$.

The thrust force of many types of existing missiles (without throttleable engines) is uncontrolled and directed along the $\mathbf{x}$ axis of the missile's body (see Figure 9.2). Missiles with thrust vector control are able to change the direction of thrust. Their autopilots change the actuator angle and, as a result, influence the components of the thrust vector. Thrust is used to control the flight of these missiles.

Gravity, which was neglected in the previous chapters, significantly influences the aerial vehicle range capability and should be included in more rigorous models than considered earlier. Usually, the gravity term is presented by the vertical coordinate of the ESF coordinate system that

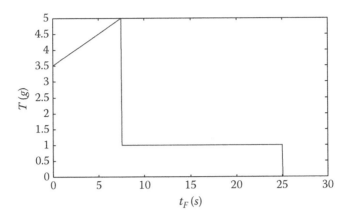

FIGURE 9.1 Thrust acceleration.

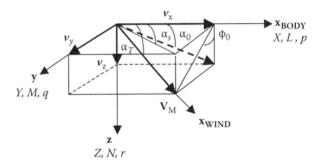

FIGURE 9.2 Coordinate systems used in aerial vehicle dynamics.

equals g. However, for the IRBM and ICBM missiles the gravity force G is distributed along three ESF coordinates as:

$$G_E = magG \frac{R_E}{\sqrt{R_E^2 + R_N^2 + (R_U + RE)^2}}, \quad G_N = magG \frac{R_N}{\sqrt{R_E^2 + R_N^2 + (R_U + RE)^2}}$$

$$G_U = magG \frac{R_U + RE}{\sqrt{R_E^2 + R_N^2 + (R_U + RE)^2}}, \quad magG = g \frac{RE^2}{R_E^2 + R_N^2 + (R_U + RE)^2}$$

(9.4)

where $RE = 6.378137*10^6$ is the Earth's equatorial radius, R_E, R_N, and R_U are the vehicle coordinates. For small altitudes $magG = g$, $G_U = g$, and $G_E = G_N = 0$.

Drag and lift belong to the so-called aerodynamic forces. Drag acts along the velocity vector (see the wind axis in Figure 9.2) and impedes the aerial vehicle's motion. It reduces vehicle speed so that reduces its acceleration capability. Lift is directed perpendicularly up with respect to drag and is the main force controlling the flight of an aerial vehicle. The lift and drag forces are presented as:

$$Lift = C_L QS, \quad Drag = C_D QS \tag{9.5}$$

where S is a reference area; C_L and C_D are the lift and drag coefficients, respectively; Q is the dynamic pressure that depends on the atmospheric pressure *PRESS* and *Mach* number:

$$Q = 0.7 \ PRESS \ (Mach)^2 \tag{9.6}$$

In the body coordinate system (see Figure 9.2; the **x** axis is directed along the aerial vehicle's body), instead of equation (9.5), the normal and axial forces generated by lift and drag are considered.

9.3   REFERENCE SYSTEMS AND TRANSFORMATIONS

A reference frame (coordinate axes) determines the origin and direction of measurement of the motion states of a dynamic model. The origin is the point from which the states are measured. The axes of the reference frame define the directions of measurement. Common reference frames in simulation are body frames, navigation frames, and inertial frames. The inertial frame is the nonaccelerating reference frame used for calculating the Newtonian equations of motion. The navigation frame is generally located at a convenient position in space; for simulation of aerial vehicles, the navigation frame may be located on the Earth's surface at a given latitude and longitude. The navigation frame may be fixed, rotating, accelerating, or moving with respect to the inertial frame. In practice, it is difficult to define a reference frame that is not accelerating with respect to inertial space. For example, an Earth-fixed reference frame is suitable for some low fidelity situations. However, in high fidelity situations the rotation and movement of the Earth needs to be accounted for in the definition of the inertial frame. The body-fixed frame has its position and orientation fixed to the vehicle body. The body carried frame has its position fixed to the vehicle body and its orientation fixed to the navigation frame. Different simulations (or phases of a single simulation) may require different inertial reference frames for the fidelity requirements. The choice of reference frames affects the numerical error incurred in the simulation. This suggests that the reference frames used by dynamic models could be chosen to reduce numerical errors.

As mentioned in Chapter 1, the flight dynamic problems require a number of reference frames for specifying relative positions, velocities, and accelerations. The equations of motion can be written with respect to any reference plane, the choice usually being a matter of convenience and accuracy requirements. The motion states of aerospace vehicles are commonly expressed using the navigation frame and the body frame. The vehicle position is commonly expressed as the position of the body frame with respect to the navigation frame, and the vehicle velocity is commonly expressed as the velocity of the body frame with respect to the navigation frame. The body frame is a convenient reference frame to express many forces and moments generated on the body.

Three orthogonal reference systems are used in the six-degree-of-freedom simulation model described below: the Earth-fixed reference system (ESF), the vehicle body system, and the seeker reference system. In Chapter 1, we described the north-east-down (NED) vehicle-carried coordinate system. As mentioned earlier, in many applications the ESF origin is near enough to the vehicle that Earth curvature is negligible, so that the NED axes are parallel to the ESF axes.

The orientation of any reference frame relative to another can be characterized by three angles (the Euler angles), which are the consecutive rotations about the **z**, **y**, **x** axes, respectively, which carry one reference frame into coincidence with the other (see Figure 9.3).

The Euler angles ψ, θ, and ϕ in Figure 9.3 correspond to the following order of rotation [4]: rotation about the $\bar{z}$ axis through angle ψ; rotation about the new position of the $\bar{y}$ axis through angle θ, putting the $\bar{x}$ axis into coincidence with the **x** axis; rotation about the **x** axis through angle ϕ.

The sequence of rotations that carry the NED frame into coincidence with the vehicle body frame are known as the body Euler angles transformation. The transformation matrices are given by:

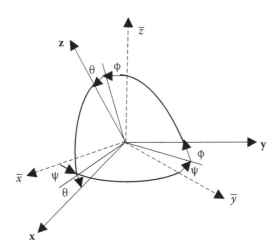

FIGURE 9.3 Euler angles.

$$L_1(\psi) = \begin{bmatrix} \cos\psi & \sin\psi & 0 \\ -\sin\psi & \cos\psi & 0 \\ 0 & 0 & 1 \end{bmatrix}, \quad L_2(\theta) = \begin{bmatrix} \cos\theta & 0 & -\sin\theta \\ 0 & 1 & 0 \\ \sin\theta & 0 & \cos\theta \end{bmatrix},$$

$$L_3(\phi) = \begin{bmatrix} 1 & 0 & 0 \\ 0 & \cos\phi & \sin\phi \\ 0 & -\sin\phi & \cos\phi \end{bmatrix} \quad (9.7)$$

so that the transformation from the NED to the vehicle body system can be described by:

$$L_{EB} = L_3(\phi)L_2(\theta)L_1(\psi) = \begin{bmatrix} \cos\theta\cos\psi & \cos\theta\sin\psi & -\sin\theta \\ \sin\phi\sin\theta\cos\psi & \sin\phi\sin\theta\sin\psi & \sin\phi\cos\theta \\ -\cos\phi\sin\psi & +\cos\phi\cos\psi & \\ \cos\phi\sin\theta\cos\psi & \cos\phi\sin\theta\sin\psi & \cos\phi\cos\theta \\ +\sin\phi\sin\psi & -\sin\phi\cos\psi & \end{bmatrix}$$

$$(9.8)$$

Taking into account that the vector of the angular velocity of the vehicle body frame relative the NED frame with coordinates $\dot\phi, \dot\theta, \dot\psi$ equals the difference between the angular velocities of the vehicle body frame (p, q, r) and the NED frame and assuming the angular velocity of the NED frame to be equal to zero, we can use the transformations (9.7) to present the rate of change of the Euler angles between the Earth axes and the vehicle body axes in terms of the body rotational rates $p, q,$ and r, the so-called roll, pitch, and yaw rates, respectively [7]:

$$\begin{bmatrix} \dot\phi \\ \dot\theta \\ \dot\psi \end{bmatrix} = \begin{bmatrix} 1 & \sin\phi\tan\theta & \cos\phi\tan\theta \\ 0 & \cos\phi & -\sin\phi \\ 0 & \sin\phi\sec\theta & \cos\phi\sec\theta \end{bmatrix} \begin{bmatrix} p \\ q \\ r \end{bmatrix} \quad (9.9)$$

If the vehicle is assumed to be roll stabilized (many simulation models are built under this assumption), i.e., $p = 0$, the simplified Euler rate equation (9.9) is used.

The Euler angles are obtained by integrating these rates and include initial values of the angles at the initiation of the simulation process, i.e.,

$$\phi(t) = \int_0^t \dot{\phi}(t)dt + \phi_0, \quad \theta(t) = \int_0^t \dot{\theta}(t)dt + \theta_0, \quad \psi(t) = \int_0^t \dot{\psi}(t)dt + \psi_0 \quad (9.10)$$

The above relations relate to the case of nonrotating spherical Earth, i.e., the rotation of the NED frame relative to the Earth-centered inertial (ECI) frame can be neglected. A more precise model should take into account the Earth rotation and its oblateness effect [7].

9.4   UNMANNED AERIAL VEHICLES DYNAMICS

Usually, missiles and aircraft are treated as a rigid body with six degrees of freedom. This is, of course, an idealization of actual flight dynamics, but avoids the complexities that a consideration of elastic forces would introduce. Assuming a rigid body, constant mass and inertia, and taking the origin of the body-fixed **x, y,** and **z** coordinates system at the vehicle center of gravity, the standard body axis six-degree-of-freedom equation of motion for a wide class of unmanned aerial vehicles can be presented as [4]:

$$\dot{v}_x = rv_y - qv_z + X + G_x + T_x$$

$$\dot{v}_y = -rv_x + pv_z + Y + G_y + T_y$$

$$\dot{v}_z = qv_x - pv_y + Z + G_z + T_z$$

$$\dot{p} = -L_{pq}pq - L_{qr}qr + L + L_T \qquad (9.11)$$

$$\dot{q} = -M_{rp}rp - M_{r^2p^2}(r^2 - p^2) + M + M_T$$

$$\dot{r} = -N_{pq}pq - N_{qr}qr + N + N_T$$

where v_x, v_y, and v_z are the components of velocity along the **x, y,** and **z** axes (see Figure 9.2), respectively; p, q, and r denote the roll, pitch, and yaw rates, respectively; G_x, G_y, and G_z are the gravity components; X, Y, and Z model accelerations produced by the aerodynamic forces; L, M, and N model angular accelerations produced by the aerodynamic moments; T_x, T_y, and T_z model propulsion system forces; and L_T, M_T, and N_T model the moments produced by the propulsion system. All variables of the right part of equation (9.11) have units of acceleration.

The coefficients L_{pq}, L_{qr}, M_{rp}, $M_{r^2p^2}$, N_{pq}, and N_{qr} are obtained as a result of simplification of a more general form of moment (m_x, m_y, and m_z) equations [4]:

$$I_{xx}\dot{p}-(I_{yy}-I_{zz})qr+I_{yz}(r^2-q^2)-I_{xz}(pq+\dot{r})+I_{xy}(rp-\dot{q})=m_x$$

$$I_{yy}\dot{q}-(I_{zz}-I_{xx})rp-I_{xz}(r^2-p^2)-I_{xy}(qr+\dot{p})+I_{yz}(pq-\dot{r})=m_y \quad (9.12)$$

$$I_{zz}\dot{r}-(I_{xx}-I_{yy})pq+I_{xy}(q^2-p^2)-I_{yz}(rp+\dot{q})+I_{xz}(qr-\dot{p})=m_z$$

where I_{xx}, I_{yy}, and I_{zz} denote the mass moments of inertia about the **x**, **y**, and **z** axes; I_{xy}, I_{xz}, and I_{yz} denote the mass product of inertia about the **x** and **y**, **x** and **z**, **y** and **z** axes, respectively.

It is desirable for the **x**, **y**, and **z** axes to coincide with the principal axes of inertia, so that the product-of-inertia terms vanish. For symmetry about the **xz** plane $I_{yz}=0$ and $I_{xy}=0$, so that instead of equation (9.12) we have:

$$I_{xx}\dot{p}-(I_{yy}-I_{zz})qr-I_{xz}(pq+\dot{r})=m_x$$

$$I_{yy}\dot{q}-(I_{zz}-I_{xx})rp-I_{xz}(r^2-p^2)=m_y \quad (9.13)$$

$$I_{zz}\dot{r}-(I_{xx}-I_{yy})pq+I_{xz}(qr-\dot{p})=m_z$$

In the case of a cruciform configuration, which is symmetrical about both the **xy** and **xz** planes, $I_{xz}=0$ as well, so that equation (9.13) is reduced to:

$$I_{xx}\dot{p}-(I_{yy}-I_{zz})qr=m_x$$

$$I_{yy}\dot{q}-(I_{zz}-I_{xx})rp=m_y \quad (9.14)$$

$$I_{zz}\dot{r}-(I_{xx}-I_{yy})pq=m_z$$

Based on the expressions for the lift and drag (9.5) and (9.6), the aerodynamic forces coefficients C_x, C_y, and C_z are modeled as nondimensional quantities and are scaled to units of force, so that:

$$\begin{bmatrix} X \\ Y \\ Z \end{bmatrix} = \frac{QS}{m} \begin{bmatrix} C_x \\ C_y \\ C_z \end{bmatrix} \quad (9.15)$$

where m in the mass of the vehicle.

The system of equations (9.13) can be presented in the form of the moment equations of (9.11), where the aerodynamic moments acting on the body are modeled as:

$$L = \frac{QSl}{I_{xx}I_{zz} - I_{xz}^2}(C_l I_{zz} + C_n I_{xz})$$

$$M = \frac{QSl}{I_{yy}}C_m \qquad (9.16)$$

$$N = \frac{QSl}{I_{xx}I_{zz} - I_{xz}^2}(C_n I_{xx} + C_l I_{xz})$$

where C_l, C_m, and C_n model nondimensional aerodynamic moment coefficients, rolling, pitching, and yawing, and l is a reference length; the cross-axis inertia symmetry term couples the roll-yaw moment equations; the coefficients L_{pq}, L_{qr}, M_{rp}, $M_{r^2p^2}$, N_{pq}, and N_{qr} in equation (9.11) equal:

$$L_{pq} = \frac{I_{xz}(I_{xx} - I_{yy} - I_{zz})}{I_{xx}I_{zz} - I_{xz}^2}, \quad I_{qr} = \frac{I_{zz}(I_{zz} - I_{yy}) - I_{xz}^2}{I_{xx}I_{zz} - I_{xz}^2}, \quad M_{pr} = \frac{I_{xx} - I_{zz}}{I_{yy}},$$

$$M_{r^2p^2} = \frac{I_{xy}}{I_{yy}}, \quad N_{qr} = \frac{I_{xz}(I_{zz} - I_{xx} - I_{yy})}{I_{xx}I_{zz} - I_{xz}^2}, \quad N_{pq} = \frac{I_{xx}(I_{xx} - I_{yy}) - I_{xz}^2}{I_{xx}I_{zz} - I_{xz}^2} \quad (9.17)$$

The above equations (9.11), (9.12), and (9.17) are written for configurations possessing the **xz** plane of symmetry, so that $I_{yz} = 0$ and $I_{xy} = 0$. Both geometrical and mass symmetry are assumed here, although it is possible that slight mass asymmetries may exist in a real vehicle configuration.

If the mass distribution is such that $I_{yy} = I_{zz}$ (for missiles with circular body cross sections), the above expressions can be simplified. In the case of a cruciform configuration, which is symmetrical about both the **xy** and **xz** planes $I_{xz} = 0$ as well, so that equations (9.11), (9.16), and (9.17) can be further simplified.

The gravitational forces are modeled as

$$\begin{bmatrix} G_x \\ G_y \\ G_z \end{bmatrix} = g\begin{bmatrix} -\sin\theta \\ \cos\theta\sin\phi \\ \cos\theta\cos\phi \end{bmatrix} \qquad (9.18)$$

where the above angles are the Euler angles of rotation by which the $\bar{x}, \bar{y}, \bar{z}$ space-fixed system of coordinates with its origin at the center of gravity comes into coincidence with the **x, y, z** body-fixed system of coordinates [4].

It is easy to conclude that equation (9.18) corresponds to the transformation of the vector $(0, 0, g)$. The angles α_0, α_s, and ϕ_0 are determined as (see Figure 9.1):

$$\alpha_0 = \tan^{-1}(v_z/v_x), \quad \alpha_s = \tan^{-1}(v_y/v_x), \quad \phi_0 = \tan^{-1}(v_y/v_z) \quad (9.19)$$

and the total angle α_T, determined as the angle between the vehicle **x** axis and the magnitude of its velocity vector $V_M = \sqrt{v_x^2 + v_y^2 + v_z^2}$ (see Figure 9.2), can be expressed as:

$$\tan^2 \alpha_T = (v_y^2 + v_z^2) / v_x^2 = \tan^2 \alpha_0 + \tan^2 \alpha_s \quad (9.20)$$

For small angles of attack and sideslip $\alpha_0 \approx v_z/v_x$, $\alpha_s \approx v_y/v_x$ and $v_x \approx V_M$. Assuming insignificant changes of speed (i.e., $\dot{v}_x \approx 0$), the accelerations $\dot{v}_y$ and $\dot{v}_z$ can be presented as $\dot{v}_y \approx v_x \dot{\alpha}_s$ and $\dot{v}_z \approx v_x \dot{\alpha}_0$. Under the above assumptions, the force equations of (9.11) can be simplified to give:

$$V_M(q\alpha_0 - r\alpha_s) = X + G_x + T_x$$

$$V_M(\dot{\alpha}_s + r - p\alpha_0) = Y + G_y + T_y \quad (9.21)$$

$$V_M(\dot{\alpha}_0 - q + p\alpha_s) = Z + G_z + T_z$$

These equations are widely used in the preliminary studies to test the autopilot design and guidance laws.

The aerodynamic coefficients C_x, C_y, and C_z related to the aerodynamic forces and C_l, C_m, and C_n related to the aerodynamic moments are typically modeled as functions of the pitch-plane angle of attack α_0, the yawn-plane sideslip angle α_s, the aerodynamic roll angle ϕ_0 (see Figure 9.2), Mach number, body rates (p, q, and r), $\dot{\alpha}_0, \dot{\alpha}_s$, the aerodynamic control surface deflections in pitch, yaw, and roll ($\delta P, \delta Y, \delta R$), center-of-gravity changes, and whether the main propulsion system is on or off.

According to equation (9.11), the body axis accelerations A_x, A_y, and A_z, the components of $A = (A_x, A_y, A_z)$ are:

$$\begin{bmatrix} A_x \\ A_y \\ A_z \end{bmatrix} = \begin{bmatrix} X + G_x + T_x \\ Y + G_y + T_y \\ Z + G_z + T_z \end{bmatrix} \quad (9.22)$$

Based on equation (9.15), this equation can be transformed in:

$$\begin{bmatrix} A_x \\ A_y \\ A_z \end{bmatrix} = \frac{QS}{m} \begin{bmatrix} C_x(\alpha_0, \alpha_s, \delta P, \delta Y, \delta R) \\ C_y(\alpha_0, \alpha_s, \delta P, \delta Y, \delta R) \\ C_z(\alpha_0, \alpha_s, \delta P, \delta Y, \delta R) \end{bmatrix} + \begin{bmatrix} G_x \\ G_y \\ G_z \end{bmatrix} + \begin{bmatrix} T_x \\ T_y \\ T_z \end{bmatrix} \qquad (9.23)$$

The functions C_x, C_y, and C_z are linearized and presented as:

$$C_x = C_{x0} + C_{xa_0}\alpha_0 + C_{xa_s}\alpha_s + C_{xV}V + C_{x\delta P}\delta P + C_{x\delta Y}\delta Y + C_{x\delta R}\delta R$$

$$C_y = C_{y0} + C_{ya_0}\alpha_0 + C_{ya_s}\alpha_s + C_{yV}V + C_{y\delta P}\delta P + C_{y\delta Y}\delta Y + C_{y\delta R}\delta R \quad (9.24)$$

$$C_z = C_{z0} + C_{za_0}\alpha_0 + C_{za_s}\alpha_s + C_{zV}V + C_{z\delta P}\delta P + C_{z\delta Y}\delta Y + C_{z\delta R}\delta R$$

where the meaning of the coefficients of the Taylor's first-order approximation is obvious.

Analogous approximation can be done for C_l, C_m, and C_n [see equation (9.16)]:

$$C_l = C_{l0} + C_{la_0}\alpha_0 + C_{la_s}\alpha_s + C_{lV}V + C_{l\delta P}\delta P + C_{l\delta Y}\delta Y + C_{l\delta R}\delta R$$

$$C_m = C_{m0} + C_{ma_0}\alpha_0 + C_{ma_s}\alpha_s + C_{mV}V + C_{m\delta P}\delta P + C_{m\delta Y}\delta Y + C_{m\delta R}\delta R \quad (9.25)$$

$$C_n = C_{n0} + C_{na_0}\alpha_0 + C_{na_s}\alpha_s + C_{nV}V + C_{n\delta P}\delta P + C_{n\delta Y}\delta Y + C_{n\delta R}\delta R$$

A more precise approximation also includes terms with coefficients depending on $\phi_0, \dot{\alpha}_0, \dot{\alpha}_s, p, q$, and r [4,7,16]. The aerodynamic coefficients are an important part of a typical aerodynamic database for a vehicle that returns the aerodynamic forces and moments for a given vehicle orientation ($\alpha_0, \alpha_s, \varphi_0$, Mach, and altitude), the aerodynamic control surface deflections ($\delta P, \delta Y, \delta R$), and the mode of the propulsion system. The aerodynamic coefficients are presented here in a general form. In practice, depending on a type of aircraft or missile, the stage of autopilot design and the requirements to accuracy, many terms of equations (9.24) and (9.25) can be excluded from consideration. The linearized small disturbance rigid body equations of motion can be found in [4,7,16,26,27,29].

Although the system (9.11) describes the dynamics of all aerial vehicles, the specific features of different types of aerial vehicles, their structure configuration, and the type of propulsion require additional equations to determine some parameters of the mentioned system and to describe the propulsion forces in details.

The helicopter dynamics look more complicated than that of fixed-wing aircrafts or missiles. A helicopter's main rotor is employed to generate lift and propulsive force while an anti-torque rotor controls the vehicle's yaw rate. The main rotor, the most important and complex subsystem, is a fundamental source of lift and forward motion for the aircraft. It produces the forces and moments required to fly and control the aircraft transferring aerodynamic forces and moments from the rotating parts (blades) to the fuselage. The rotation of its central hub with an angular speed Ω is generated by the torque created by one or more engines located in the fuselage. Each blade, attached by a set of hinges to the rotor's central hub produces the flapping motion (out of the rotor disk plane), lagging motion (back and forth in the disk plane), and pitching of feathering motion (along its longitudinal axis). By controlling the pitch angle of each blade (using the so-called *swash-plate* actuator), the pilot indirectly controls the amplitude and orientation of the load that the rotor applies to the fuselage. The interaction between the helicopter's rotor dynamics and aerodynamics (the control actions directly influence the aerodynamic loads on blades, which, in turn, influence the distribution of the airflow around the rotor) makes the helicopter simulation model rather complicated.

Various types of helicopters differ by number of rotors (e.g., dual contra-rotating rotors, quadrotor, and tandem rotors helicopters) and the type of propulsion engines. Electric engines used in many small UAVs simplify the construction of helicopters since the lift can be controlled by changing the rotor speed. They do not require complex mechanical control linkages for rotor actuation. This also simplifies their dynamic models.

Below we consider the basic helicopter model with one main rotor and modify the equations (9.11) and (9.13) by summing the forces and moments generated from the helicopter's components such as the main rotor, tail rotor, fuselage, horizontal stabilizer, and vertical stabilizer. The force terms are represented by X, Y, and Z while the moment terms in roll, pitch, and yaw directions are represented by L, M, and N, respectively; the subscripts M, T, F, H, and V represent main rotor, tail rotor, fuselage, horizontal stabilizer, and vertical stabilizer.

$$\dot{v}_x = rv_y - qv_z + \frac{1}{m}(X_M + X_T + X_H + X_V + X_F) - g\sin\theta$$

$$\dot{v}_y = -rv_x + pv_z + \frac{1}{m}(Y_M + Y_T + Y_H + Y_V + Y_F) + g\cos\theta\sin\phi \qquad (9.26)$$

$$\dot{v}_z = qv_x - pv_y + \frac{1}{m}(Z_M + Z_T + Z_H + Z_V + Z_F) + g\sin\theta\cos\phi$$

$$I_{xx}\dot{p} - (I_{yy} - I_{zz})qr = L_M + L_F + Y_M h_M + Y_T h_T + Y_V h_V + Y_F h_F + Z_M y_M$$

$$I_{yy}\dot{q} - (I_{zz} - I_{xx})rp = M_M + M_T + M_F - X_M h_M - X_T h_T - X_V h_V - X_H h_H \quad (9.27)$$

$$+ Z_M l_M + Z_H l_H + Z_T h_T$$

$$I_{zz}\dot{r} - (I_{xx} - I_{yy})pq = N_M + N_F - Y_M l_M - Y_T l_T - Y_V l_V - Y_F l_F$$

where l and h indicate the moment arms of the corresponding forces (their components; e.g., h_M and h_T are the main rotor hub and tail rotor height above CG, respectively; l_T and l_H is the tail rotor hub and stabilizer location behind CG, respectively) with respect to the center of gravity of the helicopter (CG).

The aerodynamic force coefficients can be calculated as described in [2,10,11,17,18,21,25]. Similar to missiles and other aircraft dynamic models [see equation (9.24)], the helicopter aerodynamic coefficients related to the aerodynamic forces and moments are presented as functions of the translational (v_x, v_y, and v_z) and angular (p, q, and r) velocities and the control actuation vector components.

The main rotor thrust and torque equations can be derived following [2,10,15,18,21]. The main thrust is a function of the rotor geometric parameters, the aerodynamic parameters of the blade and the operational parameters (collective pitch θ_0 and rotor speed Ω). The spatial orientation of the main rotor thrust T_M (i.e., the total aerodynamic force generated by the rotor) is controlled by a time-varying pitch command $\theta(t)$ a vertical displacement (*collective control*) or a longitudinal/lateral tilt (*longitudinal/lateral cyclic control*) of the nonrotating swash-plate [18]:

$$\theta(t) = \theta_0 + \theta_{1c}(t)\cos\Omega t + \theta_{1s}(t)\sin\Omega t \quad (9.28)$$

where θ_0, $\theta_{1s}(t)$, and $\theta_{1c}(t)$ are the collective blade pitch, the lateral cyclic pitch, and the longitudinal cyclic pitch, respectively.

The main rotor *collective pitch* control, by moving the swash-plate up and down the rotor shaft, affects the pitch angle of all blades and simultaneously changes the angle of attack of each of the blades to achieve a higher (smaller) lift. It does not induce any tilt to the swash-plate and, as the main rotor blades sweep through the air; the resulting amount of upward thrust (generally) increases with the increase of this angle.

By *cyclic pitch* control, the main rotor can be tilted as a disc to control its lateral (*lateral cyclic pitch*) and longitudinal (*longitudinal cyclic pitch*) motions. Cyclic pitch forces the blade to have a certain cyclic pitch angle that is a function of the rotational angle of the main rotor with respect to the fuselage. The so-called flapping, the oscillatory motion of the main rotor blades about the hinges, controlled by the cyclic pitch creates an uneven lift distribution. Blade flapping compensates the dissymmetry of lift. As the advancing blade flaps up due to the increased lift, the retreating blade flaps down due to the decreased lift. The change in angle of attack on each blade brought about by this flapping action tends to equalize the lift over the two halves of the rotor disc.

It is worthwhile to mention that a mathematical model of the blade, participating in indicated above types of motion, presents a complicated dynamic problem. Simplified models assume that the aerodynamic forces (lift, drag, and pitching moment) on the blade are determined only by local airflow, which depends on the overall helicopter motion, the rotation of the blade, the rotor induced flow, the rigid body motion, and the elastic deformations of the blade. The aerodynamic forces are determined by the equations similar to equations (9.5), (9.6), and (9.15) and usually the tabulated lift, drag, and pitch coefficients are determined based on wind tunnel test data.

As mentioned in [15], the complexity of helicopter flight dynamics makes modeling itself difficult, and without a good model of the flight-dynamics, the flight-control problem becomes inaccessible to most useful analysis and control design tools.

In the coupled airframe/rotor dynamics, the longitudinal and lateral blade flapping a_{1s} and b_{1s} are described, respectively, by two coupled first-order differential equations [15]:

$$\dot{a}_{1s} = -\frac{a_{1s}}{\tau_f} - q + A_{b1s}b_{1s} + A_{lat}\delta_{lat} + A_{lon}\delta_{lon}$$

$$\dot{b}_{1s} = -\frac{b_{1s}}{\tau_f} - p + B_{a1s}a_{1s} + B_{lat}\delta_{lat} + B_{lon}\delta_{lon}$$

$$(9.29)$$

where τ_f is the rotor time constant; A_{b1s}, A_{lat}, A_{lon}, B_{a1s}, B_{lat}, and B_{lon}, are the rotor flapping terms (see details in [15,18,21]); δ_{lon} and δ_{lat} are the longitudinal and lateral cyclic inputs, respectively.

Taking into account that the main rotor is coupled to the airframe dynamics through the roll and pitch angular dynamics and the longitudinal translational dynamics v_x and v_y, in equations (9.26):

$$X_M = X_{v_x}v_x + X_{a1s}a_{1s}$$

$$Y_M = Y_{v_y}v_y + Y_{b1s}b_{1s}$$

$$L_M = L_{v_x}v_x + L_{v_y}v_y + L_{a1s}a_{1s} + L_{b1s}b_{1s}$$

$$M_M = M_{v_x}v_x + M_{v_y}v_y + M_{a1s}a_{1s} + M_{b1s}b_{1s}$$

$$(9.30)$$

where the terms $X_{v_x}, Y_{v_y}, L_{v_x}, L_{v_y}, M_{v_x}, M_{v_y}, X_{als}, Y_{bls}, L_{als}, L_{bls}, M_{als}$, and M_{bls} are obtained from the corresponding analytical expressions and by experiments [15,18,21].

The heave dynamics can be approximately presented as [15]:

$$\dot{v}_z = Z_{v_z} v_z + Z_{col} \delta_{col} \qquad (9.31)$$

where δ_{col} is the collective input; Z_{v_z} and Z_{col} are the rotor flapping terms.

There exist various analytical expressions for X_M, Y_M, Z_M, L_M, M_M, and N_M (see [2,10,17,18,21,25]). All of them approximately describe the main rotor dynamics. We chose those that fit better for the later discussed autopilot design procedure.

The primary role of the tail rotor is to generate horizontal thrust varying by the collective pitch of the tail rotor blades in order to counteract the main rotor torque. It also produces the unbalanced horizontal force, which acts as a drifting force in the y direction. Its thrust T_T and torque Q_T can be computed using the same procedures as for the main rotor thrust and torque with no flapping effect included. The resulting forces and moments in equations (9.26) and (9.27) are:

$$Y_T = -T_T = -\rho C_T \Omega_T^2 R^4, \quad M_T = -Q_T = -\rho C_Q \Omega_T^2 R^5, \quad X_T = Z_T = 0 \quad (9.32)$$

where ρ is the atmosphere density, R and Ω_T are the rotor radius and speed, C_T and C_Q are the thrust and torque coefficients, respectively [17,18,21].

The remaining components have less significant contributions and are ignored in simpler models. The fuselage produces drag forces and moments that are the functions of its geometric shape. Usually, the drag of the fuselage is measured in a wind tunnel or estimated by the projected blocking area of the fuselage:

$$X_F = -\frac{\rho}{2} S_{Fx} C_D^F u_a^2, \quad Y_F = -\frac{\rho}{2} S_{Fy} C_D^F v_a^2, \quad Z_F = -\frac{\rho}{2} S_{Fz} C_D^F w_a^2 \quad (9.33)$$

where C_D^F is the fuselage drag coefficient; S_{Fx}, S_{Fy}, and S_{Fz} are the effective frontal, side, and vertical drag areas of the fuselage, respectively; u_a, v_a, and w_a are the velocity components producing pressure in the indicated areas; X_F, Y_F, and Z_F can be determined using the standard aerodynamic procedure [see also equations (9.5) and (9.6)] and substituted in equation (9.26).

The horizontal and vertical stabilizer fins create the restoring moments in the pitching and the yawing directions, respectively. They act similar to the stabilizer used in fixed-wing aircraft. The forces X_H, Z_H and X_V, Y_V in equations (9.26) and (9.27) created by the horizontal and vertical stabilizer can be presented in the form similar to equation (9.33) and calculated using the standard aerodynamic approach [17,18]. The stabilizing effect can be reflected indirectly by the rotor time constant in equation (9.29).

9.5   AUTOPILOT AND ACTUATOR MODELS

An autopilot task is to control the motion of the aerial vehicle. Maximizing vehicle performance requires choosing the appropriate autopilot structure for each stage of flight. Usually, during the midcourse phase of the missile flight, where a long flyout is required and the terminal phase, where terminal homing maneuvers are necessary, autopilots that control the missile acceleration are used. At the end of terminal homing during a guidance integrated fuse maneuver, the missile attitude may be controlled to improve the lethality of the warhead. Contemporary aircraft autopilots control an aircraft in the roll and pitch axis; control is also possible in the yaw axis, as well as an autopilot-controlled landing. Specifics of UAV autopilots are that they should be able to direct UAVs to waypoints (the landing operation can be considered as a waypoint operation as well). The flight of existing UAVs are controlled by the operator by roll and heading, pitch and altitude, and also speed commands. The future generations of UAVs will follow waypoints mostly without any operator's participation, so that their motion will be similar to a missile's motion, and their autopilots should have many features similar to missile's autopilots.

The missile dynamics model (9.11) [(9.12)–(9.16), (9.18), (9.21)–(9.25)] enables us to determine the desired values of controlled parameters. Their comparison with the real measured values of these parameters creates the error signals, which are used by autopilot controllers. Usually, instead of one autopilot, three autopilots are designed and used in practice: a pitch autopilot, a roll autopilot, and a yaw autopilot. Each of them is designed for individual channels, pitch, roll, and yaw, ignoring coupling between them. The effect of coupling is taken into account in the sophisticated design by creating interaction between the autopilots (i.e., by creating coupled autopilot channels).

The pitch rate dynamics were considered in equations (9.11), (9.16), and (9.21). Ignoring roll-yaw dynamics and considering only the components $C_{za_0}\alpha_0$, $C_{z\delta P}\delta P$, $C_{ma_0}\alpha_0$, and $C_{m\delta P}\delta P$ in equations (9.18), and (9.19), we can obtain the transfer function characterizing the relationship between A_z and δP. As shown in [4], for tail-controlled missiles it has the form (5.12), i.e.,

$$\frac{A_z(s)}{\delta P(s)} = \frac{-Bs^2 + K(AE - BC)}{s^2 + AKs - C} \tag{9.34}$$

where

$$A = QSC_{z\alpha_0}/m, \quad B = QSC_{z\delta P}/m, \quad C = QSlC_{m\alpha_0}/I,$$

$$E = QSlC_{m\delta P}/I, \quad K = 1/V_M$$

The corresponding expressions for the pitch rate and the angle of attack are [4]:

$$\frac{q(s)}{\delta P(s)} = \frac{Es + K(AE - BC)}{s^2 + AKs - C}, \quad \frac{\alpha_0(s)}{\delta P(s)} = \frac{-BKs + E}{s^2 + AKs - C} \quad (9.35)$$

The expressions (9.34) and (9.35) are used in the pitch autopilot design. As mentioned earlier, for tail-controlled missiles the transfer function (9.34) is a nonminimum phase. As the elevator δP deflects, the fin force accelerates the missile in the wrong direction. However, this force creates a pitching moment that rotates the missile. As the missile rotates, the body force accelerates the missile in the correct direction. One of the possible autopilot structures is given in Figure 9.4 (A_z is a real missile acceleration; A_{z0} is the guidance law acceleration command; τ_1 is a time constant of the actuator, which is usually modeled as a first-order lag).

The pitch control law given in [4] has the form:

$$\delta P_c(s) = W_{P1}(s)e_z(s) + W_{P2}(s)q(s) \quad (9.36)$$

where $W_{P1}(s)$ and $W_{P2}(s)$ are the transfer functions with respect to the error between the measured and desired acceleration $e_z(s)$ and with respect to the measured pitch $q(s)$.

These transfer functions are determined to guarantee the autopilot stability with the desired response and ability to operate over a broad range of aerodynamic parameters (i.e., over a broad range of aerodynamic conditions). More sophisticated autopilots have time-varying parameters to compensate changes in missile dynamics. (In the examples of the previous chapters we considered guidance laws for various ω_z, which characterized changes in missile dynamics for low and high altitudes.)

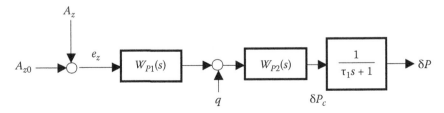

FIGURE 9.4   Pitch control.

The yaw control law can be chosen similar to the pitch control law, only in this case we operate with the components $C_{ya_s}\alpha_s, C_{y\delta Y}\delta Y, C_{na_s}\alpha_s$, and $C_{n\delta Y}\delta Y$. Some realizations of the roll control law are based on measured roll position and roll rate [4]. A constant p decouples the pitch and yaw channels [see equation (9.11)].

Over the past decades many techniques were applied to autopilot design. They include classical multivariable, modern, and optimal approaches. Integrated design methodologies that combine synthesis of guidance, estimation, and autopilot control systems have the potential for increasing missile efficiency. Testing the guidance law should be linked with a certain structure of autopilot with appropriate parameters.

Since many examples in the book relate to tail-controlled missiles, we consider here only the widely used fin actuator model. Usually, the fin actuator dynamics is modeled with a first- or second-order differential equation. The nonlinearities related to position and rate limits, as well as mechanical backlash of electromechanical actuators, should be included in the model. The autopilot pitch, yaw, and roll fin commands $(\delta P, \delta Y, \delta R)$ are distributed to the four fins producing real deflections δ_i $(i = 1,\ldots,4)$. The above-mentioned nonlinearities relates to δ_i $(i = 1,\ldots,4)$. The relationship between the actuator commands $\delta P, \delta Y, \delta R$, and individual fin deflections depends upon whether the missile has an "+" or "x" tail, i.e., whether the control surfaces are in line with wings or in planes midway between the wings.

For an "+" and "x" tail we have, respectively [4]:

$$\delta P = \frac{\delta_2 - \delta_4}{2}, \quad \delta Y = \frac{\delta_1 - \delta_3}{2}, \quad \delta R = \frac{\delta_1 + \delta_2 + \delta_3 + \delta_4}{4}$$

and

$$\delta P = \frac{-\delta_1 + \delta_2 + \delta_3 - \delta_4}{4}, \quad \delta Y = \frac{\delta_1 + \delta_2 - \delta_3 - \delta_4}{4}, \quad \delta R = \frac{\delta_1 + \delta_2 + \delta_3 + \delta_4}{4}$$

$$(9.37)$$

To obtain the unique solution of equation (9.37) with respect to the actual fin deflections δ_i $(i = 1,\ldots,4)$, the additional condition, the so-called squeeze mode (SM) condition:

$$\delta SM = \frac{\delta_1 - \delta_2 + \delta_3 - \delta_4}{4}$$

$$(9.38)$$

should be satisfied. The actual fin deflection should be chosen to make the axial force resulting from deflections as small as possible [4]. The restrictions related to the fin deflections are transformed into the auto-pilot limits (δP, δY, δR) that, in turn, impose constrains on the missile acceleration.

In contrast to missiles, where autopilots realize motion in accordance with the guidance commands that serve as the missile's brain, autopilots and control systems of aircraft realize the commands of their pilots.

A pilot of a fixed-wing aircraft uses rudder pedals, which move the rudder to control yaw, and a yoke (or a joystick) to control the attitude of the aircraft in pitch by moving elevators, when moved backward or forward, and in roll by moving ailerons, when deflected left or right. The aircraft autopilot can maintain or change the aircraft flight conditions. It can hold the aircraft attitude, altitude, and flight trajectory by controlling the control surface deflection of the aircraft and control the aircraft speed (throttle control). The three-axis autopilot controls the aircraft in pitch, roll, and yaw.

Usually, a pilot of a helicopter uses the cyclic stick, the collective lever, and the anti-torque pedals. Depending on the complexity of the helicopter, the cyclic and collective may be linked together. On most helicopters, the *cyclic,* which changes cyclically the pitch of the rotor blades, is similar to a joystick in a conventional aircraft. If the pilot pushes the cyclic forward, the rotor disk tilts forward, and the rotor produces a thrust vector in the forward direction. If the pilot pushes the cyclic to the right, the rotor disk tilts to the right and produces thrust in that direction, causing the helicopter to move sideways in a hover or to roll into a right turn during forward flight, much as in a conventional aircraft.

The collective pitch control, or *collective lever,* changes the pitch angle of all the main rotor blades collectively (i.e., all at the same time), so that increasing collective causes a climb while decreasing collective causes a descent of the helicopter. The anti-torque pedals serve a similar purpose as the rudder pedals in an airplane. They control heading by changing the pitch of the tail rotor blades, increasing or reducing the thrust produced by the tail rotor and causing the helicopter to yaw in the direction of the applied pedal, to turn left or right.

As in aircraft, the helicopter throttle should keep the rotor speed within allowable limits. It is necessary for the rotor to generate enough lift for flight. Turbine engine helicopters and some piston helicopters use electro-mechanical control systems to maintain rotor speed. Recently becoming very popular, electric engine helicopters are the most suitable for rotor speed control.

In the existing UAVs, the pilot commands are generated by an operator, a remote pilot, from a ground control station. The UAV servo systems generate inputs used by pilots, i.e., the UAV's brain activity is mostly outside of the vehicle (avoidance algorithms allow the UAV to demonstrate a certain intellect).

The guidance laws discussed in the previous chapters set the acceleration commands for the UAV autopilots to perform flights without any operator (realistically, with his minimal interference). Since these inputs are different from the inputs used by the current aircraft autopilots, the UAV autopilots should be designed for these inputs.

The autopilot design is easier for fixed-wing aircrafts than for rotary-wing aircrafts because the fixed-wing aircrafts have relatively simple, symmetric, and decoupled dynamics. That is why the fixed-wing aircraft have been favored as the platform for the UAV.

Similar to missile design, the fixed-wing UAV autopilot designer must devise a roll autopilot to provide stabilization and pitch and yaw autopilots to provide the longitudinal and lateral motions relative to the stabilized position. The design procedure is similar to missile design and pitch control law [see equations (9.34)–(9.36)] is similar to the described above (see Figure 9.4). Only the parameters of the transfer functions (9.34) and (9.35) are different, and δP is the elevator deflections. Knowledge of the transfer function of the yaw channel (its input is the rudder deflection) enables one to design control of the aircraft in yaw. The relationship between thrust T and the throttle-angle displacement δ_{th} is approximated by the first-order transfer function:

$$\frac{T(s)}{\delta_{th}(s)} = \frac{k_{th}}{(\tau_{th}s + 1)} \tag{9.39}$$

where k_{th} is a gain constant and τ_{th} is a time constant.

Details of autopilot systems design can be found in [4,6,7,16,26,27,29]. The system of equations (9.11)–(9.15), (9.21)–(9.24) describing the aircraft dynamics is basic to design a controller that would meet certain requirements.

Here we touch briefly on specifics related to design of the helicopter autopilots. Because of complicated helicopter dynamics [see equations (9.26) and (9.27)] and difficulty in determining reliably some of its dynamic parameters, the frequency approach widely used in the engineering practice to describe the input-output relationship was used in [15].

The yaw dynamics with an additional yaw rate feedback is given by [15]:

$$\frac{r(s)}{\delta_{ped}(s)} = \frac{k_r(s + k_{1r})}{s^2 + k_{2r}s + k_{3r}} \tag{9.40}$$

where δ_{ped} is the anti-torque pedal input; k_r, k_{1r}, k_{2r} and k_{3r} are constant parameters.

The acceleration Y_T is determined by equation (9.32), where C_T is a function of δ_{ped}.

The second-order nature of the response is also seen from the frequency response of the rolling and pitching rates p and q to the longitudinal δ_{lon} and lateral δ_{lat} cyclic inputs [15]. The mentioned frequency responses and the corresponding transferred functions can be obtained for the functioning helicopters with existing autopilots and/or for specific flight conditions (hover, forward flight, vertical, sideward, or rearward flight), since in the general case the helicopter dynamics (9.26) and (9.27) are unstable and should be stabilized by feedback using q, p, v_x, v_y, φ, and θ.

The aerodynamic transfer functions for the simplified equations (9.29) and (9.30) [see equations (9.26) and (9.27)] are very useful on the initial design stage. Solving equation (9.29) with respect to a_{1s} and b_{1s} we obtain the relationship between the acceleration components X_M and Y_M and the longitudinal δ_{lon} and lateral δ_{lat} cyclic inputs:

$$a_{1s}(s) = \frac{1}{\Delta(s)}[(s + b_{1s}/\tau_f)(-q(s) + A_{lat}\delta_{lat}(s) + A_{lon}\delta_{lon}(s))$$

$$- A_{b1s}(p(s) - B_{lat}\delta_{lat}(s) - B_{lon}\delta_{lon}(s))]$$

$$b_{1s}(s) = \frac{1}{\Delta(s)}[(s + a_{1s}/\tau_f)(-p(s) + B_{lat}\delta_{lat}(s) + B_{lon}\delta_{lon}(s))$$

$$- B_{a1s}(q(s) - A_{lat}\delta_{lat}(s) - A_{lon}\delta_{lon}(s))]$$

(9.41)

and

$$X_M(s) = X_{v_x}v_x(s) + X_{a1s}a_{1s}(s)$$

$$Y_M(s) = Y_{v_y}v_y(s) + Y_{b1s}b_{1s}(s)$$

(9.42)

where:

$$\Delta(s) = (s + a_{1s}/\tau_f)(s + b_{1s}/\tau_f) - A_{b1s}B_{a1s}$$

(9.43)

The above expressions enable us to build controllers that realize the guidance law components $a_{M1}(t)$ and $a_{M2}(t)$ [see equations (8.8), (8.17), and (8.18)]. They should be used in the pitch and roll autopilot design.

The acceleration command $a_{M3}(t)$ generated by the guidance law should be created by the collective δ_{col} and cyclic (longitudinal δ_{lon} and lateral δ_{lat} cyclic) controls [see equations (9.26), (9.27), (9.41)–(9.43); to compensate for negative effect of other forces additional terms can be added in the final stage of design]. The acceleration collective component can be developed similarly to the considered earlier missile pitch control [see equations (9.34)–(9.36) and Figure 9.4].

Autopilot design is beyond the scope of this book. Control theory offers various approaches and structures to build high quality controllers. For different stages of flight, different flight modes, some parameters of controllers should be changed or various autopilots should be applied. We only discussed a general approach that is compatible with the guidance laws considered in the book.

9.6 SEEKER MODEL

As a device used in a moving object, especially in a missile, a seeker locates a target by detecting some kind of emission (light, heat, or radio waves). IR imaging sensors are used in optical seekers; an antenna is used as the sensor in radar seekers. Since weaponized UAVs are able to launch missiles, the seeker systems become a part of the UAV systems, which also contain video cameras. Seekers can be useful in detecting obstacles also. A new generation of small missiles promises to improve precision strike and bring new capabilities to UAVs. It accommodates a variety of precision seekers that are accurate to within a meter of their target. A semiactive laser seeker guides the vehicle to a target being illuminated by a laser; a millimeter-wave seeker finds targets through fog and rain; and imaging infrared and shortwave infrared seekers go after heat sources such as engines.

During the terminal phase of flight, the target tracking is performed by a seeker that detects a target and tracks it within its field-of-view. Usually, seekers are mounted on gimbals. Mostly they are equipped with two mutually perpendicular yaw and pitch gimbals. Sometimes a third gimbal, the so-called roll gimbal, is added. Seekers are always stabilized (i.e., their axes remain fixed in space). To achieve stabilization and pointing control, the gimbals are controlled by torque motors using signals from rate gyros, as well as sensor information of the target position.

Analogous to the Euler body angles defined for the vehicle body frame with respect to the NED frame, the Euler angles between the vehicle body axes and the seeker axes can be defined. As indicated earlier, the $\mathbf{x}$ axis of the vehicle body coordinate system is coincident with the longitudinal axis of the vehicle; the $\mathbf{y}$ axis points out of the right side of the vehicle's body; and the $\mathbf{z}$ axis is orthogonal to both the $\mathbf{x}$ and $\mathbf{y}$ axes and defined by the right-hand rule, so that positive is defined as down. The $\mathbf{x}_s$ axis for the seeker onboard the vehicle coincides with the boresight of the seeker. The $\mathbf{y}_s$ and $\mathbf{z}_s$ axes are referred to as the yaw and pitch seeker axes. When the seeker boresight axis $\mathbf{x}_s$ coincides with the vehicle body $\mathbf{x}$ axis and the seeker gimbal angles are zero, the $\mathbf{y}_s$ and $\mathbf{z}_s$ axes coincide with the vehicle body $\mathbf{y}$ and $\mathbf{z}$ axes, respectively. The rotational sequence from the $\mathbf{x}$, $\mathbf{y}$, $\mathbf{z}$ axes to the $\mathbf{x}_s$, $\mathbf{y}_s$, $\mathbf{z}_s$ axes L_{BS} is yawn, pitch, and zero roll (ψ_s, θ_s, 0). This corresponds to such a seeker's platform that it tracks the target in azimuth and elevation.

For this case:

$$L_{BS} = L_2(\theta_s)L_1(\psi_s) = \begin{bmatrix} \cos\theta_s \cos\psi_s & \cos\theta_s \sin\psi_s & -\sin\theta_s \\ -\sin\psi_s & \cos\psi_s & 0 \\ \sin\theta_s \cos\psi_s & \sin\theta_s \sin\psi_s & \cos\theta_s \end{bmatrix} \quad (9.44)$$

Taking into account that the seeker angular velocity vector in the NED frame equals the sum of the seeker angular velocity vector with respect to the vehicle and the vehicle angular velocity vector in the NED, this relationship can be redefined in terms of the seeker body rates (p_s, q_s, r_s), the seeker Euler angle rates ($\dot{\psi}_s, \dot{\theta}_s, 0$), and the vehicle body rates (p, q, r), i.e.,

$$p_s\mathbf{x}_s + q_s\mathbf{y}_s + r_s\mathbf{z}_s = \dot{\psi}_s\mathbf{z} + \dot{\theta}_s(-\sin\psi_s\mathbf{x} + \cos\psi_s\mathbf{y}) + p\mathbf{x} + q\mathbf{y} + r\mathbf{z} \quad (9.45)$$

and solved for the seeker Euler angle rates (here we assume the unit coordinate vectors).

Using equation (9.44) and assuming zero vehicle body and seeker roll rate ($p = p_s = 0$), equation (9.45) can be written in the seeker coordinate system:

$$\begin{bmatrix} q_s \\ r_s \end{bmatrix} = \begin{bmatrix} -\sin\psi_s & \cos\psi_s & 0 \\ \sin\theta_s \cos\psi_s & \sin\theta_s \sin\psi_s & \cos\theta_s \end{bmatrix} \begin{bmatrix} -\dot{\theta}_s \sin\psi_s \\ q + \dot{\theta}_s \cos\psi_s \\ r + \dot{\psi}_s \end{bmatrix} \quad (9.46)$$

The seeker Euler angle rate equations follow immediately from equation (9.46):

$$\dot{\psi}_s = \frac{r_s - q\sin\theta_s \sin\psi_s}{\cos\theta_s} - r \quad (9.47)$$

and

$$\dot{\theta}_s = q_s - q\cos\psi_s \quad (9.48)$$

The seeker Euler angles are obtained by integrating the rate equations.

Typically, the line-of-sight error is sensed in two orthogonal components measured along the $\mathbf{y}_s$ and $\mathbf{z}_s$ axes. The elevation error $\theta_{es} = \tan^{-1}(-R_{z_s}/R_{x_s})$, measured along $\mathbf{z}_s$ axis, and the azimuth error $\alpha_{es} = \tan^{-1}(R_{y_s}/R_{x_s})$, measured

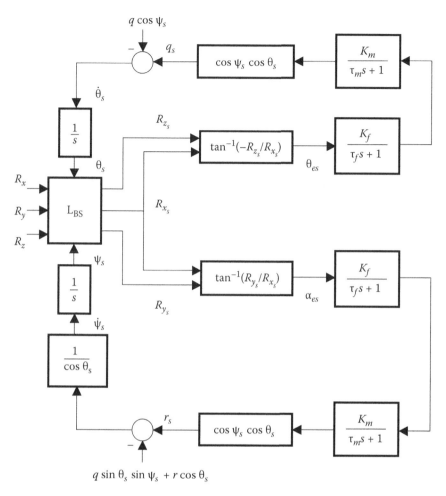

FIGURE 9.5   Seeker dynamics loop.

along the $\mathbf{y_s}$ axis, are used as the input signals to the seeker head control system that, after filtering these signals, produces torques about the axes perpendicular to the sensed error axes, which causes a gyroscopic precession of the seeker head to reduce the LOS error (the seeker Euler angle rates tend to zero). A block diagram of the seeker dynamics loop is presented in Figure 9.5.

Here the motor and filter are characterized by the first-order transfer functions with the time constants τ_m and τ_f and gains K_f and K_m, respectively; the purpose of cosines of the gimbal angles $\cos\psi_s \cos\theta_s$ is to reduce the torque of the motor as the gimbal angles increase. The closed-loop dynamics correspond to equations (9.47) and (9.48).

One of the factors that degrade performance of radar-guided missiles is the radome, which is designed to protect the missile from the airflow and reduce drag. A nonhemispheric radome causes a refraction of the incoming electromagnetic wave, thus giving a wrong indication of the target location. The radome reflection has a destabilizing effect on the missile guidance system, especially at high altitudes [30,32]. The radome effect couples the missile LOS angles to the body dynamics through the gimbal angles and causes the aberration of the measured LOS angles. One of the ways of compensating the radome effect is described in [23,32]. It combines filtering with utilization of nondestructive dither on the acceleration command signal. The LOS angles measured by the seeker are not equal to the true LOS angles. They depend nonlinearly on the horizontal and vertical gimbal angles ψ_s and θ_s, respectively. The first-order approximation gives the additional error terms $\rho_\psi \psi_s$ and $\rho_\theta \theta_s$, respectively. Following [14], in the simulation model the radome slope coefficients can be described by random processes:

$$\dot{\rho}_\psi = w_{\rho\psi}, \quad \dot{\rho}_\theta = w_{\rho\theta} \tag{9.49}$$

where $w_{\rho\psi}$ and $w_{\rho\theta}$ are zero-mean Gaussian white noise stochastic processes.

Seekers may saturate, because they operate only within their field-of-view limits. The limits on ψ_s and θ_s should be incorporated into the simulation model. The detailed seeker model including the stabilization loop enables us to obtain more accurate estimates of the line-of-sight and its derivative. It should be included in the very sophisticated models. This is beyond the scope of this book, which is focused primarily on guidance problems. For the purpose of testing guidance laws, a simplified seeker model can be considered [5]. It includes the unit determining the LOS and LOS rate and a filter. The seeker dynamics can be presented by the first- or second-order (for some types of seekers) differential equation. Noise and errors because of the radome effect can be incorporated directly into the LOS rate expression.

9.7   FILTERING AND ESTIMATION

Information required by guidance laws is obtained based on measurements provided by various sensors. As known, any measurement is accompanied with noise that distorts, at a certain degree, the result of measurement. Special measures are used to increase the accuracy. Until recently, most of measurements have been typically carried out using analog equipment. Now we are living in the digital era, when digital devices are dominating. The so-called weapon control system (WCS, a computerized program) operates missiles during the launch and midcourse phases, and microprocessors (controllers) guide missiles during the homing stage. Even if some sensors and simple filters remain analog, the information from them inputs as digital. The dialog between UAVs and their operators is in the digital language. That is why we describe digital filters below that are used in the guidance process and that can be easily incorporated in the simulation model.

The α, β and α, β, γ filters are widely used in target tracking [1]. Tracking radar systems are used to measure the target's relative position in range, azimuth angle, elevation angle, and velocity. The α, β filter produces, on the n-th observation, smoothed estimates for the position and velocity $x_s(n) = x_s(n, n)$ and $\dot{x}_s(n) = \dot{x}_s(n,n)$, respectively, as well as a predicted position for the $(n + 1)$-th observation $x_p(n + 1, n)$, i.e.,

$$x_s(n) = x_p(n) + \alpha(x_m(n) - x_p(n)) \tag{9.50}$$

$$\dot{x}_s(n) = \dot{x}_s(n-1) + \frac{\beta}{T}(x_m(n) - x_p(n)) \tag{9.51}$$

and

$$x_p(n+1) = x_p(n+1,n) = x_s(n) + T\dot{x}_s(n) \tag{9.52}$$

where α, β are the filter gains, T is the sampling period, x_m is the measured position sample, and initial conditions are defined as:

$$x_s(1) = x_p(2) = x_m(1), \quad \dot{x}_s(1) = 0, \quad \dot{x}_s(2) = (x_m(2) - x_m(1))/2$$

The recommendations concerning the choice of the parameters α and β:

$$\beta = \frac{\alpha^2}{2 - \alpha} \tag{9.53}$$

are motivated by the goal to reduce the measurement noise and minimize the tracking error (i.e., so the filter should be able to track maneuvering targets).

The fading memory filters are a subclass of the α, β filters. The filter parameters depend upon the so-called smoothing factor $0 \leq \xi \leq 1$ and are given by:

$$\alpha = 1 - \xi^2, \quad \beta = (1 - \xi)^2 \tag{9.54}$$

where heavier smoothing corresponds to larger values of the smoothing factor.

Most of the guidance laws considered earlier require some knowledge of the target's acceleration. The target's acceleration cannot be estimated accurately enough using only angle measurements made by imaging sensors on the vehicle; usually range information is required. It can be supplied either by off-board passive sensors or by onboard active sensors.

The α, β, γ filter produces, on the n-th observation, smoothed estimates for the position, velocity, and acceleration $\ddot{x}_s(n) = \ddot{x}_s(n,n)$, as well as a predicted position for the $(n + 1)$-th observation, i.e.,

$$x_s(n) = x_p(n) + \alpha(x_m(n) - x_p(n)) \tag{9.55}$$

$$\dot{x}_s(n) = \dot{x}_s(n-1) + T\ddot{x}_s(n-1) + \frac{\beta}{T}(x_m(n) - x_p(n)) \tag{9.56}$$

$$\ddot{x}_s(n) = \ddot{x}_s(n-1) + \frac{2\gamma}{T^2}(x_m(n) - x_p(n)) \tag{9.57}$$

$$x_p(n+1) = x_s(n) + T\dot{x}_s(n) + \frac{T^2}{2}\ddot{x}_s(n) \tag{9.58}$$

where initial conditions are:

$$x_s(1) = x_p(2) = x_m(1), \quad \dot{x}_s(1) = \ddot{x}_s(1) = \ddot{x}_s(2) = 0, \quad \dot{x}_s(2) = \frac{(x_m(2) - x_m(1))}{2},$$

$$\ddot{x}_s(3) = \frac{x_m(1) + x_m(3) - 2x_m(2)}{T^2}$$

The analog of equation (9.53) for the α, β, γ filter is given as:

$$2\beta - \alpha(\alpha + \beta + \gamma / 2) = 0 \tag{9.59}$$

and for the fading memory α, β, γ filters analog of equation (9.54) is:

$$\alpha = 1 - \xi^3, \quad \beta = 1.5(1 - \xi)^2(1 + \xi), \quad \gamma = (1 - \xi)^3 \tag{9.60}$$

As in the case of the α, β filters, $\xi = 0$ means that there is no smoothing.

The Kalman filter is considered as a more sophisticated filtering and estimation tool than the above-described filters. It is known to be optimal in the white Gaussian noise environment. The optimal filtering problem is formulated in the following way. For the system of difference equations:

$$x(n+1) = A_n x(n) + B_n w(n), \quad x_m(n) = H_n x(n) + v(n) \qquad (9.61)$$

where $w(n)$ and $v(n)$ are independent Gaussian random processes with zero means and covariances Q_n and R_n, respectively, based on the measurements $x_m(k)$ $(k = 1, 2,...,n)$ find the estimates $x(n, n)$ of $x(n)$ that minimize the sum of squares of the measurement errors

$$\sum_{i=1}^{n} (x_m(i) - H_i x(i,i))^T R_i (x_m(i) - H_i x(i,i))$$

(here $w(n)$ and $v(n)$ are scalars describing the process and measurement noises, respectively; A_n, B_n, and H_n are the matrices of appropriate dimensions.

The solution of the filtering problem is given as (see, e.g., [13]):

$$x(n+1,n+1) = A_n x(n,n) + P(n+1,n)H_{n+1}^T$$

$$[H_{n+1}P(n+1,n)H_{n+1}^T + R_{n+1}]^{-1}[x_m(n+1) - H_{n+1}A_n x(n,n)] \qquad (9.62)$$

where the matrix $P(n, n)$ is the solution of the matrix Riccati equation:

$$P(n+1,n) = A_n P(n,n) A_n^T + B_n Q_n B_n^T \qquad (9.63)$$

$$P(n+1,n+1) = P(n+1,n) - P(n+1,n)H_{n+1}^T$$
$$[H_{n+1}P(n+1,n)H_{n+1}^T + R_{n+1}]^{-1}H_{n+1}P(n+1,n) \qquad (9.64)$$

$P(n, n-1)$ and $P(n, n)$ are interpreted as covariance matrices representing errors in the state estimates before and after an update, respectively.

The Kalman filter equation can be presented in a form close to equations (9.50), (9.51), and (9.55)–(9.57) that describe the α, β and α, β, γ filters. By introducing the state prediction vector $x(n + 1, n)$ that satisfy the state prediction equation:

$$x(n+1,n) = A_n x(n,n) \qquad (9.65)$$

and the filter gain K_n:

$$K_n = P(n,n-1)H_n^T[H_n P(n,n-1)H_n^T + R_n]^{-1} \qquad (9.66)$$

the equation (9.62) can be rewritten as:

$$x(n+1,n+1) = x(n+1,n) + K_{n+1}[x_m(n+1) - H_{n+1}x(n+1,n)] \qquad (9.67)$$

It looks similar to equations (9.50), (9.51), and (9.55)–(9.57), which were obtained based on an intuitive approach. In contrast to the α, β and α, β, γ filters, the Kalman filter gain is time-varying. It depends on $P(n, n)$, i.e., the solution of equations (9.63) and (9.64). The solution of equations (9.63) and (9.64) depends on the initial conditions $P(0, 0)$. This presents the separate and most difficult part of the filtering problem. Nevertheless, Kalman filters are very popular and widely used in practice. The interpretation of $P(n, n-1)$ and $P(n, n)$ as covariance matrices helps choosing $P(0, 0)$. The justification of this interpretation is based on the following.

Subtracting equation (9.65) from equation (9.61) yields the state prediction error:

$$e(n+1,n) = A_n e(n,n) + B_n w(n) \qquad (9.68)$$

so that the state prediction covariance $P(n+1, n) = E[e(n+1, n)\, e(n+1, n)^T]$ satisfies equation (9.63); $P(n, n) = E[e(n, n)\, e(n, n)^T]$, called the updated covariance, and $e(n, n)$ is the updated error in the state estimates [see equation (9.67)].

Subtracting equation (9.67) from the state equation (9.61) and acting in a manner similar to equation (9.68), we can obtain equation (9.64), which can be rewritten as:

$$P(n+1,n+1) = P(n+1,n) - K_{n+1}S(n+1)K_{n+1}^T \qquad (9.69)$$

where:

$$S(n+1) = H_{n+1}P(n+1,n)H_{n+1}^T R_{n+1} \qquad (9.70)$$

is called the measurement prediction covariance; $S(n+1) = E[e_m(n+1, n)\, e_m(n+1, n)^T]$; $e_m(n+1, n)$ is the measurement prediction error:

$$e_m(n+1,n) = x_m(n+1) - H_{n+1}x(n+1,n) = H_{n+1}e(n+1,n) + v(n+1) \qquad (9.71)$$

The widespread tracking models (nearly constant velocity $\ddot{x}(t) = w_0(t)$ and near constant acceleration $\dddot{x}(t) = w_0(t)$ models [1]; $w_0(t)$ is the zero-mean white noise process) are analyzed based on the above-given expressions. For these models, the matrices $A_n = A$, $B_n = B$ and $H_n = H$ are:

$$A_n = \begin{bmatrix} 1 & T \\ 0 & 1 \end{bmatrix}, \quad B_n = \begin{bmatrix} 0 \\ 1 \end{bmatrix}, \quad H_n = \begin{bmatrix} 1 \\ 0 \end{bmatrix}$$

(9.72)

for the nearly constant velocity model

and

$$A_n = \begin{bmatrix} 1 & T & 0.5T^2 \\ 0 & 1 & T \\ 0 & 0 & 1 \end{bmatrix}, \quad B_n = \begin{bmatrix} 0 \\ 0 \\ 1 \end{bmatrix}, \quad H_n = \begin{bmatrix} 1 \\ 0 \\ 0 \end{bmatrix}$$

(9.73)

for the nearly constant acceleration model

In the discretized state equations with the sampling period T, the discrete time process noise relates to the continuous time zero-mean white noise process $w_0(t)$ with the spectral density Q_0 as:

$$w(n) = \int_0^T e^{A_c(T-\tau)} B w_0(nT + \tau) d\tau$$

(9.74)

where A_c is the state matrix of the equations $\dot{x}(t) = w_0(t)$ and $\ddot{x}(t) = w_0(t)$, respectively. Then $Q = E[w(n)w(n)^T]$ equals:

$$Q = \begin{bmatrix} \dfrac{T^3}{3} & \dfrac{T^2}{2} \\ \dfrac{T^2}{2} & T \end{bmatrix} Q_0 \quad \text{for the nearly constant velocity model} \quad (9.75)$$

and

$$Q = \begin{bmatrix} \dfrac{T^5}{20} & \dfrac{T^4}{8} & \dfrac{T^3}{6} \\ \dfrac{T^4}{8} & \dfrac{T^3}{3} & \dfrac{T^2}{2} \\ \dfrac{T^3}{6} & \dfrac{T^2}{2} & T \end{bmatrix} Q_0 \quad \text{for the nearly constant acceleration model} \quad (9.76)$$

Based on the steady-state solution of the Riccati equations (9.63) and (9.64), for these two models, the relationship between the parameters of the α, β and α, β, γ filters and the steady-state parameters of the corresponding Kalman filters is established [1]. The position gain α, velocity gain coefficient β, and acceleration gain γ are the functions of the so-called target maneuverability index

$$\lambda = \frac{\sigma_w T^2}{\sigma_v},$$

where σ_w and σ_v, the variances of the process and measurement noise, characterize the motion and observation uncertainty, respectively.

The described target state estimators are important subsystems in advanced vehicle guidance systems. Target state estimators are required for two reasons. First, the measurements provided by onboard seekers such as line-of-sight angles and its rates, as well as range and range rate, are often contaminated by noise and are not in a form usable by the guidance laws. Second, advanced guidance laws require additional information about the target such as its acceleration, which cannot be provided by the onboard sensors. The proportional navigation guidance law has been used most widely in the homing phase. However, there exist many situations (highly maneuvering targets) where the PN law performs unsatisfactorily. This fact has given rise to various modifications of the proportional navigation guidance law; some of them were considered in the previous chapters. The augmented proportional navigation guidance law, as well as other guidance laws considered in the previous chapters, need information about the target acceleration for their implementation. The optimal guidance laws are based on information about the target acceleration and the predicted intercept point or time-to-go, so that their implementation needs the estimates of the corresponding parameters. Even in the case of the classical PN law, the LOS rate and closing velocity need to be estimated, since the corresponding measurements are noisy. In the example of Chapter 3, the Kalman filter was used to produce a smoothed LOS rate estimate for use in the PN law.

In general, the use of Kalman filters can be rigorously justified, when the dynamics is considered to be linear. Moreover, although various elegant expressions have been obtained, the application of the above-described Kalman filters requires certain skills and experience. The 9-DOF and 6-DOF filters are used for estimating the position, velocity, and acceleration of the target. There exist algorithms that operate with both models and that determine when to switch from one model to another.

Although Kalman filters are based on the theory developed for linear models, they are applied also for nonlinear models. The kinematics of the target are modeled usually in the NED coordinate system, but the target position measurements are assumed to be made in the spherical system consisting of range, azimuth, and elevation angles (see (1.4) and (1.5)). The transformation from the spherical coordinate system to the Cartesian coordinate system is nonlinear. The so-called extended Kalman filters are used to increase accuracy of estimations. The approach used in the extended Kalman filter is based on the linearization of nonlinear functions by utilizing a first-order Taylor series expansion (ignoring the second- and higher-order terms). The resulting approximate measurement equation becomes linear, but the measurement matrix should be calculated in every iteration. Despite the lack of the rigorous mathematical justification (in contrast to the Kalman filter), the extended Kalman filter is widely used in attitude estimation (e.g., Euler angles). The Kalman filters (original and extended) require complete prior covariance information on the initial state, process noise, and measurement noise.

In numerous applications, essential statistic information concerning the process and measurement noise may either be missing or may be poorly defined. The filter operation is further affected by modeling errors, linearization approximations, and, as a result, the covariance matrix, computed from the Riccati equation of the extended Kalman filter, may not resemble the true covariance matrix of errors in the estimated state.

The so-called unscented filters, developed as an improvement of the extended Kalman filters, are also used in attitude estimation [28]. However, the above-mentioned filters and their tuning require an extensive experimental data.

The filters included in the simulation model used for testing guidance laws should contain well-tested parameters. The accuracy and robustness of state estimators have been some of the limiting factors for improving the guidance performance against maneuvering targets. The above-considered filters must be tuned to the most stressing threat that is expected. If the threat is less stressing, the performance will be worse than optimal compared to a Kalman filter tuned to a less stressing threat. This leads to the consideration of adaptive estimation techniques that give robust performance over a wide class of maneuvering targets.

The term *target* is absolutely clear in the case of missile guidance. In the case of UAV guidance, targets can be obstacles, objects of surveillance, and so on. Filters are used to get reliable information from measurements produced by appropriate devices. Besides the target parameter measurements, Kalman filters are used to get a vehicle's altitude estimates from accelerometer and gyro readings. A survey of numerous estimation techniques can be found in [3].

9.8 KAPPA GUIDANCE

The Kappa algorithm dominates midcourse guidance in the endoatmosphere and can be interpreted as maximization of the terminal missile velocity, i.e.,

Kappa guidance is an optimal guidance law that maximizes missile speed at the beginning of the terminal phase of flight [9]. This requirement is essential against a target at a far distance or at a low altitude. When engaging a target at long range or at low altitude, missile velocity is the prime factor. Kappa guidance is applied also for targets at close distances. When engaging a close-in target, the time line is most important, because the missile must destroy the target before it reaches within the minimum range of intercept.

The Kappa algorithm is based on knowledge of the predicted intercept point. As indicated in the introduction of Chapter 3, the accuracy of prediction or estimation significantly influences the accuracy of the engagement. Because the Kappa guidance law is obtained as the solution of the terminal optimal problem, it requires complete information (current and future) about the missile and target, as well as the external conditions during the engagement. However, the missile dynamic equations including drag and lift can only be implemented approximately, and it is impossible to estimate analytically the influence of incomplete information on the outcome of the engagement. Moreover, the predicted intercept point (PIP) and time-to-go are important parameters that dominate the accuracy of the optimal solution. These parameters can only be estimated, and it is difficult to evaluate the influence of the errors on the engagement final results.

It is difficult to reflect all factors when formulating the optimal guidance problem, so that many optimal problems discussed in the literature have not been implemented in practice. All optimal problems are also rather complex for real-time onboard implementation. In many publications, target maneuvers were either neglected or assumed to be well-defined, mostly constant. In guidance laws that explicitly include the target maneuver, the estimation of this variable, which cannot be measured directly, becomes critical. Formally, the optimal problem that generates the Kappa algorithm ignores a target behavior. Indirectly, it is taken into account in the estimate of the predicted intercept point. The optimal problem is accompanied by many assumptions. However, and it is very important, the Kappa midcourse guidance algorithm is successfully used in the SM2 (U.S. Standard Missile 2) missiles. That is why it deserves to be included in the simulation software and is considered below.

Let the vector r_{PIP} indicate the position of the predicted intercept point and the vector r_M characterize the current missile position. Then the PN law (2.24) can be rewritten as:

$$a_c = \frac{N}{t_{go}^2}(r_{PIP} - r_M - v_M t_{go}) - \frac{N}{t_{go}}(v_{PIP} - v_M)$$

where the missile velocity v_M terms were added with opposite signs to the left part of equation (2.24) and v_{PIP} is the missile terminal velocity at the intercept point.

The above equations can be presented in a more general form:

$$a_c = \frac{K_1}{t_{go}^2}(r_{PIP} - r_M - v_M t_{go}) - \frac{K_2}{t_{go}}(v_{PIP} - v_M) \qquad (9.77)$$

where K_1 and K_2 are some coefficients.

Assuming that the predicted intercept point can be evaluated during the missile flight and that v_{PIP} is a desired terminal missile velocity, the problem of finding the optimal guidance law can be formulated as a problem of finding the optimal values K_1 and K_2 that maximize the terminal value of v_M. As shown in [9], the optimal values are:

$$K_1 = \frac{w^2 r^2 (\cosh(wr) - 1)}{wr \sinh(wr) - 2(\cosh(wr) - 1)} \qquad (9.78)$$

and

$$K_2 = \frac{w^2 r^2 - wr \sinh(wr)}{wr \sinh(wr) - 2(\cosh(wr) - 1)} \qquad (9.79)$$

where

$$w^2 = \frac{D_0 L_\alpha (T/L_\alpha + 1)^2}{m^2 v^4 (2C_{L_\alpha} + T/L_\alpha)} \qquad (9.80)$$

and

$$D_0 = C_{D0}QS, \quad Lift = QSC_{L_\alpha}\alpha = L_\alpha\alpha$$

D_0 is the drag component stipulated by the component C_{D0} of $C_D = C_{D0} + C_{L_\alpha}\alpha^2$ assuming that the so-called drag curve is parabola [17] and L_α is the lift factor [see equation (9.5)]; T is thrust, m is the mass, and v is the speed of the missile.

The first term of equation (9.77) is called the proportional term and presented as:

$$a_{c1} = \frac{K_1}{t_{go}^2}(r_{tgo} - v_M t_{go}) \qquad (9.81)$$

where r_{tgo} is the range-to-go vector.

The second term of equation (9.77) is called the shaping term and presented as:

$$a_{c2} = \frac{K_2}{t_{go}}(v_{PIP} - v_M) \qquad (9.82)$$

The desired terminal velocity vector $v_{PIP} = (V_{PIPN}, V_{PIPE}, V_{PIPD})$ is given as:

$$V_{PIPN} = V_M \cos\mu_v \cos\mu_h$$

$$V_{PIPE} = V_M \cos\mu_v \sin\mu_h \qquad (9.83)$$

$$V_{PIPD} = -V_M \sin\mu_v$$

where the angles μ_v and μ_h are the vertical and horizontal trajectory shaping angles.

For tactical ballistic missiles $\mu_v \approx 45°$, for cruise missiles $\mu_v \approx -75°$.

A slightly different presentation of the proportional and shaping terms can be found in [22].

9.9   LAMBERT GUIDANCE

The so-called Lambert guidance presents the guidance law based on the solution of the Lambert problem, a terminal problem to determine initial conditions of the given system of ordinary differential equations that would satisfy the terminal (at the moment t_F) conditions. In contrast to terminal problems usually considered in the optimal theory, the Lambert problem deals with a part of initial conditions, assuming that the other part is known. The problem can be reformulated as a control problem by including δ-functions in a class of admissible controls.

The Lambert problem can be stated as follows: given the initial position $r_0 = (x_{10}, x_{20}, x_{30})$ of a body in the gravity field described by the Newton's law of universal gravitation [see also equation (9.4)]:

$$\ddot{x}_s(t) = -\frac{gm}{r^{1.5}(t)}x_s(t) \qquad (s = 1, 2, 3) \qquad (9.84)$$

where g is the gravitational acceleration, m is the body's mass, $x_s(t)$ are its coordinates in the Earth-centered coordinate system:

$$r(t) = \sqrt{\sum_{s=1}^{3} x_s^2(t)}$$

find the initial velocity orientation $V_0 = (\dot{x}_{10}, \dot{x}_{20}, \dot{x}_{30})$ so that at the moment t_F the body will be at the location $r(t_F) = (x_1(t_F), x_2(t_F), x_3(t_F))$ with the desired coordinates x_{sF}, i.e., $x_s(t_F) = x_{sF}$ ($s = 1, 2, 3$).

The solution of the formulated problem in the case of linear differential equations does not present any difficulty, since there exists the analytical expression describing the relationship between $r(t_F)$, r_0, and V_0. For the nonlinear system (9.84) such an expression cannot be obtained and various computational algorithms were offered to solve this problem. For a planar trajectory, the existing expression for $\|V_0\|$ is a function of the initial flight path angle and the expression for t_F is a function of $\|V_0\|$ and the initial flight path angle (the symbol $\| \ \|$ denotes the Euclidean norm). Formally, the number of unknown parameters remains the same, but the computational procedure looks more attractive since the iterations to solve two nonlinear algebraic equations are conducted only with respect to one parameter, the value of a flight pass angle [31]. Computational programs for two- and three-dimensional cases can be found in [31].

The existing computational methods enable us to develop algorithms dealing directly with $\dot{x}_{10}, \dot{x}_{20}$, and $\dot{x}_{30}$ (e.g., by solving for the given t_F the minimization problem

$$\min_{\dot{x}_{s0}} \|r_F - r(t_F)\|^2$$

where r_F is the desired position vector and $r(t_F)$ is the solution of the system (9.84) for the initial conditions (r_0, V_0^i); i is the step of iterations; $s = 1, 2, 3$.

The Lambert problem relates to a body motion in a gravity field and its solution, an instant impulse, is more applicable to steer spacecraft rather than missiles.

Applying the Lambert problem to missile guidance we assume that the moving body is a launched missile, t_F is the flight time and the target position satisfies the terminal condition r_F. The solution of the Lambert problem gives the missile launch direction and velocity required to reach the target at the time t_F.

Since some existing guidance algorithms use the predicted intercept point and the time to intercept t_F, the Lambert problem was modified to obtain guidance laws for moving targets [31]. In this case, the Lambert terminal problem is solved for the predicted $r(t_F)$, i.e., the terminal condition becomes fuzzy. The obtained velocity vector $V(t)$ determines the so-called velocity-to-be-gained vector ΔV_M, the difference between the required velocity $V(t)$ and the current missile velocity $V_M(t)$, i.e.,

$$\Delta V_M = V(t) - V_M(t) \tag{9.85}$$

If a current acceleration vector is $a_M(t_i)$, then the direction of the acceleration at the moment t_{i+1} should be aligned with the velocity-to-be-gained vector $V_M(t)$, i.e.,

$$a_{Ms}(t_{i+1}) = a_M(t_i)\frac{\Delta V_{Ms}}{\|\Delta V_M\|} \qquad (s = 1, 2, 3) \qquad (9.86)$$

Formally, if the missile's thrust acceleration vector is aligned with the velocity-to-be-gained vector, then the desired velocity can be achieved (after that the engine should be cut off) and the missile should fly ballistically to the target. However, this is possible only and only if at the moment when the engine is cut off the missile position, the future intercept point and the time of intercept satisfy the Lambert equations.

The mathematically rigorous Lambert problem requires the known initial and final points and the time of flight t_F. It can be used for missiles that fly ballistically to hit stationary targets. However, for moving targets this approach lacks rigorous justification.

The use of a Lambert guidance looks reasonable for offensive ballistic missiles, in which the boost phase brings them on a gravity field trajectory, which can be calculated in advance. However, the guidance law (9.85) requires a more complicated thrust vector controlled (TVC) boosting motor with the cut-off mode. In the case when the interceptor is a defensive missile, the time of flight and the final point (related to the procedure - predicted intercept point/time of intercept) are interconnected and unknown so that, as mentioned above, a modified rigorous mathematical problem became nonrigorous. For interceptors, the minimal time of intercept is the most important factor. But a gravity field free trajectory takes more time than a forced trajectory, and Lambert guidance deals with such trajectories. That is why even if the interceptor operates successfully following Lambert guidance (there exists no proof of the convergence of the offered algorithms) its performance (time of intercept) can be improved by using another guidance law.

9.10   SIMULATION MODELS OF UNMANNED AERIAL VEHICLES

In the previous sections, we described the main aerial system elements that should be included in the simulation model. Depending on the accuracy requirements and the guidance laws under consideration, some of the elements may be not needed in the model, some equations can be simplified.

The simulation model should analyze the performance of guidance laws in a realistic simulation environment, which accounts for the effects of drag and flight control system dynamics on the vehicle's performance.

Analysis of the vehicle's performance criteria should be used as the measure of effectiveness and basis of comparison.

Modern threats have become faster, stealthier, and more maneuverable. Successfully engaging such threats will require a system approach that implements a combination of advanced sensor processing algorithms, guidance algorithms, and control processing techniques. Missile defense interceptor flight control system design requirements are generally driven by high-maneuver rates that are needed for terminal homing. These requirements must be met while retaining stability and robustness throughout a large possible engagement envelope. The engagement envelope or kinematic boundary is of paramount importance. The kinematic boundary represents the maximum range at which the missile will achieve a hit, when there is no noise in the system. It can, therefore, be used as a criterion to compare the performance of guidance laws. Among other significant features of guidance systems performance are the miss distance, the time of intercept, maximum rate of turn, and maximum lateral acceleration. The comparative analysis of guidance laws is more restrictive. It includes some of these features (the engagement envelope and miss distance, the time of intercept), as well as specific features, such as the missile terminal speed and impact angle.

The proportional navigation guidance law and the Kappa and Lambert guidance laws can be used as the baseline against which the other guidance laws will be tested. The candidate guidance laws include the guidance laws considered in this book and their combination (i.e., hybrid guidance laws, as well as other guidance laws considered in the literature). The guidance laws should be tested against nonmaneuvering and maneuvering targets.

The UAV performance should be judged by analyzing how close to the prescribed flight pass the UAV operates, how efficiently it avoids obstacles, and so on. Since the existing UAVs are guided mostly by operators (i.e., the use of the guidance law generated by an operator based on his knowledge and ability), it is possible to test the offered guidance laws by comparing them with the operator guided flights based on the detailed information about these flights.

The simulation model should be built based on the module principle, which is the most efficient way to create simulation models that can be enhanced in the future or simplified, if it is necessary. Separate modules serve as building bricks. If needed, some of them can be deleted without damaging the structure or the new ones can be added to make the structure more sophisticated.

Normally the design of the models could be developed using a computer aided software engineering (CASE) tool supporting an object-oriented methodology such as the Unified Modeling Language, an object-based methodology such as HOOD or a structured design methodology such as Yourdon. These tools generate the model stubs, and the models would then typically be implemented in C++, ADA, or C. Simulink and Matlab developed software based on six degrees of freedom dynamics and simplified aerodynamics that can be used in the aerospace industry.

Building a simulation model is an art. Despite the existing design tools, deep knowledge of specific problems enables one to create more sophisticated programs than by using the existing "general use" tools. Below we describe the structure of hypothetical simulation models that, in our opinion, in the best way meets the research requirements. The missile simulation model is focused on analyzing various intercept problems in the endoatmosphere. The UAV simulation model is focused on analyzing various flight paths. They can be realized by using, for example, Visual Fortran.

9.10.1   6-DOF Simulation Models

The missile simulation model should properly reflect two stages of the missile flight, its midcourse and homing stages. As mentioned earlier, in certain scenarios the mission requirements call for the payload to impact the target from a specific direction. These requirements are of importance during the so-called endgame, the final part of the homing stage.

The simulation process usually starts from the midcourse phase. It means that the missile prehistory (its uncontrolled boost stage) should be presented by the missile position and velocity at the beginning of the midcourse stage (see Figure 9.6). The launch parameters, which define the direction of the missile flight, are determined by the predicted intercept point (PIP). The initial unguided boost stage is strictly programmable depending on the target position at the launch time and some other external measurable factors.

Typically, there exist tables that enable us to find, depending on the predicted intercept point, the position and velocity of the missile up to about 6 s after its launch (i.e., at the beginning of the midcourse phase).

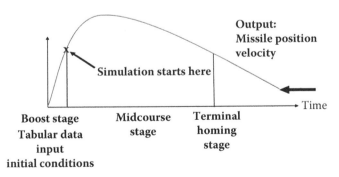

FIGURE 9.6   Stages of the simulation process.

They have been obtained from various experiments combined with analytical analysis. For missiles launched from ships, corrections should be calculated for ship movement, ballistic wind influence, and launch cell parallax. This data and related equations should be incorporated into the model.

As indicated earlier, during the midcourse phase the missile is guided by the weapon control system. The most sophisticated simulation models should be able to model the main WCS operations that include: (1) algorithms to determine the predicted intercept point and time-to-go; (2) filtering that provides the midcourse guidance with inputs characterizing the target and missile position, velocity and, if needed, acceleration; and (3) midcourse guidance commands. The predicted intercept point and time-to-go are used also in some versions of the PN law in the terminal guidance phase. As mentioned, the difficulties in predicting target position stem from uncertainty in interception time because of unpredictability of the target's future behavior. Factors contributing to uncertainties in the intercept point are (1) random and systematic errors in the defense detection and tracking system's measurement of the position and velocity and estimate of acceleration of the attacking missile, (2) lack of knowledge of the attacking missile's target, and (3) intentional trajectory shaping and intentional evasive maneuvers. A reasonable assumption is that at long ranges the missile need only travel approximately in the right direction. Hence, at long range the time to intercept needs to be estimated only roughly. As interception approaches, the need for accuracy increases. The time-to-go can be estimated roughly as $t_{go} = -\dot{r}/r$. In practice, a more detailed scenario-specific analysis is used to determine PIP and the time-to-go. This topic is beyond the scope of this book.

Absence of the PIP module in the simulation model would require a certain homework to determine the missile initial position and velocity at the beginning of the midcourse stage. If the simulation process deals only with the terminal stage, the missile position and velocity should be determined separately based on specifics of the scenario under consideration.

The filtering module should contain a set of equations discussed earlier. During the midcourse, the interceptor receives frequent updates on its position from off-board tracking sensors, so that formally the estimates of the position and velocity (as well as acceleration for some guidance algorithms) not only of the target but also the missile are needed. In the less sophisticated simulation models, the filtering operations are applied to the relative position and velocity of the missile with respect to the target. This relates also to the terminal stage. The filtering module realizes signal processing algorithms to provide smoothed data that is used in the guidance law. It is very important to model the errors expected in tracking using infrared sensors, surface- and air-based radars based on the analysis of these sensors. Then the obtained filtering results are more realistic than under assumption of Gaussian white noise.

The target module consists of a 3-DOF point-mass presentation of the target motion. The target model should be capable of executing a maneuver at a given time. The user should be able to adjust the time and extent of this maneuver. From an almost infinite variety of possible maneuvers, it is important to choose the most representative maneuvers for various types of missiles and the intercept scenarios.

In the previous chapters, we considered the so-called step and weave maneuvers. Inside these two main classes, special types of maneuvers are specified. For example, analyzing boost-phase intercept systems (attacking ballistic missiles should be disabled in their boosting phase, during the first few minutes of flight) it is reasonable to model a sudden increase or decrease in the target's missile angle of attack and a switch back maneuver, in which the target switches from a positive to negative angle of attack. Maneuvers like these (*lunge* maneuvers) might be performed either to shape the attacking missile trajectory or to try to evade an anticipated interceptor. In the simulation model, a lunge maneuver should be executed during the last few seconds before the predicted intercept time. Another kind of maneuver, a *jinking* maneuver, is a periodic maneuver usually a sinusoidal modulation of the acceleration that would produce a fish tail-like motion of the missile during the last few seconds before the predicted intercept. Such maneuvers are likely to be within the capabilities of the attacker's missiles. Cruise missiles can perform various maneuvers. The most typical are *diving* maneuvers, when a cruise missile significantly changes its altitude, and weave maneuvers in a horizontal plane, when it approaches the target at a very low altitude.

Optimal control and game theories were used to formulate precisely and solve the problem of optimal pursuit and evasion. Unfortunately, this approach cannot work because it is difficult to build an analytical model that matches well to reality and can be used in practice. Deterministic optimal problems require ideal information, which cannot be obtained, about the target and missile flight parameters. However, the optimal approach does enable an evaluation of the best possible scenario for the evader that can be used for comparison with strategies that can be realized in practice. The optimal evader's maneuver, if a priori information concerning the missile guidance system is available, is described in [31]. An intuitive evasive maneuver, which is formulated as the reverse proportional navigation, is given in [12]. It is expected (although it is to be proved) that the target would be able to avoid the pursuing missile by turning its velocity vector reversely proportional to the LOS rate. Based on analysis of the optimal maneuvers for various scenarios, practical periodic target evasive maneuvers were considered in [30,31]. Random target maneuvers were discussed in Chapter 7.

Evasive maneuver design parameters include magnitude, weave period (for weaving evasion), initiation time, and duration. The maximum achievable maneuver magnitude period combination is a function of initiation and duration times and may vary within the established flight envelope of interest. Offensive missile design information including airframe configuration, mass properties parameters, aerodynamic, and propulsion parameters need to be used. Based on this information, flight performance can be evaluated and the maximum achievable maneuver, as well as the region within the vehicle's flight profile where the maneuver is most likely to occur, can be determined. Missile Datcom, a widely used semiempirical datasheet component buildup method for preliminary design of missile aerodynamics and performance, can be used to build a dynamic model [see, e.g., equation (5.65)] of an offensive missile and establish the maximum achievable maneuver magnitude. The optimal weaving frequency can be determined as described in Chapter 5 (see Figure 5.13).

The midcourse guidance module contains algorithms realizing the guidance law under consideration. Based on smoothed data from the filtering module, the commanded acceleration is calculated in the NED coordinate system and transferred to the missile model. Taking into account that the missile trajectory of the model, formally deterministic and generated by the guidance law, differs from the real one, it is reasonable to use in the model as the missile data the deterministic signal plus noise depending on the missile data accuracy requirements. Usually, the calculation of the commanded acceleration is accompanied by the calculation of the gravity acceleration [see equation (9.4)]. However, depending on taste, the separate missile gravity module can be created that calculates the acceleration in the NED coordinates according to equation (9.4) or in the missile body coordinate system according to equation (9.18). The WCS operates with a certain frequency (usually, 4–10 Hz). The acceleration commands are transferred with this frequency to the missile model that operates with a significantly higher frequency.

Typically the missile model includes: (1) the thrust module, (2) the aerodynamics module, (3) the missile dynamics module, (4) the autopilot module, (5) the fin/actuator module, (6) the seeker module, (7) the missile trajectory module, and (8) the coordinate transformation module.

The thrust module contains data to obtain a certain thrust profile. One of its components is the pressure table. It is used to calculate the second term in equation (9.4). The aerodynamics module contains the aerodynamic forces and aerodynamic moment coefficients [see equations (9.15), (9.16), and (9.23)–(9.25)] used in the missile dynamics module. The missile dynamics module models missile dynamics from the aerodynamic forces (aerodynamic module), thrust, and gravity (trust module) [see equations (9.6), (9.11)–(9.16), (9.21)]. It contains tables of mass, center of gravity, and moment of inertia that are used to determine missile dynamics [see equations (9.11) and (9.17)].

The autopilot module calculates the required roll, pitch, and yaw within admissible limits to realize the guidance law. The fin/actuator module receives the fin commands from the autopilot module, translates these commands from roll, pitch, and yaw commands to the required input to each actuator. The fins configuration must match the aerodynamic data used in the missile aerodynamic module. As mentioned earlier, the autopilot equations (9.34)–(9.36) correspond only to one channel of control and a certain type of autopilot. They are given as an illustration without detailed consideration (see details, e.g., in [4,31]). It is desirable for the simulation model to be oriented on a concrete type of autopilot used in specific missiles. The real missile acceleration is described by equation (9.23). Based on equation (9.11) it is possible to determine the components of the missile velocity that correspond to the determined fin deflection. The missile trajectory module integrates the equations of motion [see equations (1.1), (1.2), and (1.20)]. Preliminary, the missile velocity or acceleration vectors in the missile body coordinate system are transferred into the NED coordinate frame by the L_{BE} operator; $L_{BE} = L_{EB}^{-1} = L_{EB}^{T}$ [see equation (9.8)]. This operation is produced in the coordinate transformation module. If the system (9.11) is used and the missile position $r_{M,k-1}$ at t_{k-1} is known, then at a moment t_k:

$$v_M = L_{EB}^T V_M, \quad r_{M,K} = \int_{t_{k-1}}^{t_k} v_M dt + r_{M,k-1} \tag{9.87}$$

The less precise expression follows from:

$$\ddot{r}_M = L_{EB}^T A, \quad v_{M,k} = \int_{t_{k-1}}^{t_k} \ddot{r}_M dt + v_{M,k-1} \tag{9.88}$$

where $v_{M,k-1}$ and $v_{M,k}$ are the missile velocity in the NED frame at t_{k-1} and t_k, respectively.

For small time increments Δ, we can use the approximation:

$$v_{M,k} = \ddot{r}_{M,k}\Delta + v_{M,k-1}, \quad r_{M,k} = \ddot{r}_{M,k-1}\Delta^2/2 + v_{M,k-1}\Delta + r_{M,k-1} \tag{9.89}$$

where the notations are obvious.

During the homing stage, the target information is received by a seeker. However, since we use a 3-DOF point-mass presentation of the target motion, this motion is presented initially in the NED coordinate frame. Then using the transformations $L_{EB}L_{BS}$ from the coordinate transformation module, the target position r_T, velocity $\dot{r}_T$, and acceleration $\ddot{r}_T$ (if needed) vectors are transformed into the seeker coordinate system, altogether with r_M and $\dot{r}_M$, and used in the seeker's module. The relative position of the target with respect to the missile is used to compute the actual line-of-sight. Band limited white noise is added to the LOS components to reflect the influence of noise on missile performance. The corrupted LOS vector is created by including noise, as described in Chapter 7, as well as the radome boresight errors [see equation (9.49)]. The estimated LOS rate is produced by a filter. The LOS and LOS rate vectors in the seeker coordinate frame are transformed into the missile body coordinate frame, where all operations are performed similar to the midcourse phase.

The effectiveness of the new missile guidance laws considered in the previous chapters should be tested by comparing them with four commonly used guidance laws: the "pure" PN guidance [see equation (2.23)], the "predictive" PN guidance that requires knowledge of the time-to-go [see equation (2.24)], the APN guidance [see equation (2.28)], and the Kappa guidance. These laws should be contained in the guidance reference module. It is also useful to create the management module that would control all operations between all of the above-mentioned modules.

The UAV simulation model should properly reflect all stages and modes of the UAV flight stages. Since usually the UAV's launch and landing are very specific operations, they require algorithms not covered in this book. They should be developed and tested separately.

As in the case of a missile model, the simulation process usually starts from the UAV's position, which the vehicle reaches several seconds after the launch. It means that the UAV prehistory (its uncontrolled launch stage) should be presented by the UAV position and velocity (see Figure 8.1). Launchers are used for many UAVs, mini UAVs can be launched by hand so that the launch operation and the initial part of the UAV trajectory, similar to the initial part of the missile trajectory, is not controllable. The launch parameters, which define the direction of the UAV flight, are determined by the first waypoint generated by the prescribed trajectory model.

Similar to the missile simulation model a certain homework is needed to determine the UAV initial position and velocity based on specifics of the scenario under consideration.

At a certain degree, the prescribed trajectory model is analogous to the target model in the missile simulation structure (see Figure 9.7). The desired UAV's trajectory is presented by a sequence of waypoints, and each waypoint presents a dummy target for the UAV. In contrast to the target model, in addition to waypoints, the recommended UAV speed can also be indicated for each part of the trajectory. The UAV moves toward the waypoint as a missile in accordance with the guidance laws discussed in Chapter 8. After the first waypoint is reached (within an admissible

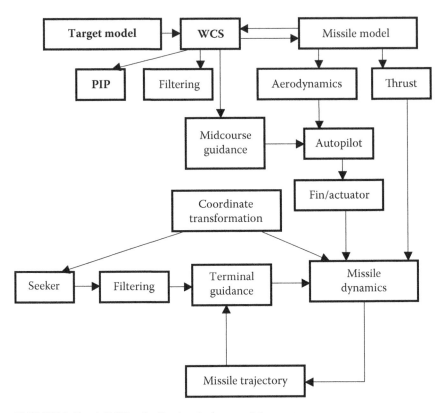

FIGURE 9.7   6-DOF missile simulation model structure.

accuracy), the prescribed trajectory model generates the next waypoint, and the guidance law directs the UAV to this point, and so on.

The simulation models should include the operations necessary to determine all components of the guidance law: (1) equations to determine the LOS, the LOS rate, and the closing velocity; (2) filtering that provides the guidance with inputs characterizing the UAV position, velocity and, if needed, acceleration; and (3) guidance commands.

The filtering module should contain a set of equations discussed earlier. It realizes signal processing algorithms to provide smoothed data that are used in the guidance law. It is very important to model the errors expected from GPS, INS, and other devices. Then the obtained filtering results are more realistic than under assumption of Gaussian white noise.

The guidance module contains algorithms realizing the guidance law under consideration. Based on smoothed data from the filtering module, the commanded acceleration is calculated. Taking into account that the UAV trajectory of the model, formally deterministic and generated by the guidance law, differs from the real one, it is reasonable to use in the model as the UAV data the deterministic signal plus noise depending on the UAV data accuracy requirements. Usually, the calculation of the commanded acceleration is accompanied by the calculation of the gravity acceleration [see equation (9.18)].

Usually, the INS filtered information is updated at a 10 *Hz* rate, while the GPS filtered information is updated at a 1 *Hz* rate, so that the guidance commands can input with a 1–10 *Hz* frequency the UAV model that operates with a significantly higher frequency.

Typically the UAV model (see Figure 9.8) includes: (1) the thrust module, (2) the aerodynamics module, (3) the UAV dynamics module, (4) the autopilot module, (5) the actuator module, (6) the obstacle module, (7) the UAV trajectory module, and (8) the coordinate transformation module.

The thrust module contains data generating a certain thrust profile and relationship between the thrust and speed, which is very important for propeller engines. The aerodynamic module contains the aerodynamic forces and aerodynamic moment coefficients [see equations (9.15), (9.16), (9.23)–(9.25), and (9.30)–(9.33)] used in the UAV dynamics module [see equations (9.26) and (9.27)]. The UAV dynamics module models UAV dynamics from the aerodynamic forces (aerodynamics module), thrust and gravity (trust module) [see equations (9.3), (9.11)–(9.16), (9.21), (9.26), (9.27), (9.29)–(9.33)]. It contains tables of mass, center of gravity, and moment of inertia that are used to determine UAV dynamics [see equations (9.11), (9.17), (9.26), and (9.27)]. The autopilot module calculates the roll, pitch, and yaw within admissible limits to realize the guidance law. The actuator module receives the commands from the autopilot module, translates these

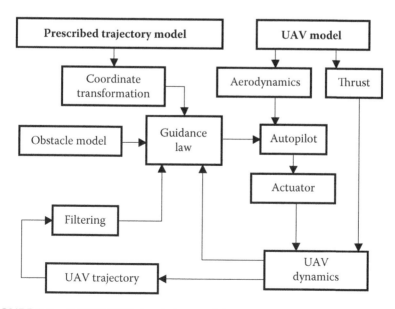

FIGURE 9.8 6-DOF UAV simulation model structure.

commands from roll, pitch, and yaw commands to the required input to each actuator. As mentioned earlier, the autopilot equations (9.39)–(9.42) correspond, as an illustration, only to several separate channels of control. The real UAV acceleration is described by equation (9.23). As indicated earlier, the dynamic equations describing fixed-wing UAVs have many common features with the equations describing missile dynamics. However, the dynamics of rotating-wing UAVs are more complicated and the 6-DOF models of such UAVs are less accurate than the 6-DOF models of missiles and fixed-wing UAVs. Based on equations (9.11), (9.26), and (9.27) it is possible to determine the components of the UAV velocity and then to determine the UAV trajectory by integrating the equations of motion [see equations (1.1), (1.2), and (1.20)].

As in the case of the missile simulation model, the UAV velocity or acceleration vectors in the vehicle body coordinate system are transferred into the NED coordinate frame by the L_{BE} operator; $L_{BE} = L_{EB}^{-1} = L_{EB}^{T}$ [see equation (9.8)]. The trajectory parameters are calculated by equations (9.87)–(9.89).

The obstacle model can be presented by equations describing the boundary of a certain region, which is very close or belongs to the predetermined (represented by the waypoints) UAV's trajectory. Such an approach will simplify the testing procedure of the obstacle avoidance algorithm, since it does not require the model to include any devices that detect obstacles. The obstacle model determines the minimal admissible distance between the UAV and the "dangerous" region. When this distance becomes equal or less than an admissible bound, which characterizes the resolution of an obstacle detector, the obstacle avoidance algorithm starts working instead of the guidance algorithm. The guidance algorithm resumes working and the UAV moves toward the waypoint when the distance exceeds the admissible bound.

It is of importance for the future generation of UAVs to be able to operate successfully in the National Air Space System with manned commercial and military aircraft. The UAV must not interfere with manned aircraft operations and must be relied upon to strictly observe the "Right of Way Rules" developed for manned aviation platforms. Special simulation scenarios should be developed to test these rules.

Of course, the described structure of the 6-DOF simulation models (see Figure 9.7 and Figure 9.8) is only a possible realization of the unmanned vehicle simulation models. The concise discussion above was focused on the main components that should be present in sophisticated 6-DOF simulation models.

9.10.2   3-DOF SIMULATION MODELS

The 3-DOF simulation models are significantly simpler than the described 6-DOF simulation models. However, they can be successfully used to test the discussed guidance laws. All operations are performed in the NED coordinate system (below the indices 1, 2, and 3 will indicate N, E, and D coordinates, respectively). There are no dynamic models of a seeker and autopilot. The flight control system is represented by the transfer function similar to the planar case considered in the previous chapters [see, e.g., equations (3.100), (5.12), (5.15), and (5.34)]. However, in the 3-DOF simulation model the differential equations corresponding to the above-mentioned transfer function should describe the relationship between the coordinates of the commanded and actual accelerations (i.e., the dimension of the system of differential equations is three times higher than in the planar models).

The main difficulty in building the 3-DOF model relates to the presentation of the total vehicle acceleration. The missile's thrust force is directed along to the missile's body; for missiles without throttleable engines the controlled part of the commanded acceleration acts orthogonal to the body; drag forces are directed along to the missile velocity vector. Without the knowledge of the angle of attack, it is impossible to combine properly the corresponding components of the missile acceleration. However, the 3-DOF missile model contains insufficient information to determine the angle of attack. Knowledge of the missile velocity vector enables us to determine the component of the missile acceleration orthogonal to this vector. Assuming small angles of attack, more precisely zero angle of attack, the above-mentioned components can be presented along and orthogonal to the velocity vector. Such models exist. However, their accuracy is not high enough, especially in the case of highly maneuvering targets.

The approximate values of the angle of attack can be obtained from the missile aerodynamic data. The aerodynamic module should contain the following regression models describing the relationship between the angle of attack α_T and the lift C_L, normal C_N, and axial C_A force coefficients [the expressions for normal C_N and axial C_A forces are analogous to equation (9.5)]:

$$\alpha_T = k_{00} + k_{01}C_L + k_{02}C_L^2$$

$$\alpha_T = k_{10} + k_{11}C_N + k_{12}C_N^2$$

$$C_A = k_{20} + k_{21}\alpha_T + k_{22}\alpha_T^2 \qquad (9.90)$$

which can be built based on the missile aerodynamic data available from experiments or generated, for example, by Missile Datcom (see Appendix C).

Here and below we will use the k-coefficients without any additional explanation. It is assumed that they are known or can be calculated.

The coefficients k_{sl} ($s = 0, 1, 2$; $l = 0, 1, 2$) are determined for a set of Mach numbers $Mach(i)$ ($i = 1, 2,..., n_i$) and altitudes Alt (j) ($j = 1, 2,..., n_j$) based on the missile aerodynamic data, so that a certain mesh $k_{sl}(i, j)$ ($s = 0, 1, 2$; $l = 0, 1, 2$) with known values in its nodes (i, j) should be created. For a concrete Mach number $Mach$ belonging to $[i_0, i_0 + 1)$ and a concrete altitude Alt belonging to $[j_0, j_0 + 1)$, the regression coefficients can be calculated by using various interpolation formulas [20], for example,

$$k_1 = k_{sl}(i_0, j_0) + \frac{k_{sl}(i_0 + 1, j_0) - k_{sl}(i_0, j_0)}{Mach(i_0 + 1) - Mach(i_0)}(Mach - Mach(i_0))$$

$$k_2 = k_{sl}(i_0, j_0 + 1) + \frac{k_{sl}(i_0 + 1, j_0 + 1) - k_{sl}(i_0, j_0 + 1)}{Mach(i_0 + 1) - Mach(i_0)}(Mach - Mach(i_0))$$

$$k_{sl} = k_1 + \frac{Alt - Alt(j_0)}{Alt(j_0 + 1) - Alt(j_0)}(k_2 - k_1) \tag{9.91}$$

Based on equations (9.90) and (9.91), the angle of attack and the drag generated axial force can be calculated. Formally, the angle of attack could be calculated from the first equation of (9.90) and (9.5), assuming that the lift is created by the orthogonal (to the velocity) component of the commanded acceleration. However, in this case we ignore acceleration limits imposed by the autopilot. That is why the first equation of (9.90) can be used only to calculate the initial condition $\alpha_T(0)$ of the computational procedure determining the angle of attack.

In the 3-DOF simulation model the autopilot module contains operations related to the computation of the angle of attack and the commanded acceleration affecting the missile trajectory. If $v_M = (V_{M1}, V_{M2}, V_{M3})$ is the missile velocity vector, then the unit velocity vector $e_M = (e_{M1}, e_{M2}, e_{M3})$ has components:

$$e_{Mi} = V_{Mi} / \sqrt{V_{M1}^2 + V_{M2}^2 + V_{M3}^2} \qquad (i = 1, 2, 3) \tag{9.92}$$

The projection a_L of the guidance law commanded acceleration $a_c = (a_{c1}, a_{c2}, a_{c3})$ on the velocity vector is

$$a_L = a_c e_M = \sum_{i=1}^{3} a_{ci} e_{Mi},$$

so that the coordinates of the vector-projection $a_L = (a_{L1}, a_{L2}, a_{L3})$ are

$$a_{Li} = a_L e_{Mi} \qquad (i = 1, 2, 3) \tag{9.93}$$

and the acceleration $a_{cN} = (a_{cN1}, a_{cN2}, a_{cN3})$ normal to the velocity vector equals

$$a_{cNi} = a_{ci} - a_L e_{Mi} \qquad (i = 1, 2, 3) \tag{9.94}$$

The unit vector $e_{LN} = (e_{LN1}, e_{LN2}, e_{LN3})$ orthogonal to the velocity vector is given by:

$$e_{LNi} = a_{cNi} / \sqrt{a_{cN1}^2 + a_{cN2}^2 + a_{cN3}^2} \qquad (i = 1, 2, 3) \qquad (9.95)$$

For the given angle of attack $\alpha_T = \alpha_T(0)$, the unit vector $e_B = (e_{B1}, e_{B2}, e_{B3})$ along to the body axis can be presented as:

$$e_B = e_M \cos\alpha_T + e_{LN} \sin\alpha_T \qquad (9.96)$$

Acting analogously to equations (9.93) and (9.94), we obtain the components of the commanded acceleration normal to the missile body $a_{cBN} = (a_{cBN1}, a_{cBN2}, a_{cBN3})$:

$$a_{cBNi} = a_{ci} - a_B e_{Bi} \qquad (i = 1, 2, 3) \qquad (9.97)$$

where

$$a_B = a_c e_B = \sum_{i=1}^{3} a_{ci} e_{Bi}.$$

The autopilot acceleration limit a_{lim} (pitch, raw/yawn) can be presented by half-empirical expressions $a_{lim} = f(Q)$. They reflect the fact that during the flight the missile is subjected to varying pressure depending on the altitude of the missile. This affects its fins displacement (which is limited), since less deflection is required when the missile is flying in dense atmosphere, and more deflection is needed in a rare atmosphere. If $a_{cBN} = \sqrt{a_{cBN1}^2 + a_{cBN2}^2 + a_{cBN3}^2} > a_{lim}$, then:

$$a_{cBNi} = a_{cBNi} \frac{a_{cBN}}{a_{lim}} \qquad (i = 1, 2, 3) \qquad (9.98)$$

Based on the commanded acceleration (9.98) and expressions (9.5) and (9.6), the coefficient C_N can be calculated, and from the second equation of equation (9.90) the new value of the angle of attack $\alpha_T(1)$ is determined. If the difference between $\alpha_T(0)$ and $\alpha_T(1)$ is small enough, $\alpha_T = \alpha_T(1)$. Otherwise, a certain computational procedure $\alpha_T(j + 1) = \alpha_T(j) + \Delta$ can be applied to make two consecutive $\alpha_T(j)$ close enough, where j is a step of iterations and Δ is an increment. The new value of the angle of attack $\alpha_T(j + 1)$ is used in equation (9.96), and the above-described operations should be repeated. Assuming the initial value of the angle of attack to be positive and evaluating its first difference (equivalent to the derivative for discrete time), it is possible to operate with positive and negative angles of attack. There is no rigorous proof that the computational procedure converges. However, the tests performed on a 3-DOF simulation model show that for the quite accurate regression models (9.90) and an appropriate search procedure (see, e.g., [8]), which we leave the reader to choose, the required iterations number only in the tens.

The missile dynamics module sums up all components of acceleration (thrust $a_{thrust} = Te_B$, normal component of the guidance law a_{cBN}, gravity, and drag generated axial component a_{axial}). The drag generated axial component of acceleration is determined by computing the axial force coefficient C_A of equation (9.90), for a given angle of attack, and then using the expressions similar to equations (9.5) and (9.15).

The total acceleration $a_{MT} = (a_{MT1}, a_{MT2}, a_{MT3})$ [see also equations (9.22) and (9.23)]:

$$a_{MT} = a_{cBN} + \left(T - \frac{QS}{m} C_A \right) e_B + G \qquad (9.99)$$

serves the input of the system of differential equations describing the flight-control system dynamics:

$$\dot{x}_{1i} = x_{2i}, \quad \dot{x}_{2i} = x_{3i},$$

$$\dot{x}_{3i} = -\frac{\omega_M^2}{\tau} x_{1i} - \frac{\omega_M^2 + 2\zeta\omega_M}{\tau} x_{2i} - \frac{2\zeta\omega_M\tau+1}{\tau} x_{3i} + \frac{\omega_M^2}{\tau} a_{MTi}$$

$$\ddot{r}_j = x_{3j-2,i} - \frac{1}{\omega_z^2} x_{3j,i} \qquad (i, j = 1, 2, 3) \qquad (9.100)$$

Here ω_M, ζ, ω_z, and τ are the flight control system natural frequency, damping, airframe zero frequency, and the actuator time constant, respectively. The indicated parameters are functions of time and depend on the dynamic pressure, missile aerodynamic characteristics, its variable mass, and other factors [see equation (9.34)]. Numerical integration of equation (9.100), as well as differential equations considered earlier, is performed using the Runge-Kutta method described in Appendix D. The use of numerical integration routines in simulation introduces numerical error, which can be propagated during the course of the simulation. To keep results accurate over long simulation runs, controlling the numerical errors are essential. Higher order methods, such as the Runge-Kutta series, express the derivative function as a power series to calculate a more accurate estimate of the incremental term. As shown in the previous chapters, the most "sensitive parameter" for tail-controlled missiles is the airframe zero frequency that should be changed depending on the missile altitude. The first-order unit describing the actuator dynamics and included in equation (9.98) can be placed

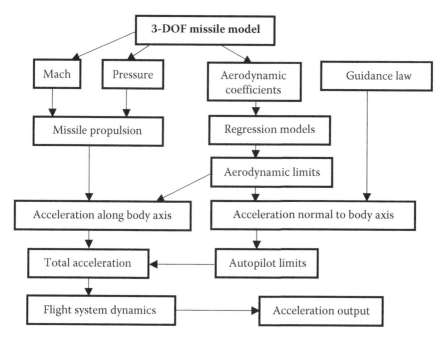

FIGURE 9.9   3-DOF missile model structure.

before the autopilot limiter (as a possible modification of the missile dynamics module).

The structure of the described 3-DOF missile model is shown in Figure 9.9. In addition to the modules considered earlier, this structure includes the aerodynamic limit unit. In reality, this unit does not exist. It simply reflects the fact that the angle of attack in the regression models (9.90) is limited. The models are obtained based on the aerodynamic data for the trim angles of attack, so that equation (9.90) are valid for definite values of these angles below the upper limit, which depends on Mach number and altitude. Although the 3-DOF model does not describe missile dynamics precisely, and the angle of attack is determined only approximately, its relative simplicity makes it an effective tool of guidance laws performance analysis.

As in the case of the 6-DOF simulation model, the missile trajectory [equation (9.88)] and velocity [equation (9.89)] are obtained by integrating the actual missile acceleration [equation (9.100)]. Both models should calculate the closing velocity. Its negative value indicates the end of the simulation process. The range between the missile and the target at this moment presents the miss distance, and the time of flight is the estimate of the time of intercept.

It was mentioned before and it is essential to underline that the miss distance is the most important but not the only important parameter characterizing missile performance. The miss distance should be considered altogether with the engagement envelope. The time of intercept and the missile terminal velocity (speed and impact angle) are also important factors in evaluating missile performance. It means that the comparative analysis of guidance laws should be based on the vector criterion including the above-mentioned components.

Since it is impossible to be certain what specific missile threats we will face in the future or from where these threats will come, the long-term strategy is to strengthen and maximize the flexibility of missile defense capabilities. The development of new guidance laws, which are the "brains" of missiles, and testing them using sophisticated simulation models is an important component of this strategy.

For UAVs, the 3-DOF models can be very efficient to test guidance laws since their flight conditions are more predictable than that of missiles and their trajectories consist of parts with insignificant changes of altitudes, so that aerodynamic conditions are rather stable. For predetermined trajectories, it is possible to evaluate approximate values of the angles of attack and the corresponding aerodynamic coefficients and incorporate them in the model.

When many system parameters are not precisely determined, a simpler model can produce more plausible results than a complicated one. That is why the 3-DOF models can be useful especially for rotating-wing vehicles with complex aerodynamics. If the parametric errors in the missile model can result in erroneous evaluation of the ability to hit maneuvering targets, in the case of stationary waypoints some errors in the estimate of the UAV acceleration mainly influence the time of flight between the waypoints rather than the UAV's ability to follow the prescribed trajectory. The input-output relationship between the commanded and real UAV's acceleration presented by the transfer function of the UAV flight control system and obtained analytically or experimentally usually provides reliable information about dynamic properties of the vehicle.

The 3-DOF model contains significantly less parameters than the 6-DOF model and it is easier to examine their influence on the UAV performance.

However, the main deficiency of the 3-DOF models is that the use of transfer functions is equivalent to operating with linear models (i.e., the assumption that the commanded acceleration can be "ideally" realized). The real acceleration is presented by equation (9.23), and it is necessary to include the acceleration limits in the model based on information about the components of equation (9.23) for concrete aerial vehicles under consideration. Moreover, these limits should be indicated for all possible flight scenarios. Without such limits the results of simulations cannot be considered as reliable to judge the efficiency of the tested guidance laws.

REFERENCES

1. Bar-Shalom, Y., and Fortmann, T. E. *Tracking and Data Association*. Boston, MA: Academic Press, 1988.

2. Bramwell, A., Done, G., and Balmford, D. *Helicopter Dynamics*. New York, NY: Elsevier, 2000.

3. Cloutier, J., Evers, J., and Feeley, J. An Assessment of Air-to-Air Missile Guidance and Control Technology, *IEEE Control Systems Magazine* 9 (1989): 27–34.

4. Cronvich, L. Aerodynamic Consideration for Autopilot Design. In *Tactical Missile Aerodynamics*. Edited by M. Hemsch and J. Nielsen J. Progress in Aeronautics and Astronautics. Vol. 124. Washington, DC: AIAA, 1986.

5. Ekstrand. B. Tracking Filters and Models for Seeker Applications, *IEEE Transactions on Aerospace and Electronic Systems* 37, no. 3 (2001): 965–77.

6. Etkin, B., and Reid, L. *Dynamics of Flight: Stability and Control*. New York, NY: John Wiley & Sons, 1966.

7. Etkin, B. *Dynamics of Atmospheric Flight*. Mineola, NY: Dover, 2005.

8. Fletcher, R. *Methods of Optimization*. New York, NY: John Wiley & Sons, 2000.

9. Grey, J., and Hecht, N. *A Derivation of Kappa Guidance*, 1–14. Dahlgren, VA: Naval Surface Warfare Center, 1989.

10. Johnson, W. *Helicopter Theory*. Princeton, NJ: Princeton University Press, 1980.

11. Koo, T., and Sastry, S. Differential Flatness Based Full Authority Helicopter Control Design. In *Proceedings of the 38th Conference on Decision and Control*, Phoenix, AZ, 1982–1988, 1999.

12. Kuo, V. Evasive Maneuver Against a Rogue Aircraft in Air Traffic Management, *AIAA Guidance, Navigation, and Control Conference*, AAIA 2003-5512, Monterey, CA, 2002.

13. Lee, R. *Optimal Estimation, Identification, and Control*. Cambridge, MA: The MIT Press, 1964.

14. Lin, J. M., and Chau. Y. F. Radome Slope Compensation Using Multiple-Model Kalman Filters, *Journal of Guidance, Control and Dynamics* 18 (1994): 637–40.

15. Mettler, B., Tischler, M., and Kanade, T. System Identification Modeling of a Small-Scale Unmanned Rotorcraft for Flight Control Design, *Journal of the American Helicopter Society* 47 (2002): 50–63.

16. Nelson, R. *Flight Stability and Automatic Control*. New York, NY: McGraw-Hill Co., 1998.

17. Nielsen, J. N. *Missile Aerodynamics*. Mountain View, CA: Nielsen Engineering & Research, Inc., 1998.

18. Padfield, G. *Helicopter Flight Dynamics: The Theory and Application of Flying Qualities and Simulation Modeling*. Oxford, UK: Blackwell, 1996.

19. Pamadi, B. *Performance, Stability, Dynamics, and Control of Airplanes*. Washington, DC: AIAA Education Series, 1998.

20. Phillips, G. M. *Interpolation and Approximation by Polynomials*. New York, NY: Springer Verlag, 2003.

21. Prouty, R. W. *Helicopter Performance, Stability and Control*. Boston, MA: PWS Engineering, 1986.

22. Serakov, D., and Lin, C.-F. Three-Dimensional Mid-Course Guidance State Equations. In *Proceedings of American Control Conference*, San Diego, CA, 6 (1999): 3738–42.

23. Shneydor, N. A. *Missile Guidance and Pursuit*. Chichester, UK: Horwood Publishing, 1998.

24. Spencer, J. *The Ballistic Missile Threat Handbook*. Washington, DC: The Heritage Foundation, 2002.

25. Tischler, M. System Identification Requirements for High-Bandwidth Rotorcraft Flight Control System Design, *Journal of Guidance and Control* 13, no. 5 (1990): 835–41.

26. Tischler, M. System Identification Methods for Aircraft Flight Control Development and Validation. In *Advances in Aircraft Flight Control*. New York, NY: Taylor & Francis, 1996.

27. Stevens, B., and Lewis, F. *Aircraft Control and Simulation*. New York, NY: Wiley Interscience, 1992.

28. Wan, E., and van der Merwe, R. The Unscented Kalman Filter. In *Kalman Filtering and Neural Networks*. Edited by S. Haykin. New York, NY: Wiley, 2001.

29. Wise, K., and Broy, D. Agile Missile Dynamics and Control, *Journal of Guidance, Control and Dynamics* 21 (1998): 441–49.

30. Yanushevsky, R. Analysis of Optimal Weaving Frequency of Maneuvering Targets, *Journal of Spacecraft and Rockets* 41, no. 3 (2003): 477–79.

31. Zarchan, P. *Tactical and Strategic Missile Guidance, Progress in Aeronautics and Astronautics*. Vol. 124, Washington, DC: AIAA, 1999.

32. Zarchan, P., and Gratt, H. Adaptive Radome Compensation Using Dither, *Journal of Guidance, Control and Dynamics* 22 (1999): 51–57.

10 Integrated Design

10.1 INTRODUCTION

The development of new aerial vehicle systems starts from the formulation of their operational requirements. An operational requirement presents a document, which describes the tactical need for the vehicle system and the area of its use. The operational requirements are then translated into performance specifications, which are given to contractors. In the case of a vehicle guidance system, the performance specifications may define the specific type of guidance to be employed. The tactical problem is the basis for the operational requirement and tactical considerations are the over-riding considerations in all stages of the design of a vehicle guidance system.

The first problem faced by the designer of an aerial vehicle guidance system is that of translating the related tactical problem into specifications for the guidance system design. After that, the mathematical model (i.e., the mathematical expressions) that governs its behavior is developed. The design process starts usually with simplified equations of vehicle motion, aerodynamic, kinematic, and inertial coupling being ignored. Then cross-coupling between the subsystems are taken into consideration, and the three-dimensional aerodynamic model is needed for the 6-DOF simula-tion. As an aid to the process of design, simulation of the system accom-panies the design process. As the design progresses, complete simulation may give way to partial simulation by substituting some of the completed elements of the system (parts constituting actual "hardware") for the mathematical expressions previously employed. When the guidance sys-tem has been developed, the behavior of the equipment is proved by flight tests. Data collected during these tests provide the designer of the guid-ance system with additional information to conclude whether the design system meets the functional requirements formulated at the beginning of the design process or the system needs some corrections or maybe, in the worst case, it should be redesigned.

In Chapter 5, we presented a block diagram of an interceptor's main subsystems (see Figure 5.1) and described briefly their functions and the subsystems interconnection.

The traditional approach for missile guidance and control systems has been to design these subsystems separately, then to integrate them together, and after that to verify their performance. If the overall system performance is unsatisfactory, individual subsystems are redesigned to improve the whole system performance.

Integrated design of the flight vehicle systems is an emerging trend within the aerospace industry. Currently, there are major research initiatives within the aerospace industry, the department of defense, and NASA to attempt interdisciplinary optimization of the whole vehicle design, while preserving the innovative freedom of individual subsystem designers. Integrated design of guidance, control, and fuze/warhead systems represents a parallel trend in the missile technology. There has been an increasing interest in integrated synthesis of missile guidance and control systems in recent literature [3,4,7–11]. Proponents of the integrated approach state that missile performance can be enhanced by utilizing methods exploiting the synergism between guidance and control (autopilot) subsystems. More cautious advocates of integrated missile design believe that because the traditional approach can lead to modifications subsequently made to each subsystem in order to achieve the desired weapon system performance, this approach can result in excessive design iterations, and may not always exploit any synergism existing between the missile guidance, autopilot, and fuze/warhead subsystems. That is why the methods for achieving tighter integration between the missile guidance, autopilot, and fuze/warhead subsystems have the potential to enhance missile performance and should be developed and tested in practice.

The integrated design of missile guidance and control systems is considered as a first step toward the development of integrated missile design methodologies. As indicated in the literature [9–11], integrated guidance and control systems are expected to result in significant improvements in missile performance, leading to lower weight and enhanced lethality. Both of these factors will lead to a more effective, lower-cost weapon system.

Figure 10.1 presents a wording form of Figure 3.1 and Figure 5.4 considered in the previous chapters. We analyzed the models in Figure 3.1 and Figure 5.4 assuming the guidance law is known. In traditional flight control systems, the guidance law uses the relative missile/target states to generate acceleration commands. More precisely, they were examined analytically in the case of the proportional navigation guidance law. The modifications of the proportional navigation (PN) law were offered based on the principle of parallel navigation, without any connection with the autopilot design. The guidance laws efficiency was tested for various parameters of airframe that, as indicated, depend on the altitude of missile motion.

The autopilot design presents one of the most important parts of missile design. The autopilot control system makes the real missile acceleration follow the commanded acceleration created by the guidance law. It receives the guidance commands and issues the relevant aerodynamic (e.g., fin), thrust-vector, or divert control commands necessary to achieve

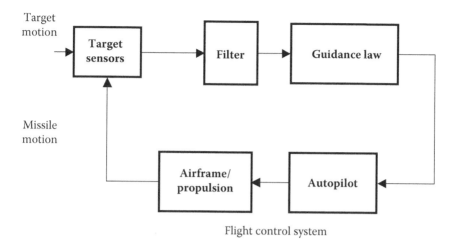

FIGURE 10.1 Traditional missile guidance and control design architecture.

the commanded acceleration. The autopilot tracks the acceleration commands by changing the missile attitude to generate angle of attack and angle of sideslip using fin deflections and/or moments generated using the reaction jet thrust.

Usually the autopilot designers couple three autopilots: a roll autopilot provides roll-stabilization while pitch and yaw autopilots provide controlled maneuvers in any desired direction relative to the stabilized position. As mentioned in Chapter 9, to maximize overall missile performance for each phase of flight the appropriate autopilot command structure should be chosen. This may include designing different autopilots for the boost, midcourse, and terminal phases [12].

The problems connected with the autopilot design were stimulus for control theory in the 1940s. The development of nonlinear control theory is indebted to the nonlinearity of autopilots (e.g., a limited fin angular position).

The autopilot serves as a controller of a nonlinear time-varying plant, the missile airframe. Without any doubt, technical advances in the various interceptor elements (e.g., airframe, actuation, sensor, and propulsion systems) enhance missile performance. However, any advances should be tested for the guidance law implemented in practice. It looks natural to believe that the decrease of the autopilot time constant will result in improving missile performance. However, as we have shown, for fin-controlled missiles guided by the proportional navigation law at high altitudes this can significantly decrease missile performance.

Traditional architecture separates guidance and flight control functions. Guidance laws are developed separately and tested for existing functioning autopilots, and vice versa autopilots are designed independently using

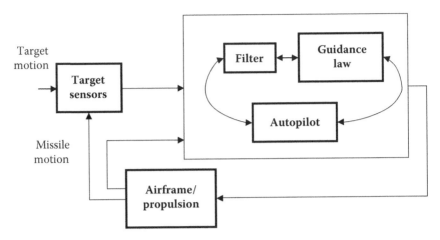

FIGURE 10.2   Integrated missile guidance and control design architecture.

methods of classical or modern control theory [1,2,5,6,12,14] and tested
for existing functioning guidance laws. The models of the airframe considered in Chapters 1–8 are too simple compared with the detailed model
given in Chapter 9, which is used in the autopilot design.

Integrated guidance and control laws are supposed to combine guidance
and control functions. Integrated missile design operates with a detailed
missile model and considers the target states relative to the missile as a
part of a generalized model. The guidance and control laws are obtained as
the solution of a certain optimal problem, which must guarantee the internal stability of the missile dynamics. The architecture of the integrated
guidance and control system is given in Figure 10.2.

Below we describe two basic models of the integrated missile and control system presented in the literature [9–11]. The integrated guidance-
control law for homing missiles is obtained as a solution of finite-interval
optimal control problems. The model of [11] includes filter design as a
part of the integrated design process comprising the design of a guidance
filter, guidance law, and autopilot. The model of [9,10] looks "more mod-
est" and operates only with guidance law and fin deflections as control
actions. Since the integrated guidance and control laws are linked with
optimal problems, the Bellman equations for various performance indices
are examined.

Special performance indices are considered that make it possible to
simplify the Bellman functional equations and solve the optimal problems
dealing with the Lyapunov equations. The integrated laws obtained based
on the modern control theory procedure are compared with the laws using
the classical control theory approach. In contrast to the previous chapters, where the term *missile guidance system* combined the guidance and

control units, here we will consider them separately. To emphasize that we will use the term *missile guidance and control system.*

Since the term *guidance law* was not applied to aerial vehicles operated by pilots and there are no publications related to the integrated UAV design, we consider only the problem of integrated missile design. However, the analysis of this problem and recommendations provided are relevant to all unmanned aerial vehicles.

10.2   INTEGRATED GUIDANCE AND CONTROL MODEL

The integrated guidance and control model is presented usually in the form:

$$\dot{x}(t) = f(x,t) + B(x,t)u(t) + D(x,t)w(t), \quad x(t_0) = x_0$$

$$y(t) = c(x,t) + D_1(x,t)w(t) \tag{10.1}$$

where $x(t)$ is the $m \times 1$ state-vector, $u(t)$ is the $n \times 1$ control vector, $w(t)$ is the $p \times 1$ vector of disturbances, $y(t)$ is the $l \times 1$ vector of output variables; $f(x, t)$, $c(x, t)$, $B(x, t)$, $D(x, t)$, and $D_1(x, t)$ are vector-functions and matrices of appropriate dimensions.

Since we operate with the nonlinear system, the traditional approach uses the linearization technique and sequentially uses the solutions of the linearized problems. First a set of linearized models at a large number of flight conditions is developed and a control law is designed for each linearized model using an appropriate synthesis technique [5].

An alternate approach to control design is based on the so-called extended linearization concept that requires a nonlinear system to be factored (when it is possible) in a way that it has a "linear-looking" structure, the so-called state dependent coefficient form [3,4,9–11].

For example, the system:

$$\begin{bmatrix} \dot{x}_1 \\ \dot{x}_2 \end{bmatrix} = \begin{bmatrix} x_1^2 - x_1 + x_1 x^2 + u_1 \\ x_1^2 x_2 - x_2 + x_2^2 + u_2 \end{bmatrix}$$

can be parameterized as:

$$\begin{bmatrix} \dot{x}_1 \\ \dot{x}_2 \end{bmatrix} = \begin{bmatrix} x_1 - 1 & x_1 x_2 \\ x_1 x_2 & x_2 - 1 \end{bmatrix} \cdot \begin{bmatrix} x_1 \\ x_2 \end{bmatrix} + \begin{bmatrix} u_1 \\ u_2 \end{bmatrix}$$

However, this parameterization is not unique. There exist infinite numbers of possible representations in the state dependent coefficient form and, unfortunately, there is no criterion that would justify our choice as the best.

For the nonlinear system (10.1), the extended linearization representation has the following form:

$$\dot{x}(t) = A(x,t)x(t) + B(x,t)u(t) + D(x,t)w(t), \quad x(t_0) = x_0$$

$$y(t) = C(x,t)x(t) + D_1(x,t)w(t) \tag{10.2}$$

where $A(x, t)$ and $C(x, t)$ are matrices of appropriate dimensions.

We rewrite the system (9.11), (9.14), and (9.22) assuming missile airframe x–z axis symmetry and that the mass distribution is such that $I_{yy} = I_{zz}$:

$$
\begin{aligned}
\dot{v}_x &= rv_y - qv_z + A_x \\
\dot{v}_y &= -rv_x + pv_z + A_y \\
\dot{v}_z &= qv_x - pv_y + A_z \\
I_{xx}\dot{p} &= m_x \\
I_{yy}\dot{q} - (I_{zz} - I_{xx})rp &= m_y \\
I_{zz}\dot{r} - (I_{xx} - I_{yy})pq &= m_z
\end{aligned}
\tag{10.3}
$$

where

$$
\begin{bmatrix} A_x \\ A_y \\ A_z \end{bmatrix} = \begin{bmatrix} X + G_x + T_x \\ Y + G_y + T_y \\ Z + G_z + T_z \end{bmatrix}
\tag{10.4}
$$

The v_x, v_y, and v_z are the components of velocity along the x, y, and z axes (see Figure 9.2), respectively; p, q, and r denote the roll, pitch, and yaw rates; A_x, A_y, and A_z are the body axis accelerations; G_x, G_y, and G_z are the gravity components; X, Y, and Z model accelerations produced by the aerodynamic forces; T_x, T_y, and T_z model propulsion system forces; m_x, m_y, and m_z model angular accelerations produced by the aerodynamic moments.

Differentiating the expressions for the angle of attack α_0 and the sideslip angle α_s [see equation (9.19)]:

$$\tan \alpha_0 = v_z/v_x, \quad \tan \alpha_s = v_y/v_x$$

i.e.,

$$\dot{\alpha}_0 = \frac{\dot{v}_z v_x - v_z \dot{v}_x}{v_x^2} \cos^2 \alpha_0, \quad \dot{\alpha}_s = \frac{\dot{v}_y v_x - v_y \dot{v}_x}{v_x^2} \cos^2 \alpha_s \quad (10.5)$$

and substituting the components of the derivative of the missile velocity vector V_M from equation (10.3), taking into account:

$$V_M^2 = v_x^2 + v_y^2 + v_z^2 = v_x^2 + v_x^2 \tan^2 \alpha_s + v_x^2 \tan^2 \alpha_0 = \Lambda v_x^2 \quad (10.6)$$

where

$$\Lambda = 1 + \tan^2 \alpha_s + \tan^2 \alpha_0 \quad (10.7)$$

the expressions (10.5) can be presented in the following form:

$$\dot{\alpha}_0 = \frac{A_z \Lambda \cos^2 \alpha_0}{V_M} - \frac{A_x \Lambda \cos \alpha_0 \sin \alpha_0}{V_M} + q - (p + r \tan \alpha_0) \tan \alpha_s \cos^2 \alpha_0 \quad (10.8)$$

and

$$\dot{\alpha}_s = \frac{A_y \Lambda \cos^2 \alpha_s}{V_M} - \frac{A_x \Lambda \cos \alpha_s \sin \alpha_s}{V_M} - r - (p + q \tan \alpha_s) \tan \alpha_0 \cos^2 \alpha_s \quad (10.9)$$

The last three equations of (10.3) can be solved for $\dot{p}, \dot{q}$, and $\dot{r}$:

$$\dot{p} = \frac{m_x}{I_{xx}}$$

$$\dot{q} = \frac{m_y}{I_{yy}} + \frac{I_{zz} - I_{xx}}{I_{yy}} rp \quad (10.10)$$

$$\dot{r} = \frac{m_z}{I_{zz}} + \frac{I_{xx} - I_{yy}}{I_{zz}} pq$$

To meet the requirements of the extended linearization representation, the expressions for the aerodynamic forces and moments are presented as [see also equations (9.23)–(9.25)]:

$$
\begin{bmatrix} X \\ Y \\ Z \\ m_x \\ m_y \\ m_z \end{bmatrix} = V_M^2 \begin{bmatrix} c_{X\alpha 0} & c_{X\alpha s} & c_{X\delta P} & c_{X\delta Y} & c_{X\delta R} \\ c_{Y\alpha 0} & c_{Y\alpha s} & c_{Y\delta P} & c_{Y\delta Y} & c_{Y\delta R} \\ c_{Z\alpha 0} & c_{Z\alpha s} & c_{Z\delta P} & c_{Z\delta Y} & c_{Z\delta R} \\ c_{m_x\alpha 0} & c_{m_x\alpha s} & c_{m_x\delta P} & c_{m_x\delta Y} & c_{m_x\delta R} \\ c_{m_y\alpha 0} & c_{m_y\alpha s} & c_{m_y\delta P} & c_{m_y\delta Y} & c_{m_y\delta R} \\ c_{m_z\alpha 0} & c_{m_z\alpha s} & c_{m_z\delta P} & c_{m_z\delta Y} & c_{m_z\delta R} \end{bmatrix} \cdot \begin{bmatrix} \alpha_0 \\ \alpha_s \\ \delta_P \\ \delta_Y \\ \delta_R \end{bmatrix} + V_M^2 \begin{bmatrix} c_X^0 \\ c_Y^0 \\ c_Z^0 \\ c_{mx}^0 \\ c_{my}^0 \\ c_{mz}^0 \end{bmatrix} \tag{10.11}
$$

i.e., the aerodynamic force and moment coefficients c_{kl}^s are described in a polynomial form with respect to the angle of attack α_0, the sideslip angle α_s, the pitch fin deflection δP, the yaw fin deflection δY, and the roll fin deflection δR. The possible constant terms in the polynomial representation have the upper index "0." The most significant nonzero term is the drag component c_X^0. For simplicity, here and below we indicated the dependence of the aerodynamic forces and moments on the dynamic pressure only by the factor V_M^2 of equation (9.6); it is assumed that other components of the dynamic pressure, as well as the missile mass m and reference parameters S, l in equations (9.15) and (9.16), are reflected in the coefficients c_{kl}^s.

As an example, assuming $c_{mx}^0 = 0$, the roll rate expression (10.10) is parameterized as:

$$
\dot{p} = \frac{c_{m_x\alpha 0}\alpha_0 + c_{m_x\alpha s}\alpha_s + c_{m_x\delta P}\delta P + c_{m_x\delta Y}\delta Y + c_{m_x\delta R}\delta R}{I_{xx}} \tag{10.12}
$$

The requirements to the missile aerodynamic model to be specified in a polynomial form may not be acceptable in situations where the design has to be based on nonsmooth aerodynamic data obtained from wind tunnel tests.

Using the first equation of (9.9)

$$
\dot{\phi} = p + q\sin\phi\tan\theta + r\cos\phi\tan\theta \tag{10.13}
$$

and comparing the commanded Euler roll angle rate $\dot{\phi}_c$ with $\dot{\phi}$ determined by equation (10.13), the expression for the roll error ε_ϕ can be presented as [11]:

$$\dot{\varepsilon}_\phi = -\frac{1}{\tau_\phi}\varepsilon_\phi + \frac{1}{\tau_\phi}(\dot{\phi}_c - p - q\sin\phi\tan\theta - r\cos\phi\tan\theta) \qquad (10.14)$$

where τ_ϕ is an adjustable parameter.

Tail-fin actuators are modeled as having second-order dynamics [11], so that:

$$\ddot{\delta}_P = -2\xi_a\omega_a\dot{\delta}_P + \omega_a^2(\delta_1 - \delta_P)$$

$$\ddot{\delta}_Y = -2\xi_a\omega_a\dot{\delta}_Y + \omega_a^2(\delta_2 - \delta_Y)$$

$$\ddot{\delta}_R = -2\xi_a\omega_a\dot{\delta}_R + \omega_a^2(\delta_3 - \delta_R) \qquad (10.15)$$

where δ_i ($i = 1, 2, 3$) is the commanded pitch-yaw-roll angular tail-fin position, ω_a and ξ_a represent the natural frequency and damping ratio of the tail-fin servo-actuator, respectively. The control vector $u(t) = (\delta_1, \delta_2, \delta_3)$ consists of the three tail-fin angular position commands. This is typical for tail-controlled missiles.

The expression for target-missile relative acceleration in the Earth-fixed inertial frame is similar to equation (1.20):

$$\ddot{r}_x = a_{Tx} - a_{Mx}, \quad \ddot{r}_y = a_{Ty} - a_{My}, \quad \ddot{r}_z = a_{Tz} - a_{Mz} \qquad (10.16)$$

where r_x, r_y, and r_z are the components of the range-vector.

In [11], the target acceleration model is presented as the first-order lag process:

$$\dot{a}_{Tx} = \frac{1}{\tau_T}(-a_{Tx} + w_T), \quad \dot{a}_{Ty} = \frac{1}{\tau_T}(-a_{Ty} + w_T), \quad \dot{a}_{Tz} = \frac{1}{\tau_T}(-a_{Tz} + w_T)$$
$$(10.17)$$

where τ_T represents the target maneuver time constant and $w_T(t)$ is a disturbance input.

The above-considered equations can be presented in the form (10.2) with the following state, output and control vectors:

$$
x(t) = \begin{bmatrix} r_x(t) \\ \dot{r}_x(t) \\ r_y(t) \\ \dot{r}_y(t) \\ r_z(t) \\ \dot{r}_z(t) \\ \alpha_0(t) \\ \alpha_s(t) \\ p(t) \\ q(t) \\ r(t) \\ \varepsilon_\phi(t) \\ \delta_P(t) \\ \dot{\delta}_P(t) \\ \delta_Y(t) \\ \dot{\delta}_Y(t) \\ \delta_R(t) \\ \dot{\delta}_R(t) \\ a_{Tx}(t) \\ a_{Ty}(t) \\ a_{Tz}(t) \end{bmatrix}
\qquad
y(t) = \begin{bmatrix} r_x(t) \\ r_y(t) \\ r_z(t) \\ a_{Mx}(t) \\ a_{My}(t) \\ a_{Mz}(t) \\ p(t) \\ q(t) \\ r(t) \\ \varepsilon_\phi(t) \\ \delta_P(t) \\ \delta_Y(t) \\ \delta_R(t) \\ a_{Tx}(t) \\ a_{Ty}(t) \\ a_{Tz}(t) \\ A_{x0}(t) \\ g(t) \end{bmatrix}
\qquad
u(t) = \begin{bmatrix} \delta_1(t) \\ \delta_2(t) \\ \delta_3(t) \end{bmatrix}
$$

(The components of the output vector A_{x0} and $g(t)$ are defined as pseudomeasurements to account for axial thrust-minus-drag and gravity, respectively.)

In the above-considered model, the state variables $r_x, \dot{r}_x, r_y, \dot{r}_y, r_z, \dot{r}_z, a_{Tx}, a_{Ty}$, and a_{Tz} do not depend directly upon other state variables, so that the system (10.2) consists of two separate subsystems. The dynamic equations were written separately for the guidance and control subsystems as if they are considered independently. However, this is not the case. The components of the missile acceleration in equation (10.16) written in the Earth-fixed inertial frame correspond to the components (10.4) presented in the missile body frame, so that all subsystems of the model are interconnected.

The linkage between the missile-target position coordinates and the state variables of equation (10.3) is more evident in the model considered in [9]:

$$\dot{x}_b = v_x + y_b r - z_b q$$
$$\dot{y}_b = v_y - x_b r + z_b p \qquad (10.18)$$
$$\dot{x}_b = v_z + x_b q - y_b p$$

where the missile-target position coordinates x_b, y_b, and z_b are presented in the missile body coordinate system.

However, equation (10.18) is valid under the assumption that the target velocity vector is negligible compared with the missile velocity vector. In the general case, the missile-target position coordinates r_x, r_y, and r_z in the Earth-fixed inertial frame should be transformed to the missile body frame [see equation (9.8)] and the relationship (9.9) between the Euler angles and the body rotational rates p, q, and r should be used. Of course, such a model will be more complicated.

To increase accuracy of the aerodynamic model, the aerodynamic forces and moments in equation (10.3) can be approximated by higher-order polynomials. Then instead of equation (10.11) we have:

$$V_M^{-2}\begin{bmatrix} X \\ Y \\ Z \\ m_x \\ m_y \\ m_z \end{bmatrix} = \begin{bmatrix} c_{X\alpha0} & c_{X\alpha0}^1 & c_{X\alpha s} & c_{X\alpha s}^1 & c_{X\delta P} & c_{X\delta Y} & c_{X\delta R} \\ c_{Y\alpha0} & c_{Y\alpha0}^1 & c_{Y\alpha s} & c_{Y\alpha s}^1 & c_{Y\delta P} & c_{Y\delta Y} & c_{Y\delta R} \\ c_{Z\alpha0} & c_{z\alpha0}^1 & c_{Z\alpha s} & c_{Z\alpha s}^1 & c_{Z\delta P} & c_{Z\delta Y} & c_{Z\delta R} \\ c_{m_x\alpha0} & c_{m_x\alpha0}^1 & c_{m_x\alpha s} & c_{m_x\alpha s}^1 & c_{m_x\delta P} & c_{m_x\delta Y} & c_{m_x\delta R} \\ c_{m_y\alpha0} & c_{m_y\alpha0}^1 & c_{m_y\alpha s} & c_{m_y\alpha s}^1 & c_{m_y\delta P} & c_{m_y\delta Y} & c_{m_y\delta R} \\ c_{m_z\alpha0} & c_{m_z\alpha0}^1 & c_{m_z\alpha s} & c_{m_z\alpha s}^1 & c_{m_z\delta P} & c_{m_z\delta Y} & c_{m_z\delta R} \end{bmatrix} \cdot \begin{bmatrix} \alpha_0 \\ \alpha_0^3 \\ \alpha_s \\ \alpha_s^3 \\ \delta_P \\ \delta_Y \\ \delta_R \end{bmatrix} + \begin{bmatrix} c_X^0 \\ c_Y^0 \\ c_Z^0 \\ c_{mx}^0 \\ c_{my}^0 \\ c_{mz}^0 \end{bmatrix}$$

$$(10.19)$$

(Notation of additional coefficients is obvious.)

Substituting the aerodynamic forces and moments from equation (10.19) into equation (10.3) and considering the state vector $x(t) = (p\ q\ r\ v_x\ v_y\ v_z\ x_b\ y_b\ z_b)^T$, we obtain the following components of the matrix $A(x,t) = [A_{ij}]$ in equation (10.2):

$$A_{11} = A_{12} = A_{13} = A_{17} = A_{18} = A_{19} = 0$$

$$A_{14} = \frac{1}{I_{xx}}(c_{mx}^0 + c_{m_x\alpha0}\alpha_0 + c_{m_x\alpha0}^1\alpha_0^3 + c_{m_x\alpha s}\alpha_s + c_{m_x\alpha s}^1\alpha_s^3)v_x$$

$$A_{15} = \frac{1}{I_{xx}}(c_{mx}^0 + c_{m_x\alpha0}\alpha_0 + c_{m_x\alpha0}^1\alpha_0^3 + c_{m_x\alpha s}\alpha_s + c_{m_x\alpha s}^1\alpha_s^3)v_y$$

$$A_{16} = \frac{1}{I_{xx}}(c_{mx}^0 + c_{m_x\alpha0}\alpha_0 + c_{m_x\alpha0}^1\alpha_0^3 + c_{m_x\alpha s}\alpha_s + c_{m_x\alpha s}^1\alpha_s^3)v_z$$

$$A_{21} = \frac{I_{zz} - I_{xx}}{I_{yy}}r, \quad A_{22} = A_{23} = A_{27} = A_{28} = A_{29} = 0$$

$$A_{24} = \frac{1}{I_{yy}}(c_{my}^0 + c_{m_y\alpha0}\alpha_0 + c_{m_y\alpha0}^1\alpha_0^3 + c_{m_y\alpha s}\alpha_s + c_{m_y\alpha s}^1\alpha_s^3)v_x$$

$$A_{25} = \frac{1}{I_{yy}}(c_{my}^0 + c_{m_y\alpha0}\alpha_0 + c_{m_y\alpha0}^1\alpha_0^3 + c_{m_y\alpha s}\alpha_s + c_{m_y\alpha s}^1\alpha_s^3)v_y$$

$$A_{26} = \frac{1}{I_{yy}}(c_{my}^0 + c_{m_y\alpha0}\alpha_0 + c_{m_y\alpha0}^1\alpha_0^3 + c_{m_y\alpha s}\alpha_s + c_{m_y\alpha s}^1\alpha_s^3 +)v_z$$

$$A_{31} = \frac{I_{xx} - I_{yy}}{I_{zz}}q, \quad A_{32} = A_{33} = A_{37} = A_{38} = A_{39} = 0$$

$$A_{34} = \frac{1}{I_{zz}}(c_{mz}^0 + c_{m_z\alpha0}\alpha_0 + c_{m_z\alpha0}^1\alpha_0^3 + c_{m_z\alpha s}\alpha_s + c_{m_z\alpha s}^1\alpha_s^3)v_x$$

$$A_{35} = \frac{1}{I_{zz}}(c_{mz}^0 + c_{m_z\alpha0}\alpha_0 + c_{m_z\alpha0}^1\alpha_0^3 + c_{m_z\alpha s}\alpha_s + c_{m_z\alpha s}^1\alpha_s^3)v_y$$

$$A_{36} = \frac{1}{I_{zz}}(c_{mz}^0 + c_{m_z\alpha0}\alpha_0 + c_{m_z\alpha0}^1\alpha_0^3 + c_{m_z\alpha s}\alpha_s + c_{m_z\alpha s}^1\alpha_s^3)v_z$$

$$A_{41} = A_{47} = A_{48} = A_{49} = 0, \quad A_{42} = -v_z, \quad A_{43} = v_y$$

$$A_{44} = -(c_X^0 + c_{X\alpha 0}\alpha_0 + c_{X\alpha 0}^1\alpha_0^3 + c_{X\alpha s}\alpha_s + c_{X\alpha s}^1\alpha_s^3)v_x$$

$$A_{45} = -(c_X^0 + c_{X\alpha 0}\alpha_0 + c_{X\alpha 0}^1\alpha_0^3 + c_{X\alpha s}\alpha_s + c_{X\alpha s}^1\alpha_s^3)v_y$$

$$A_{46} = -(c_X^0 + c_{X\alpha 0}\alpha_0 + c_{X\alpha 0}^1\alpha_0^3 + c_{X\alpha s}\alpha_s + c_{X\alpha s}^1\alpha_s^3)v_z$$

$$A_{51} = v_z, \quad A_{53} = -v_x, \quad A_{52} = A_{57} = A_{58} = A_{59} = 0$$

$$A_{54} = (c_Y^0 + c_{Y\alpha 0}\alpha_0 + c_{Y\alpha 0}^1\alpha_0^3 + c_{Y\alpha s}\alpha_s + c_{Y\alpha s}^1\alpha_s^3)v_x$$

$$A_{55} = (c_Y^0 + c_{Y\alpha 0}\alpha_0 + c_{Y\alpha 0}^1\alpha_0^3 + c_{Y\alpha s}\alpha_s + c_{Y\alpha s}^1\alpha_s^3)v_y$$

$$A_{56} = (c_Y^0 + c_{Y\alpha 0}\alpha_0 + c_{Y\alpha 0}^1\alpha_0^3 + c_{Y\alpha s}\alpha_s + c_{Y\alpha s}^1\alpha_s^3)v_z$$

$$A_{61} = -v_y, \quad A_{62} = v_x, \quad A_{63} = A_{67} = A_{68} = A_{69} = 0$$

$$A_{64} = -(c_Z^0 + c_{Z\alpha 0}\alpha_0 + c_{Z\alpha 0}^1\alpha_0^3 + c_{Z\alpha s}\alpha_s + c_{Z\alpha s}^1\alpha_s^3)v_x$$

$$A_{65} = -(c_Z^0 + c_{Z\alpha 0}\alpha_0 + c_{Z\alpha 0}^1\alpha_0^3 + c_{Z\alpha s}\alpha_s + c_{Z\alpha s}^1\alpha_s^3)v_y$$

$$A_{66} = -(c_Z^0 + c_{Z\alpha 0}\alpha_0 + c_{Z\alpha 0}^1\alpha_0^3 + c_{Z\alpha s}\alpha_s + c_{Z\alpha s}^1\alpha_s^3)v_z$$

$$A_{72} = -z_b, \quad A_{73} = y_b, \quad A_{74} = 1, \quad A_{71} = A_{75} = A_{76} = A_{77} = A_{78} = A_{79} = 0$$

$$A_{81} = z_b, \quad A_{83} = -x_b, \quad A_{85} = 1, \quad A_{82} = A_{84} = A_{86} = A_{87} = A_{88} = A_{89} = 0$$

$$A_{91} = -y_b, \quad A_{92} = x_b, \quad A_{96} = 1, \quad A_{93} = A_{94} = A_{95} = A_{97} = A_{98} = A_{99} = 0$$

(signs in A_{ij}, $i = 4 - 6$, $j = 4 - 6$, correspond to the direction of axial and normal forces.)

For the control vector $u(t) = (\delta_P, \delta_Y, \delta_R)^T$ the matrix $B(x,t) = [B_{ij}] = V_M^2 [B_{ij}^0]$ in equation (10.2) has the following form [the model can be easily enhanced by including tail-fin actuators dynamics similar to equation (10.15)]:

$$B_{11}^0 = \frac{1}{I_{xx}} c_{m_x\delta P}, \quad B_{12}^0 = \frac{1}{I_{xx}} c_{m_x\delta Y}, \quad B_{13}^0 = \frac{1}{I_{xx}} c_{m_x\delta R}$$

$$B_{21}^0 = \frac{1}{I_{yy}} C_{m_y \delta P}, \quad B_{22}^0 = \frac{1}{I_{yy}} C_{m_y \delta Y}, \quad B_{23}^0 = \frac{1}{I_{yy}} C_{m_y \delta R}$$

$$B_{31}^0 = \frac{1}{I_{zz}} C_{m_z \delta P}, \quad B_{32}^0 = \frac{1}{I_{zz}} C_{m_z \delta Y}, \quad B_{33}^0 = \frac{1}{I_{zz}} C_{m_z \delta R}$$

$$B_{41}^0 = C_{X \delta P}, \quad B_{42}^0 = C_{X \delta Y}, \quad B_{43}^0 = C_{X \delta R}$$

$$B_{51}^0 = C_{Y \delta P}, \quad B_{52}^0 = C_{Y \delta Y}, \quad B_{53}^0 = C_{Y \delta R}$$

$$B_{61}^0 = -C_{Z \delta P}, \quad B_{62}^0 = -C_{Z \delta Y}, \quad B_{63}^0 = -C_{Z \delta R}$$

$$B_{71}^0 = B_{72}^0 = B_{73}^0 = B_{81}^0 = B_{82}^0 = B_{83}^0 = B_{91}^0 = B_{92}^0 = B_{93}^0 = 0$$

The difference between the above-described two models used in the integrated guidance and control system is in choosing the state variables of the models. The state variables in [11] combine the state variables of two separate designs—the guidance law and autopilot. The choice of the angle of attack and the sideslip angle as the state and output variables is physically justifiable and used in the current practice of autopilots design. However, the matrix $A(x, t)$ of the extended linearization representation (10.2) is too complicated. It depends upon the Euler angles and the components of the missile velocity vector, i.e., the first three equations of (10.3) are indirectly present in the model. The model of [9] has a certain advantage over the model of [11]. As an integrated model, it looks more logical than the model in [11] because the coupling of the guidance and control is more noticeable. However, as mentioned, it is obtained under the assumption that the target velocity vector is negligible compared with the missile velocity vector.

The above models presented in the state-space form required by the modern control theory are used to formulate the integrated guidance and control problem as an optimal control problem. The authors of [9–11] offer the solution of the integrated design problem utilizing the existing solutions of linear optimal problems. Assuming first $w(t) = 0$ and considering the linearized model (10.2) and the performance index [see equations (A7)–(A17)]:

$$I = \frac{1}{2} \left(x^T(t_F) C_0 x(t_F) + \int_{t_0}^{t_F} \left(x^T(t) R x(t) + \|u(t)\|^2 \right) dt \right) \qquad (10.20)$$

we obtain the optimal solution in the form:

$$u(t) = -B^T W(t)x(t) \qquad (10.21)$$

and

$$\dot{W} + A^T W + WA - WBB^T W + R = 0, \quad W(t_F) = C_0 \qquad (10.22)$$

Instead of the Riccati equation (10.22), the so-called state dependent Riccati equation was considered in [3,4,9–11]. The state dependent Riccati equation technique is applied for the equations of motion in the extended linearized form (10.2) together with a state dependent quadratic performance index. A state dependent algebraic Riccati equation is formulated using the model (10.2) ignoring disturbances and assuming $A(x, t) = A(x)$ and $B(x, t) = B(x)$ and introducing the state dependent weight matrix $R(x)$ in the performance index. As shown in [3,4] for the performance index:

$$I = \int_{t_0}^{\infty} (x^T(t)R(x)x(t) + \|u(t)\|^2)dt \qquad (10.23)$$

the state dependent algebraic Riccati equation can be written as:

$$A^T(x)W(x) + W(x)A(x) - W(x)B(x)B^T(x)W(x) + R(x) = 0 \qquad (10.24)$$

For the performance index:

$$I = \frac{1}{2}\left(x^T(t_F)C_0(x)x(t_F) + \int_{t_0}^{t_F} \left(x^T(t)R(x)x(t) + \|u(t)\|^2 \right)dt \right) \qquad (10.25)$$

the modification of equation (10.24), similar to equation (10.22), can be obtained.

The linear quadratic dynamic games approach to the model (10.2) allows us to find the control $u(t)$ counteracting to the disturbances $w(t)$, i.e., the optimal solution for the functional

$$\min_{u(t)} \max_{w(t)} I = \frac{1}{2}\left(x^T(t_F)C_0(x)x(t_F) \right.$$

$$\left. + \int_{t_0}^{t_F} \left(x^T(t)R(x)x(t) + \|u(t)\|^2 + \gamma\|w(t)\|^2 \right)dt \right)$$

is similar to the optimal solution for the functional (10.25), where γ is a constant coefficient, which is also obtained from the Riccati equation [11]. In all above-mentioned cases the control structure is the same:

$$u(t) = -B^T(x)W(x)x(t) \qquad (10.26)$$

i.e., at each moment fin deflections depend on the current values of the state vector of the model chosen in the design process. If not all the system states are available from measurements, the procedure can be modified to synthesize state dependent estimators.

The described method of the integrated guidance and control system design based on the modification of the optimal solution for linear optimal problems by utilizing the nonlinear equations in the extended linearized form needs more rigorous mathematical justification. It looks too complicated to be used in practice despite the published literature [9–11] experimental results supporting its efficiency.

Since the control laws used in [9–11] are obtained based on the procedure of dynamic programming, below we describe more general optimal control laws that do not require the extended linearized representation of guidance-control models.

10.3 SYNTHESIS OF CONTROL LAWS

10.3.1 Minimization of Standard Functionals

First we consider the optimal problem for the system presented in more general form than equation (10.1):

$$\dot{x}(t) = f(x,u,t), \quad x(t_0) = x_0 \qquad (10.27)$$

and the generalized performance index:

$$I = V_0(x(t_F)) + \int_0^{t_F} L(x(t),u(t),t)dt \qquad (10.28)$$

where the function $L(x(t),u(t),t)$ depends simultaneously on two variables—the state vector $x(t)$ and the control vector $u(t)$; $V_0(x(t_F))$ is the function of the terminal state $x(t_F)$.

The synthesis problem of a closed-loop optimal control system consists of finding the controller equations $u(t) = \Gamma[x(t), t]$ that, together with equation (10.27), form the stable system and minimize the functional (10.28). The procedure of obtaining the Bellman functional equation is described in Appendix A. For the performance index (10.28) and the system (10.27) it can be presented as:

$$\frac{\partial V}{\partial t} + \min_{u(t)} \left\{ L(x,u,t) + \frac{\partial V^T}{\partial x} f(x,u,t) \right\} = 0 \qquad (10.29)$$

where the function $V(x(t))$ should satisfy the condition $V(t_F) = V_0(x(t_F))$ and

$$\frac{\partial V^T}{\partial x} = \left(\frac{\partial V}{\partial x_1}, \ldots, \frac{\partial V}{\partial x_m} \right).$$

Although equation (10.29) is presented in a compact form, in many practical cases it is difficult to find its solution. Assuming that $u_0(t)$ minimizes the expression in braces of equation (10.29) and substituting $u_0(t)$ in equation (10.29) we have:

$$\frac{\partial V}{\partial t} + \frac{\partial V^T}{\partial x} f(x,u_0,t) = -L(x,u_0,t) \qquad (10.30)$$

or

$$\frac{\partial}{\partial u_0} L(x,u_0,t) + \frac{\partial}{\partial u_0} \left\{ \frac{\partial V^T}{\partial x} f(x,u_0,t) \right\} = 0 \qquad (10.31)$$

where the partial derivatives is written with respect to the components of $u_0 = (u_{01},\ldots,u_{0n})^T$.

The optimal synthesis problem can be solved only if we can obtain the solution of the system of nonlinear partial differential equations (10.31) and then present u_0 as a function of x and t. Difficulties in solving this problem are demonstrated below.

Let us consider the model similar to equation (10.1):

$$\dot{x}(t) = f(x,t) + B(x,t)u(t), \quad x(t_0) = x_0 \qquad (10.32)$$

For this model and the additive functional:

$$I = V(x(t_F)) + \int_{t_0}^{t_F} (R(x(t),t) + Q(u(t),t))dt \qquad (10.33)$$

the equations (10.30) and (10.31) have the form:

$$\frac{\partial V}{\partial t} + \frac{\partial V^T}{\partial x} (f(x,t) + B(x,t)u_0) + Q(u_0,t) = -R(x,t) \qquad (10.34)$$

and

$$\frac{\partial Q^T(u_0,t)}{\partial u_0} + \frac{\partial V^T}{\partial x} B(x,t) = 0 \tag{10.35}$$

Assuming that equation (10.35) can be resolved with respect to $u_0 = (u_{01},\ldots,u_{0n})^T$, i.e.,

$$u_0 = \Gamma\left[B^T(x,t)\frac{\partial V}{\partial x}, t \right] \tag{10.36}$$

and substituting equation (10.36) in equation (10.34) we obtain:

$$\frac{\partial V}{\partial t} + \frac{\partial V^T}{\partial x} f(x,t) + \frac{\partial V^T}{\partial x} B(x,t)\Gamma\left[B^T(x,t)\frac{\partial V}{\partial x}, t \right]$$

$$+ Q\left(\Gamma\left[B^T(x,t)\frac{\partial V}{\partial x}, t \right], t \right) = -R(x,t) \tag{10.37}$$

The solution of this equation satisfying the boundary condition $V(t_F) = V_0(x(t_F))$ should determine the optimal control based on equation (10.35). It is possible to prove that if

$$Q(u,t) - Q(u_0,t) - \frac{\partial Q^T(u_0,t)}{\partial u_0}(u - u_0)$$

is a positive definite function of u that equals zero only when $u = u_0$, then equation (10.36) is the optimal control [6,13].

Let the functional (10.33) have the form:

$$I = V_0(x(t_F)) + \int_0^{t_F} \left(R(x(t),t) + \frac{1}{2}u^T(t)K^{-1}u(t) \right) dt \tag{10.38}$$

where $K = [k_{ii}]$ is a diagonal matrix with $k_{ii} > 0$.

In this case, the function

$$Q(u,t) - Q(u_0,t) - \frac{\partial Q^T(u_0,t)}{\partial u_0}(u - u_0)$$

satisfies the above-mentioned conditions of the existence of the unique optimal solution (10.36). The Bellman functional equation and the expression for the optimal control can be written as [see equations (10.30) and (10.31)]:

$$u = u_0 = -KB^T(x,t)\frac{\partial V}{\partial x} \tag{10.39}$$

and

$$\frac{\partial V}{\partial t} + \frac{\partial V^T}{\partial x}f(x,t) - \frac{1}{2}\frac{\partial V^T}{\partial x}B(x,t)KB^T(x,t)\frac{\partial V}{\partial x} = -R(x,t),$$

$$V(t_F) = V_0(x(t_F)) \tag{10.40}$$

The solution of the functional Bellman equation even in the form (10.40) is a matter of insurmountable difficulties even when we seek an approximate solution of Bellman functional equations. The analytical solution exists only for the linear-quadratic problems. The method of power series to solve equation (10.40) was offered in [1]. The solution is sought as:

$$V = \frac{1}{2}\sum_{i=1}^{m}\sum_{j=1}^{m}\gamma_{ij}x_ix_j + \frac{1}{3}\sum_{i=1}^{m}\sum_{j=1}^{m}\sum_{r=1}^{m}\gamma_{ijr}x_ix_jx_r + ... \tag{10.41}$$

where the unknown coefficients should be determined from the system of the ordinary differential equations. This approach can be useful for the models similar to equation (10.2).

10.3.2 MINIMIZATION OF SPECIAL FUNCTIONALS

For the model described by equation (10.32) we consider the functional:

$$I = V_0(x(t_F)) + \int_{t_0}^{t_F}(R(x(t),t) + Q(u(t),t) + Q_0(u_0(t),t))dt \tag{10.42}$$

where $Q(u(t), t)$ and $Q_0(u_0(t), t)$ are such functions that

$$Q(u,t) + Q_0(u_0,t) - \frac{\partial Q^T(u_0,t)}{\partial u_0}u$$

is the positive definite, with respect to $\boldsymbol{u}$, function that equals zero at $\boldsymbol{u} = \boldsymbol{u}_0$. The function $\boldsymbol{u}_0$ is an unknown optimal control. The functionals including unknown optimal controls were introduced and used for synthesis of control systems in [6].

Following [6], we will show that the optimal control for the functional (10.42) is determined from the expression:

$$\frac{\partial Q^T(\boldsymbol{u}_0,t)}{\partial u_0} = -\frac{\partial V^T}{\partial x} B(\boldsymbol{x},t) \tag{10.43}$$

where $V(x, t)$ is the solution of the Lyapunov equation (10.32) when $\boldsymbol{u} \equiv 0$:

$$\frac{\partial V}{\partial t} + \frac{\partial V^T}{\partial x} f(\boldsymbol{x},t) = -R(\boldsymbol{x},t), \quad V(t_F) = V_0(\boldsymbol{x}(t_F)) \tag{10.44}$$

The Bellman equation for the optimal problem (10.42) and (10.32) has the form:

$$\frac{\partial V}{\partial t} + \min_{u(t)}\{R(\boldsymbol{x}(t),t) + Q(\boldsymbol{u}(t),t) + Q_0(\boldsymbol{u}_0(t),t) + \frac{\partial V^T}{\partial x}(f(\boldsymbol{x},t) + B(\boldsymbol{x},t)\boldsymbol{u})\} = 0 \tag{10.45}$$

Minimization of the expression in braces immediately gives equation (10.43). Substituting equation (10.43) in the modified equation (10.45):

$$\frac{\partial V}{\partial t} + R(\boldsymbol{x}(t),t) + Q(\boldsymbol{u}_0(t),t) + Q_0(\boldsymbol{u}_0(t),t) + \frac{\partial V^T}{\partial x}(f(\boldsymbol{x},t) + B(\boldsymbol{x},t)\boldsymbol{u}_0) = 0 \tag{10.46}$$

and taking into account that

$$Q(\boldsymbol{u}_0,t) + Q_0(\boldsymbol{u}_0,t) - \frac{\partial Q^T(\boldsymbol{u}_0,t)}{\partial u_0}\boldsymbol{u}_0$$

equals zero we obtain equation (10.44).

For the "modernized" functional (10.38):

$$I = V_0(\boldsymbol{x}(t_F)) + \int_{t_0}^{t_F}\left(R(\boldsymbol{x}(t),t) + \frac{1}{2}[\boldsymbol{u}^T(t)K^{-1}\boldsymbol{u}(t) + \boldsymbol{u}_0^T(t)K^{-1}\boldsymbol{u}_0(t)]\right)dt \tag{10.47}$$

the optimal solution is:

$$u = u_0 = -KB^T(x,t)\frac{\partial V}{\partial x} \qquad (10.48)$$

i.e., it looks identical to equation (10.39).

However, if here $V = V(x, t)$ is the solution of the Lyapunov equation (10.44), for the performance index (10.38) and control law (10.39) $V(x, t)$ is the solution of a more complicated nonlinear partial differential equation (10.40), i.e., instead of the Bellman equation we operate with the Lyapunov equation (10.44) and the Lyapunov function $V(x, t)$. The existence of the Lyapunov function satisfying equation (10.44) assumes the uncontrolled ($u \equiv 0$) process described by equation (10.32) to be stable. In the case of instability, it should be initially stabilized. Later the stabilization feedback should be included in the controller structure obtained by the above-considered method.

Similar to equation (10.41), the method of power series can be used to solve the Lyapunov equation (10.44). Assuming that the components of $f(x(k)) = [f_i]$ in equation (10.32) can be presented by convergent power series in a certain domain X, i.e.,

$$f_i = \sum_{j=1}^{m} a_{ij}x_j + \sum_{j=1}^{m}\sum_{k=1}^{m} a_{ijk}x_jx_k + \sum_{j=1}^{m}\sum_{k=1}^{m}\sum_{r=1}^{m} a_{ijkr}x_jx_kx_r + \ldots \qquad (10.49)$$

and presenting the positive definite functions V_0 and R of the functional (10.46) as:

$$V_0 = \frac{1}{2}\sum_{i=1}^{m}\sum_{j=1}^{m} \rho_{ij}x_ix_j + \frac{1}{3}\sum_{i=1}^{m}\sum_{j=1}^{m}\sum_{r=1}^{m} \rho_{ijr}x_ix_jx_r + \ldots \qquad (10.50)$$

and

$$R = \frac{1}{2}\sum_{i=1}^{m}\sum_{j=1}^{m} \mu_{ij}x_ix_j + \frac{1}{3}\sum_{i=1}^{m}\sum_{j=1}^{m}\sum_{r=1}^{m} \mu_{ijr}x_ix_jx_r + \ldots \qquad (10.51)$$

where a_* and μ_* are constant or time-dependent coefficients and ρ_* are constant coefficients, we seek the solution of the Lyapunov equation in the form (10.41). It is supposed that the power series:

$$\frac{\partial V}{\partial x_i} = \sum_{j=1}^{m} \gamma_{ij}x_j + \sum_{j=1}^{m}\sum_{k=1}^{m} \gamma_{ij}x_jx_k + \sum_{j=1}^{m}\sum_{k=1}^{m}\sum_{r=1}^{m} \gamma_{ijkr}x_jx_kx_r + \ldots \qquad (10.52)$$

converges in X.

For the functional (10.46) and the diagonal matrix $K(t) = [K_i(t)]$ the optimal control law (10.47) can be written as:

$$u_i = -k_i(t) \sum_{j=1}^{m} B_{ij} \frac{\partial V}{\partial x_j}$$ (10.53)

where the coefficients of equation (10.52) satisfy the system of differential equations [6]:

$$\dot{\gamma}_{\underbrace{i...q}_{S}} - \sum_{S_1=1}^{S-1} \frac{S_1!(S-S_1)!}{(S-1)!} \sum_{s=1}^{m} \left\{ \gamma_{\underbrace{si...h}_{S_1}} a_{\underbrace{kj...q}_{S-S_1}} \right\} = -\mu_{\underbrace{i...q}_{S}}$$ (10.54)

with the boundary conditions:

$$\gamma_{i...q}(t_F) = \rho_{i...q}, \quad i,...,q = 1,2,...,m$$ (10.55)

(Here the symbol {*} denotes summation of the product in braces for each possible change of indices in γ and a.)

There exists one more very effective approach to solve the problem (10.47) of minimization of the performance index (10.47) subject to equation (10.32). Taking into account that the expression (10.44) corresponds to the derivative of V along any trajectory of the uncontrolled ($u \equiv 0$) system (10.32), i.e.,

$$\dot{x}_u(t) = f(x_u,t)$$ (10.56)

we can rewrite equation (10.44) as:

$$\frac{dV}{dt} = -R(x_u,t)$$ (10.57)

From this equation we have:

$$V(x_u(t_F),t_F) - V(x_u(t),t) = -\int_t^{t_F} R(x_u(t),t)dt$$ (10.58)

or, taking into consideration the terminal condition $V(t_F) = V_0(x(t_F))$,

$$V(x_u(t),t) = V_0(x_u(t_F)) + \int_t^{t_F} R(x_u(t),t)dt$$ (10.59)

Let $X(x, t, \sigma)$ be the solution of equation (10.56) for the initial condition $x_t = x(t)$, where $x(t)$ is the current state of the system (10.32). Then instead of equation (10.43) we can present the analytical expression of the optimal control in the form:

$$\frac{\partial Q^T(u_0, t)}{\partial u_0} = -\frac{\partial}{\partial x}\left[V_0(X(x,t,t_F)) + \int_t^{t_F} R(X(x,t,\sigma))d\sigma\right]^T B(x,t) \quad (10.60)$$

For the functional equation (10.46) we have:

$$u = u_0 = -KB^T(x,t)\frac{\partial}{\partial x}\left[V_0(X(x,t,t_F)) + \int_t^{t_F} R(X(x,t,\sigma))d\sigma\right] \quad (10.61)$$

The performance indices considered in this section enable us to obtain the optimal control algorithms that require significantly less computational operations than the performance indices examined in the previous section. In turn, the control law (10.61) has a certain advantage over others considered in this section. The computational algorithm based on equation (10.61) should include the following operations in the discrete-time:

i. At the beginning of each discrete moment of time $k = k_0, k_0 + 1, \ldots, k_F$, when the control values are determined, the current value of the state vector of equation (10.32) should be determined (or estimated).
ii. The solution $X(x, t, \sigma)$ of (10.56) on the interval $[k, k_F]$ for the initial conditions coinciding with the current (or close to it) state of equation (10.32) should be determined.
iii. Computation of the gradient of $V(x, t)$ for the current moment of time k [see equations (10.59) and (10.60)].
iv. Computation of the control actions according to equation (10.61).

Since computational operations are performed by a computer, the integration in equation (10.60) is changed to the summation. The discrete form of equation (10.60) is obvious.

Below we consider the discrete-time analogue of the problem (10.32) and (10.42):

$$x(k+1) = f(x(k)) + B(x(k+1))u(k), \quad x(k_0) = x_0 \quad (10.62)$$

and

$$I = V_0(x(k_F)) + \sum_{k_0}^{k_F} R(x(k)) + Q(u(k)) + Q_0(u_0(k)) \quad (10.63)$$

assuming that the function

$$Q(u) + Q_0(u_0) - \frac{\partial Q^T(u_0)}{\partial u_0} u$$

is positive definite with respect to u and equals zero at $u = u_0$.

The Bellman equation can be presented as:

$$V_i[x(i)] = \min_{u(i)} \{ R(x(i)) + Q(u(i)) + Q_0(u_0(i)) + V_{i+1}^T [f(x(i)) + B(x(i))u(i)] \}$$

$$(10.64)$$

$$V(x(k_F)) = V_0(x(k_F)), \quad i = k_F - 1, \quad k_F - 2,\ldots$$

Acting analogously to the continuous case, instead of equations (10.43) and (10.44) we have:

$$\frac{\partial Q^T(u_0(i))}{\partial u_0} = -\frac{\partial V_{i+1}^T}{\partial x(i+1)} B(x(i)) \qquad (10.65)$$

and

$$V_{i+1}[f(x(i)) + B(x(i))u_0(i)] - V_i(x(i)) - \frac{\partial V_{i+1}^T}{\partial x(i+1)} B(x(i))u_0(i) = -R(x(i))$$

$$(10.66)$$

The first-order approximation of $V_{i+1}[f(x(i)) + B(x(i))u_0(i)]$ as:

$$V_{i+1}[f(x(i)) + B(x(i))u_0(i)] \approx V_{i+1}[f(x(i))] + \frac{\partial V_{i+1}^T}{\partial x(i+1)} B(x(i))u_0(i) \quad (10.67)$$

makes it possible to simplify equation (10.66), so that it instead of equation (10.66) we can use the equation in the form similar to equation (10.44):

$$V_{i+1}[f(x(i))] - V_i[x(i)] = -R(x(i)), \quad V(x(k_F)) = V_0(x(k_F)) \quad (10.68)$$

As mentioned earlier, the examined optimal problems here have a certain computational advantage over the optimal problems considered in the previous section and can compete successfully with them being used in the integrated guidance and control systems design.

10.4 INTEGRATION AND DECOMPOSITION

Decomposition means the disintegration, the division of the whole system into absolutely independent or weakly interconnected subsystems, which on certain stages can be treated separately. Solutions obtained for these subsystems are used later to obtain the solution for the whole problem. During many years, this approach was considered as a natural and logical way and in many cases as the only way to resolve complex problems. The decomposition methods formed a special part of computational mathematics. They were widely developed and used especially on the initial stage of the computer era, when the time of computation was the main restricting factor.

The term *integration* means an act of combining into an integral whole. Currently, powerful computers enable us to solve problems we can only have dreamed about solving years ago. Does it mean that it is worth loading computers with huge models, containing many "fuzzy" parameters and rely upon the solutions obtained? Does it mean that we should not trust our intuition and rely upon sober sense? Computers, how powerful they might be, need time for computations. Homing missiles use their onboard computers, and for them the time of computation, as well as the computer's weight, depending on its computational ability, are very important factors.

As shown in Figure 10.1, the guidance law and control units are connected in series. It means that, formally, each of these units, at least on the initial stage, can be designed separately. However, as seen from Figure 10.1, they are interconnected also by the existing feedback loop, and that is the reason why the integrated approach to missile guidance and control has merit.

The system concept is paramount on all stages of design, because a decision in one particular design area on one specific component may radically affect other parts of the missile system design. However, it does not contradict the necessity to examine thoroughly separate elements of the system.

As mentioned earlier, the natural approach to solving a complicated problem is to consider simplified parts of the problem and solve them first; later the problem is brought closer to reality by gradually including the complicating factors of the full problem. This enables the designers to feel deeply all aspects and details of the problem.

This approach is used in the autopilot design [5]. At the earliest stages of design, the behavior of servos and airframe is approximated by linear differential equations with constant coefficients, so that well-known analytical methods can be used to evaluate (roughly) some design parameters. Despite the approximate nature of the obtained analytical solutions, they make it possible to arrive at qualitative estimates of the most important effects of the missile system parameters on the system's accuracy. As mentioned earlier, at the preliminary stage of the autopilot design, three rotational channels (roll, pitch, and yaw) are investigated separately. The flight control system must stabilize and control the attitude of a missile about the three body axes—roll, pitch, and yaw. As shown in Figure 9.2, roll is defined as being about the longitudinal axis; yaw is orthogonal to roll and is contained in the trajectory plane, pitch is orthogonal to both and completes the right-handed set. The three channels of the flight control system are similar, with pitch and yaw usually being nearly identical. Their separate consideration significantly simplifies the design procedure. Their aerodynamic coupling is taken into account later, and the control system is modified to meet additional requirements.

The item of prime importance to the designer is the accuracy of the missile system. However, accuracy requirements are linked with other important characteristics of the whole missile system. In addition, the autopilot should guarantee the system stability over the missile operational range. The desired autopilot response is required to be fast with minimum overshoot to meet structural limitation. The autopilot should also provide attenuation of high frequencies, so that it does not react to high-frequency aeroelastic behavior that can affect the sensor signals or to noise accompanying the acceleration commands [5].

The airframe parameters and structure significantly influence missile performance, and the guidance system designer should have full knowledge of the airframe response (e.g., its frequency response characteristics) to the guidance system commands. In turn, the dynamic characteristics of the airframe depend on the type of control chosen (e.g., jet, tail, canard, or wing) and the location of control devices.

The guidance designer is vitally interested in the weight admissible for the guidance equipment and the dimensions and locations of the space allocated for it. The location of some elements of the guidance system (e.g., gyros) may be quite critical. The designer should be involved in the airframe design and space allocation to place the motion-sensitive elements properly. During the preliminary design, the missile body is usually assumed rigid. However, taking elastic behavior of the structure and its effect on aerodynamics into account later, additional corrections can be made. For example, the accelerometers should be placed where the translational vibrations of the structure are minimal, and the pitch and yaw rate sensors should be placed where the rotational vibrations are minimal.

We described only some problems that should be resolved in the process of the missile guidance and control system design. Most of them relate to the autopilot design, assuming that the guidance law has been already chosen.

Proponents of the integrated approach [3,4,7–11] offer to design the autopilot and to determine the guidance law based on a certain criterion.

However, the criteria chosen [see equations (10.25) and (10.30)] are closely linked with the generalized Riccati equation [see equations (10.22) and (10.24)]. As indicated in [9–11], after transforming the missile model into the state dependent coefficient form, the next major responsibility of the designer is to select the state weighting matrices $C_0(x)$ and $R(x)$ positive semidefinite for all expected values of x. As admitted in [9], selecting matrix functions that meet these requirements for all values of the state vector is not generally practical. It is difficult to justify the choice of coefficients in the above-indicated matrices for various engagement scenarios. Moreover, there is no rigorous proof that the closed-loop, nonlinear system based on the procedure described in [9–11] will be stable. That is why the optimal approach discussed in the previous section looks more encouraging. It is more rigorous and it is easier to solve the Lyapunov equations similar to equation (10.44) than the Bellman equations (10.20), (10.24), or (10.40).

The optimal methods described in this chapter belong to a class of the analytical controller design and suffer the drawback pertaining to this class: the choice of the performance index coefficients presents a separate independent problem, and the realization of the optimal solution entails certain difficulties because the measurement of the whole state-vector is required.

The methods of analytical design of controllers based on the state-space models and optimal control theory have been extensively presented in the literature. During the last 50 years, thousands of papers and books were dedicated to this problem [14]. However, they were not widely used in the engineering practice. Engineers, dealing with real physical systems, prefer to operate with the input-output relationships and frequency characteristics of separate elements of the systems. They feel better about the system potential, when they know, for example, its bandwidth. Frequency domain methods, developed in the United States many years ago, are still very popular and widely used in the engineering practice because they are very physical. Following the traditional autopilot design procedure, as soon as the method of guidance has been determined, and the general bandwidth specifications for the main autopilot components have been developed, the design of the system then proceeds using analytical methods, modeling, and simulation.

Too much passion for mathematics, rather than physics of a process under consideration, can be more dangerous than neglecting modern mathematical tools and relying on intuition based on deep physical understanding of the process.

It is worth reminding that traditionally, based on past experience, the guidance objective is formulated in terms of line-of-sight (LOS). The widely used proportional navigation and pure pursuit guidance laws are based on certain principles, the so-called geometrical rules. As mentioned earlier, according to the pure pursuit rule, the pursuer should be directed at the target; according to the parallel navigation rule, the direction of the line-of-sight should be kept parallel to the initial LOS. People adopted these rules from nature, observing the behavior of predators. The optimal law corresponding to the proportional navigation, which is the simplest implementation of the parallel navigation rule, was obtained as a solution of a specially constructed optimal problem for an extremely simplified missile guidance model (2.44) and (2.54), or (2.37) and (3.17). The optimal solution requires information about the future target behavior. Indirectly, it can be presented by the indication of the time-to-go or/and the predicted intercept point. It can be justified for the midcourse guidance, where at least there is time to improve the situation. However, for homing guidance the mistake in determining the time-to-go or the predicted intercept point can be crucial.

Classical control theory is based on the feedback principle. The optimal control theory supplies us with an optimal control law as a function of time. Only for a special class of optimal problems the optimal solution can be presented as a function of the system state vector (i.e., present controller equations similar to the closed-loop system examined in classical control theory).

Does it mean that the optimization methods of modern control theory are useless? Not at all. The optimal solutions (10.26), (10.47), (10.53), and (10.61) help engineers to choose a rational structure for the control system. Although the measurements available on board the missile are limited, and the authors of [9–11] assumed that all the measurements required for the implementation of the integrated guidance and control law were available, analyzing the structure in Figure 10.1 and the optimal solution (10.26) we can easily conclude that the autopilot input in Figure 10.1 lacks many components of the state vector of equation (10.2). Most of these components influence the missile actual acceleration a_{Mx}, a_{My}, and a_{Mz}, so that it is logical to conclude that the structure with the missile acceleration feedback is better than the structure in Figure 10.1. We made the same conclusion in Chapter 6, supporting the necessity of the acceleration feedback as a measure to make the actual missile acceleration close to the commanded acceleration (in full accordance with the feedback principle).

Here we present Figure 6.1 in a more general form. As seen from Figure 10.3, the autopilot input is considered as the commanded acceleration, i.e., the real guidance law consists of two parts: the first component depends directly on a chosen initially guidance law; the second component presents a correction caused by the difference between the chosen guidance law and its realization. The same structure can be interpreted in a different way. We can consider it as consisting of the guidance part presented by a chosen guidance law and the autopilot that includes the acceleration

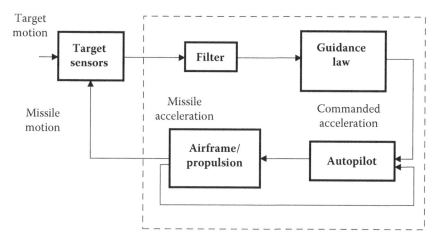

FIGURE 10.3 Integrated guidance control system.

feedback. The difference in terminology does not change the whole problem aimed to increase the missile system accuracy.

Usually the design of a product, which has its "predecessors," does not start from zero. It uses components of the previous design and improves them. Should we ignore this approach?

The structure in Figure 10.3 and the procedure of determining its components can be considered as a modernization of the existing systems. Analysis of the relationship between the miss distance and the unmanned aerial vehicle parameters (see Chapters 5 and 7) enables the designers to establish the "critical" parameters and to see the way of improving vehicle accuracy (to redesign certain components that would change these parameters). The design procedure to modernize the existing guidance and control systems will enable designers to modify the existing design procedure analyzing the influence of the vehicle main design parameters on the miss distance and consider the new guidance laws and the new guidance-autopilot structure to improve vehicle performance.

As mentioned in Chapter 9, the design process is an art and we hope the above given material presents useful information for reflection.

REFERENCES

1. Albrekht, E. On Optimal Stability of Nonlinear Systems, *Journal of Applied Mathematics and Mechanics* 25, no. 5 (1961): 836–44.
2. Blakelock, J. *Automatic Control of Aircraft and Missiles.* 2nd ed. New York, NY: Wiley-Interscience, 1991.
3. Cloutier, J., D'Souza, C., and Mracek, C. Nonlinear Regulation and Nonlinear H Control via the State-Dependent Riccati Equation Technique. In *Proceedings of the International Conference on Nonlinear Problems in Aviation and Aerospace*, Daytona Beach, FL, May 1996.

4. Cloutier, J., Mracek, C., Ridgely, D., and Hammett, K. State Dependent Riccati Equation Techniques: Theory and Applications, *ACC Workshop Tutorial* 6 (1998).

5. Cronvich, L. Aerodynamic Consideration for Autopilot Design. In *Tactical Missile Aerodynamics*. Edited by M. Hemsch and J. Nielsen. Progress in Aeronautics and Astronautics. Vol. 124, Washington, DC: AIAA, 1986.

6. Krasovskii, A. *Systems of Automatic Control of Flight and Their Analytical Design*. Moscow, Russia: Nauka, 1973.

7. Lin, C., Wang, Q., Speyer, J., Evers, J., and Cloutier, J. Integrated Estimation, Guidance, and Control System Design Using Game Theoretic Approach, 3220–24. In *Proceedings of the American Control Conference*, 1992.

8. Lin, C., Ohlmeyer, E., Bibel, J., and Malyevac, S. Optimal Design of Integrated Missile Guidance and Control, *World Aviation Conference*, AIAA-985519, 1998.

9. Menon, P., and Ohlmeyer, E. Integrated Design of Agile Missile Guidance and Control Systems. In *Proceedings of the 7th Mediterranean Conference on Control and Automation (MED99)*, Haifa, Israel, 1999.

10. Menon, P., Sweriduk, G., and Ohlmeyer, E. Optimal Fix-Interval Integrated Guidance-Control Laws for Hit-to-Kill Missiles, *AIAA Guidance, Navigation, and Control Conference*, AIAA, 2003-5579, Austin, TX, 2003.

11. Palumbo, N., Reardon, B., and Blauwkamp, R. Integrated Guidance and Control for Homing Missiles, *Johns Hopkins APL Technical Digest* 25, no. 2 (2004): 121–30.

12. Wise, K., and Broy, D. Agile Missile Dynamics and Control, *Journal of Guidance, Control and Dynamics* 21 (1998): 441–49.

13. Yanushevsky, R. A Controller Design for a Class of Nonlinear Systems Using the Lyapunov-Bellman Approach, *Transaction of the ASME, Journal of Dynamic Systems, Measurement, and Control* 114 (1992): 390–93.

14. Yanushevsky, R. *Theory of Optimal Linear Multivariable Control Systems*. Moscow, Russia: Nauka, 1973.

11 Guidance Laws for Boost-Phase Interceptors Launched from UAVs

11.1 INTRODUCTION

The considered guidance laws were developed without taking into consideration a real missile acceleration ability. The commanded acceleration is a command to the missile guidance and control system to realize the prescribed acceleration. However, the missile inability to perform these commands would demonstrate practical inefficiency of the chosen guidance law. That is why the implementation of the guidance laws should be accompanied with simulation that helps to choose proper guidance law parameters. Acceleration limits depend on types of interceptors and technology invested in their production. These limits should be taken into consideration on all stages of missile design.

Here we will consider the procedure of choosing and testing the guidance laws for the future generation of interceptors with one of their possible applications—boost-phase defense against intercontinental ballistic missiles (ICBNs). Our current national missile defense program focuses on ground-based interceptors that would destroy warheads launched by ICBMs but before they re-enter the atmosphere. Although during the so-called midcourse phase of flight, which lasts approximately several tens of minutes, the warheads follow predictable ballistic trajectories to expect high probability to destroy them, many experts believe that the defense system can be defeated by countermeasures and penetration aids, including a large number of lightweight decoys that would be difficult to distinguish from real warheads outside the atmosphere.

Boost-phase intercept systems, which would try to disable ICBMs during the first few minutes of flight, while their boosters are still burning, are considered as a possible alternative of the existing defense scheme. The boost phase of an ICBM usually lasts only several minutes, so that interceptors should be able to operate at a very high speed not achievable now by existing missiles.

The technical feasibility and required performance of boost-phase intercept systems were examined by the American Physical Society (APS)

Study group [2]. The APS report found that the interceptors that burnout in 40 to 50 seconds and reach speeds of at least 6.5–10 *km/s* would be required to defend against ICBMs launched from North Korea and Iran. Such interceptors would have to be substantially larger and capable of higher performance than existing interceptors. A 5 *km/s* interceptor would work against slow-burning (5 minutes or longer) liquid-propellant ICBMs; to defend against solid-propellant ICBMs would require interceptors that could reach speeds of about 10 *km/s* [2].

It is assumed that boost-phase intercept systems include the following important components. The detection and tracking system, which can be space, ground, sea, or aircraft based, should detect the initial ballistic missile launch and then to track the target from launch or cloud break until intercept. The target should be tracked at a high enough data rate to generate accurate guidance commands to the interceptor until its seeker can acquire the target. The interceptor consists of a boost vehicle that accelerates a kill vehicle (KV) to the burnout velocity required to engage the target when the interceptor launch point may be far from the target launch point. The kill vehicle guides to the target by the use of lateral divert engines and should destroy the target in a high speed collision. It is assumed that the KV consists of a terminal infrared (IR) seeker with possibly a laser-ranging device, lateral divert engines, a receiver, and an inertial reference unit. When the IR seeker acquires the boosting target it must provide angular rate information of sufficient accuracy to enable the KV to hit the target. The KV laser ranging device may also be used in conjunction with an IR seeker to provide range information that can be used for better guidance accuracy. The KV lateral divert engines generate the acceleration required by the guidance law. A receiver is required during the KV's midcourse phase of flight (i.e., before the KV seeker acquires the target) to receive guidance commands from the off-board tracking system and possibly range information during the homing phase of flight. An inertial reference system and possibly GPS are also required in order to determine the kill vehicle position, velocity, acceleration, and angular orientation with sufficient accuracy.

A limited interceptor's speed and the short time available to intercept the attacking missile restrict the interceptor's operational range. The interceptor's range is limited by the highest speed that is technically feasible and the time available to complete the intercept, so that boost-phase defense is possible only if interceptors can be positioned close enough to the required intercept locations, generally within 400–1000 *km*.

The innovative approach to boost-phase intercept offered in [3] involves the use of UAVs to launch interceptors. Because of UAV pay-load weight constraints, airborne interceptors would have lower burnout velocities than surface-based interceptors. But this drawback of airborne interceptors is compensated by the advantage of unmanned aircraft, their ability to penetrate much closer to ICBM launch sites than it would be possible by using surface-based interceptors, so that that the required burnout velocity of air-launched interceptors can be significantly less than that of surface-based interceptors. In addition, the use of stealthy UAVs with an IR search and track system would make less severe time lines for intercepting ICBMs. (In [2] the following time lines were determined for surface-based missiles to intercept ICBMs launched from North Korea to Alaska: for a liquid-propellant ICBM having burnout time 240 s the rocket detection time is 45 s, an interceptor is launched 30 s after the rocket has been detected, and the maximum time available to achieve intercept is 92 s; for a solid-propellant ICBM having burnout time 170 s the rocket detection time is 30 s, an interceptor is launched 30 s after the rocket has been detected, and the maximum time available to achieve intercept is 62 s.)

In this chapter we will test the guidance laws considered in the book applying them to the airborne interceptors launched from UAVs to demonstrate their advantages (less time of intercept, higher accuracy, and simplicity of implementation) over the existing widespread guidance laws. Advanced guidance algorithm development is essential and necessary for meeting lethality requirements against future advanced maneuvering threats and for defining future interceptor concepts and associated critical enabling technologies. The specific features of the boost-phase interceptors, which distinguish them from other types of interceptors are the robust ability to intercept maneuvering targets and accelerating and maneuvering boosters. This requirement translates into a need for large divert capability and relatively high acceleration capability. The problem of the development of a boost-phase intercept system is a part of a more general problem to develop a next generation interceptor and concept of operations that can defeat a wide range of threats in the boost and terminal phases of flight.

Since the kill vehicle is the most important component of the boost-phase interceptors, it is important first to choose guidance laws that would guarantee its best performance.

11.2   KILL VEHICLES FOR BOOST-PHASE DEFENSE

The flight of the kill vehicle may be divided in three stages: kill vehicle divert, kill vehicle homing, and the endgame [2]. In [2], the different proportional navigation (PN) law gains were used for the different stages; the "hybrid" PN/APN guidance scheme was also used. The two types of maneuvers (lunge and jinking) were considered, and the miss distance of 0.5 m or less was considered as the most important kill vehicle's functional requirement [2]. The kill vehicle's dynamics were modeled by a fifth-order binomial model with five time constants $\tau = 0.1\ s$, i.e., (see Figure 5.6):

$$G_2(s) = (\tau s + 1)^{-5} \qquad (11.1)$$

This model is assumed to be conservative, so that for the real model the performance results will be better [2]. The step miss for various effective navigation ratios N of the linear PN guided missile model is given in Figure 11.1 (solid, dashed, and dash-dotted lines correspond to $N = 3$, $N = 4$, and $N = 5$, respectively).

Based on Figure 11.1, it is possible to conclude that with the time-to-go about 5 s we cannot expect intercept for an 8-g lunge maneuver, since the miss distance is less than 0.5 m only for $N = 3$ and the time-to-go more than 4 s. However, if we take into account the 15-g KV's acceleration limit, then the nonlinear system describing the missile guidance model [the system in Figure 5.6 with a limited $a_c(t)$] becomes BIBO unstable with respect to t_F for the effective navigation ration $N = 3$ and $N = 4$.

To test the efficiency of guidance laws for maneuvering targets, equation (5.42) is used to determine the optimal weaving frequency of evasive maneuvers (see also Figure 5.7). The peak miss and the target weaving maneuvers for various effective navigation ratios is analyzed based on the expressions of Chapter 5 and presented in Figure 11.2 (solid, dashed, and dash-dotted lines correspond $N = 3$, $N = 4$, and $N = 5$, respectively). For the

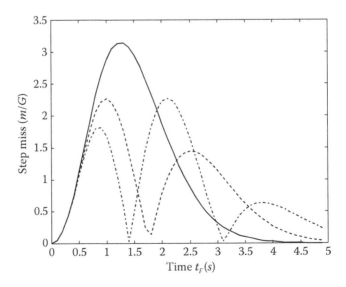

FIGURE 11.1 Step miss for the linear binomial model for various N.

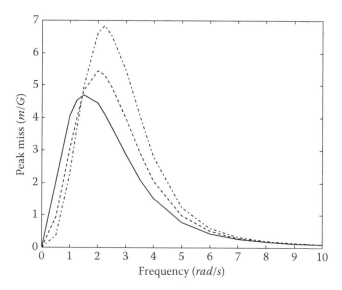

FIGURE 11.2   Peak miss for the linear fifth-order binomial model for various N and weaving maneuvers.

binomial model, the optimal target frequency is around 8 rad/s for $N = 3$, 10 rad/s for $N = 4$, and 11 rad/s for $N = 5$, i.e., its range is $1.25 - 1.75$ Hz.

Based on Figure 11.2 we can conclude that for the fifth-order model the peak miss for a 2-g jinking maneuver is more than 0.5 m, so that this binomial model and the PN guidance law gain chosen cannot guarantee the performance requirements formulated in [2].

For the first-order model, the optimal target frequency and peak miss are around 7.2 rad/s and 0.038 m/G for $N = 3$; 10 rad/s and 0.025 m/G for $N = 4$; 12 rad/s and 0.018 m/G for $N = 5$, respectively.

The maximum acceleration that can be provided by the kill vehicle's propulsion system is modeled by imposing a fixed upper limit on the commanded acceleration $a_c(t)$. According to [2], a series of simulations showed that a 15-g acceleration is adequate to assure a miss distance of 0.5 m or less for closing velocities less than or about 14 km/s. The acceleration of the kill vehicle was limited to 15-g for a lunge 8-g maneuver, and the effective navigation ratio $N = 6$ was used for the endgame simulations in [2]. Although a high $N = 6$ gain tends to increase the effect of sensor noise, in the endgames analyzed in [2] the noise was low enough relative to other factors to be acceptable. Figure 11.3 shows the results of simulations for $N = 5$, 6, and 7 (solid, dashed, and dash-dotted lines, respectively).

Based on Figure 11.3, we cannot expect intercept within 5 s. Since binomial models are not realistic, they should not be used as the worst scenario even on the initial stage of design. Figure 11.4 shows the results of simulations of the PN guidance law for the first-order model of the kill

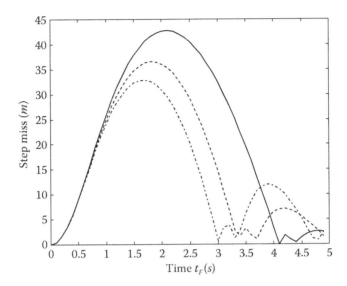

FIGURE 11.3 Step miss for the fifth-order binomial model (PN law, 8-*g* lunge maneuver, 15-*g* limit).

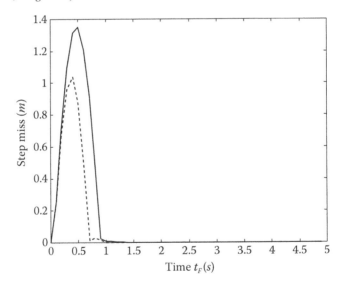

FIGURE 11.4 Step miss for the first-order model (PN law, 8-*g* lunge maneuver, 15-*g* limit).

vehicle with $\tau = 0.1$ *s* for $N = 5$ and 6 (solid and dashed lines, respectively). Although usually such a simple model gives optimistic results, the considered nonlinear model for $N = 3$ and 4 is BIBO unstable, as in the case of the binomial nonlinear model.

In the future we will consider a more realistic model than in [2] and compare the results with that of the first-order model.

11.3 DEVELOPMENT OF THE MISSILE MODEL AND SELECTION OF GUIDANCE LAW PARAMETERS

In [2], the authors adopted the admissible miss distance of 0.5 m or less. They believe that the 0.2 m diameter kill vehicle capable of aiming at the centerline of 1 m diameter booster with a high probability of a miss distance of 0.5 m or less would almost certainly collide with the offensive missile. In their simulations, they used a simple KV model with 0.1 s time constant and with the fixed 15-g acceleration limit. The endgame simulations were made for an 8-g lunge maneuver and a 2-g jinking 1 Hz maneuver.

As mentioned earlier, if the KV acceleration is limited to 15-g, the missile guidance system for the gains $N = 3$ and 4 is BIBO unstable. This explains why the unusually high $N = 6$ gain was chosen in [2]. Figure 11.5 shows the miss distance of the first-order model with $\tau = 0.1$ s for 8-g lunge and 2-g, 1 Hz jinking maneuvers (the lunge maneuver corresponds to $N = 6$, the jinking maneuver corresponds to $N = 3$). The PN law applied to this model satisfies the formulated earlier requirements. However, it is unrealistic to assume that that the created kill vehicle would have such dynamic characteristics. The considered first- and fifth-order models ignore such important dynamic characteristics of the kill vehicle as the natural frequency and damping. That is why the PN law applied to the model that takes into account these characteristics produces different results.

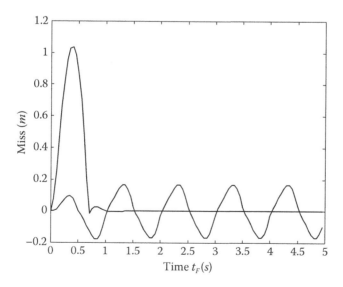

FIGURE 11.5 Miss distance for the first-order model (8-g lunge and 2-g, 1 Hz jinking maneuvers).

The parameters of the KV's flight control system depend on the kill vehicle's mass, configuration, and so on. These parameters are time-varying. However, even the model with constant parameters, which reflects properly the KV's dynamics, would allow a designer to obtain more realistic estimates than the first- and/or fifth-order models.

Although the performance of guided missile systems is assessed by their terminal effect and the generation and intelligent control of this terminal effect is one of the key requirements to the missile systems, the terminal effect depends significantly on parameters of the subsystems. The performance of a separate subsystem dictates requirements to the interconnected ones. For example, the KV's airframe parameters determine the airframe natural frequency ω_M that influences significantly KV dynamics and, as a result, influences the autopilot system requirements. Higher accuracy guidance and autopilot systems can employ smaller warheads. The seeker dynamic parameters influence the guidance system accuracy. As indicated in the previous chapter, the traditional approach to designing missile guidance and autopilot systems usually neglects interaction between these systems and treats individual missile subsystems separately. The subsystems are designed separately and then integrated before verifying their performance. The quantification of the impact of missile parameters on the miss distance is the first important step toward integrated design of missile guidance and autopilot systems. As mentioned in Chapter 5, the main factors that influence the miss distance in homing missiles are the seeker errors, airframe characteristics, autopilot lag τ, and target maneuvers. An appropriate choice of the estimation system parameters can reduce requirements to a seeker's accuracy and a guidance law effective navigation ratio N. The frequency response (5.42) and (5.43) obtained for the proportional navigation law enables us to analyze the influence of the basic guidance system parameters on the miss distance for step and weave maneuvers.

Dynamic properties of the flight control system that are close to reality are very important on the initial stage of design. It is impossible to ignore the natural frequency of the airframe. It should be taken into account that the payload will affect the natural frequency of the KV's structure, i.e., when the payload is increased, the system fundamental frequency will be decreased. The natural frequency depends on the kill vehicle configuration, number, displacement of thrusters, and so on. Although this information is unavailable, we will choose the model's parameters based on the known models of small size missiles and analyze the relationship between optimal evasive maneuver weaving frequency, peak miss, and the kill vehicle dynamic parameters (damping, natural frequency, and time lag). The new model is compared with the model considered in [2]. Then the parameters of the guidance laws will be chosen for the newly developed model.

The flight control system dynamics are presented by the third-order transfer function:

$$W(s) = \frac{1}{(\tau s + 1)\left(\dfrac{s^2}{\omega_M^2} + \dfrac{2\zeta}{\omega_M}s + 1\right)} \tag{11.2}$$

with damping ζ and natural frequency ω_M ($\zeta = 0.7$ and $\omega_M = 20$ *rad/s*), the flight control system time constant $\tau = 0.1$ *s*. (On this stage, we consider the deterministic case and ignore the influence of a filter on missile performance.)

Figure 11.6 compares the step responses of the fifth- (dash-dotted line), first-order (dashed line) binomial, and newly developed (solid line) models. It is natural to expect that the PN guidance law for the new model, with better dynamic characteristics than the fifth-order binomial model, results in the less miss step than for the fifth-order binomial model (see Figure 11.7 and Figure 11.1; in Figure 11.7 solid, dashed, and dash-dotted lines correspond to $N = 3$, 4, and 5, respectively) but worse than for the first-order model (see Figure 11.4).

The relationship between the target weaving frequency, the peak miss, and the KV main functional parameters is obtained based on equation (5.42) and presented in Figure 11.8 (dashed, solid, and dash-dotted lines correspond to $N = 3$, 4, and 5, respectively) and Table 11.1.

As seen from Table 11.1, the less KV flight control system time constant gives the less peak miss and higher the optimal target weaving frequency ω_{opt},

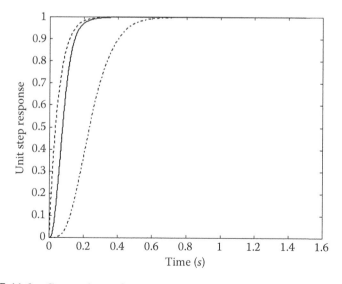

FIGURE 11.6 Comparison of step responses.

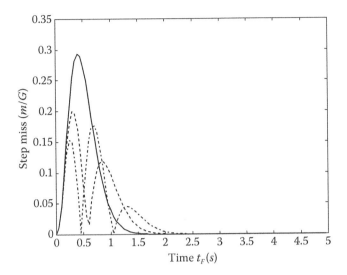

FIGURE 11.7   Step miss for the new model for various N.

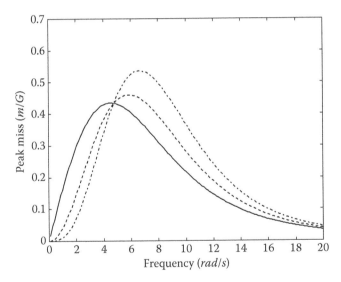

FIGURE 11.8   Peak miss for the new model for various N and weaving maneuvers.

which is more difficult to implement efficiently (with an appropriate ampli-tude) by the boosting missile. The increase of damping increases the peak miss and decreases the optimal target weaving frequency. The increase of natural frequency of the missile flight control system decreases the peak miss and increases the optimal target weaving frequency. Designers of the flight control system should focus on decreasing its time constant and find a reason-able tradeoff in choosing its natural frequency and damping. To achieve a high-speed missile response the payload must be limited.

TABLE 11.1

Influence of Flight Control System Parameters on Optimal Weaving Frequency and Peak Miss Distance for PN Guidance (1-g Weaving Maneuver, $N = 3$, Unlimited Control)

Case #	τ s	ζ	ω_M rad/s	Peak miss m	ω_{opt} rad/s
1	0.1	0.7	20	0.4346	4.5
2	0.1	0.6	20	0.3977	4.9
3	0.1	0.65	20	0.4161	4.7
4	0.1	0.75	20	0.453	4.4
5	0.1	0.8	20	0.4724	4.3
6	0.1	0.85	20	0.4916	4.1
7	0.1	0.9	20	0.5117	4.0
8	0.05	0.7	20	0.2629	6.5
9	0.075	0.7	20	0.3473	5.4
10	0.125	0.7	20	0.526	3.9
11	0.15	0.7	20	0.6224	3.5
12	0.1	0.7	25	0.3368	4.9
13	0.1	0.7	30	0.2765	5.2
14	0.1	0.7	35	0.2364	5.4
15	0.1	0.7	40	0.2076	5.6
16	0.075	0.7	25	0.2639	5.9
17	0.075	0.7	30	0.2133	6.3
18	0.075	0.7	35	0.1795	6.6
19	0.075	0.7	40	0.1557	6.9

Comparing the peak miss for the considered models, we can conclude that the peak miss for the new model is about 10 times larger than for the first-order model and about 10 times less than for the fifth-order binomial model.

To test the ability of the guidance laws considered in the previous chapters to act better than the PN and APN laws, we will consider the guidance law:

$$a_{Mc1}(t) = 3v_{cl}\dot{\lambda}(t) + N_1\dot{\lambda}^3(t) \tag{11.3}$$

Since the effective navigation ratio $N = 3$ is optimal (the energy efficient optimal value) in the accordance with the criteria (2.55) and (3.17) for the parallel navigation, we use this optimal value. The gains N_1 of the cubic terms should be chosen based on available information about the admissible noise level and acceleration limits.

As mentioned earlier, it is impossible to realize the PN law with $N = 3$ and 4 in a case of an 8-g lunge maneuver because of the insufficient 15-g missile acceleration ability. For $N = 5$ and the PN law is realizable but gives worse results than in the case $N = 6$. It is easy to calculate that the line-of-sight (LOS) rate that corresponds to the 15-g limit and the closing velocity 14 km/s for the PN guidance with $N = 5$ equals $\dot{\lambda}_0 = 15 \cdot 9.81/5 \cdot 14000 = 0.0021$. Since, as indicated in [2], the gain $N = 6$ does not make the noise level unacceptable, the upper limit of N_1 is determined from the condition that the guidance law (11.3) should not exceed the 15-g limit for $\dot{\lambda}_0$ and $N = 6$, i.e., $N_1 = 3/\dot{\lambda}_0^2 = 6.8 \cdot 10^5$.

Since the miss distance for the fifth-order binomial model is far away from the performance requirement, we will demonstrate the efficiency of the guidance (11.3) on the example of the first-order model.

11.4   ENDGAME REQUIREMENTS AND THE COMPARATIVE ANALYSIS OF EFFICIENCY OF GUIDANCE LAWS

11.4.1   PLANAR MODEL

The step miss for the guidance law (11.3) (here and earlier, it is clear from the text what kind of missile acceleration, commanded or real, is meant) for the first-order model is shown in Figure 11.9 (solid line). It is slightly worse than for the PN law with $N = 6$ (see Figure 11.5) but consumes less energy (as mentioned earlier, the PN law with $N = 3$ cannot be realized).

According to [2], "APN might reduce somewhat 15-g acceleration result, but its response to switchback maneuvers and jinking maneuvers could be counterproductive and would have to be studied" ([2], p. 237). Figure 11.9 (dashed line) shows the step miss for the guidance law:

$$a_M(t) = Nv_{cl}\dot{\lambda}(t) + N_1\dot{\lambda}^3(t) + N_3 a_T(t) \tag{11.4}$$

where $a_T(t) = 8g$, $N = 3$, $N_1 = 2.26 \cdot 10^5$, $N_3 = \begin{cases} 1 \\ 1.1 \end{cases}$ if $sign(a_T(t)\dot{\lambda}(t)) \begin{smallmatrix} \leq 0 \\ \geq 0 \end{smallmatrix}$

The step miss is significantly better than without the target acceleration term.

Figure 11.10 shows the miss distance in the newly developed model for an 8-g lunge maneuver for the guidance law (11.3) (solid line), where $N = 3$ and $N_1 = 6.8 \cdot 10^5$. For comparison, the miss distance for the PN

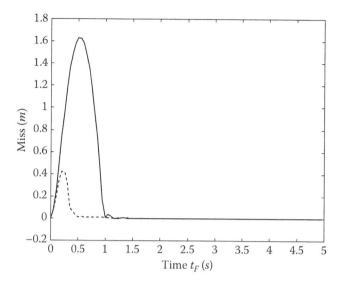

FIGURE 11.9 Miss for the first-order model with 15-g acceleration limit (8-g lunge maneuver and guidance law (11.3)).

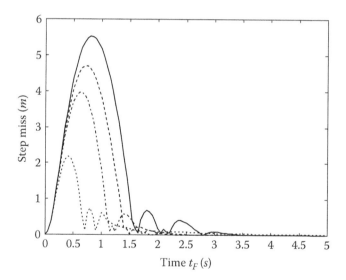

FIGURE 11.10 Miss distance for an 8-g lunge maneuver (new model).

law for $N = 5$ (dashed line) and $N = 6$ (dash-dotted line) are shown. As indicated earlier, for $N = 3$ and 4, the PN guided system under an 8-g lunge maneuver is BIBO unstable.

The dotted line in Figure 11.10 characterizes the miss distance for the guidance law (11.4), where $N = 3$ and $N_1 = 2.26 \cdot 10^5$. It shows that the additional target acceleration term significantly improves missile performance.

All above-considered guidance laws are able to hit the target if the maneuver is more than about 2 *s* to go.

REMARK: To reach the target with comparable recourses, the predator with limited inner resources distributes these resources properly by declining from the optimal strategy (minimum resources) when it is necessary. That is why we keep $N = 3$ all the time as basic, assuming that the additional terms should react efficiently only to sharp target maneuvers.

For a 2-*g* jinking maneuver we consider the same law that was used for an 8-*g* lunge maneuver [see equations (11.3) and (11.4)]. Simulation results for the first-order model are given in Figure 11.11.

Although the PN law (solid line) guarantees the peak miss less than 0.5 *m*, the guidance law (11.4) with the cubic term (dashed line) and, in addition, with the target acceleration term (dotted line) significantly decrease the peak miss.

Figure 11.12 shows the miss distance for the model (11.2) and the weaving frequency 1 *Hz*. The additional target acceleration term (dotted line) decreases the peak miss insignificantly compared to the cubic term (solid line) to satisfy the requirement 0.5 *m* or less. Inefficiency of the cubic term in this case can be explained by the existing 15-*g* acceleration limit and a significantly higher than in reality range of the LOS rate in the chosen planar model (see also Figure 11.8 and Table 11.1 for the linear case).

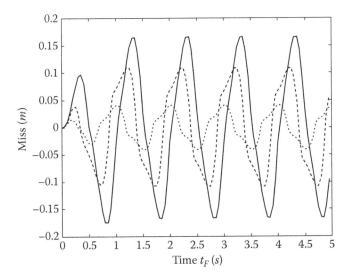

FIGURE 11.11 Miss distance for a 2-*g*, 1 *Hz* jinking maneuver (first-order model and new guidance law).

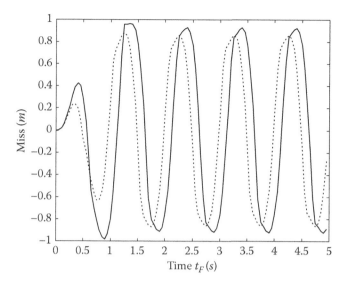

FIGURE 11.12   Miss distance for a 2-g jinking maneuver (new model and new guidance law).

As seen in Figure 11.8, for low target frequencies, less than the frequencies corresponding to the peak values, the increase of N gains decreases the peak miss. The opposite character of the gain influence can be seen for high frequencies. Table 11.2 shows that the peak frequencies for the guidance with the additional cubic term are close to the linear case without acceleration constraints. The 1 Hz (6.28 rad/s) target jinking frequency used in [2] lies within the range of frequencies of Tables 11.1 and 11.2. It will be considered as basic for testing the performance of the three-dimensional kill vehicle model against targets performing sinusoidal weave maneuvers.

As mentioned earlier, according to [2], a 15-g acceleration is adequate to assure a miss distance of 0.5 m or less for closing velocities less than or about 14 km/s. Such a statement can be reliable only if the parameters of the kill vehicle are already known. For the known parameters, it would be easier to choose appropriate guidance law parameters. However, on the initial stage of design such information is unknown. The models considered in [2] (first- and fifth-order binomial models) ignore such an important parameter as natural frequency of the airframe. The more precise model with $\omega_M = 20$ rad/s enables us to justify the use of a 1 Hz target weaving frequency to test the guidance law performance.

As shown in Chapter 5, the generalized planar model [see Figure 5.12 and equation (5.65)], which takes into account the boosting missile dynamics, allows us to evaluate more precisely the miss step and peak miss distances.

TABLE 11.2

Influence of Flight Control System Parameters on Optimal Weaving Frequency and Peak Miss Distance for the Guidance Law (11.3) (2-g Weaving Target Maneuver and 15-g Acceleration Limit)

Case #	τ s	ζ	ω_M rad/s	Peak miss m	ω_{opt} rad/s
1	0.1	0.7	20	1.02	4.5
2	0.1	0.65	20	0.9859	4.6
3	0.1	0.75	20	1.064	4.5
4	0.1	0.8	20	1.1043	4.4
5	0.1	0.85	20	1.1462	4.3
6	0.1	0.9	20	1.1940	4.2
7	0.05	0.7	20	0.6418	5.8
8	0.075	0.7	20	0.8365	5.3
9	0.125	0.7	20	1.2135	4.3
10	0.15	0.7	20	1.4104	3.8
11	0.1	0.7	25	0.7705	4.6
12	0.1	0.7	30	0.6184	5.1
13	0.1	0.7	35	0.5331	5.8
14	0.1	0.7	40	0.4658	6.3

Simulations show that the miss distances for the 8-g lunge and 2-g jinking target $a_T(t)$ maneuvers and the transfer function:

$$W_T(s) = \frac{1 - \dfrac{s^2}{15^2}}{(0.15s+1)\left(\dfrac{s^2}{5^2} + \dfrac{2 \cdot 0.8}{5}s + 1\right)} \tag{11.5}$$

are significantly less (more than 20%) than obtained for the usual KV guidance model. However, the conservative approach and using the model in Figure 5.6 is reasonable when dealing with insufficient information about the KV model.

High values of the effective navigation ratio used in [2] can be explained by a low (less than 2) ratio between the target acceleration and the missile acceleration limit. This presents an essential difficulty for designers of the guidance and control system. Moreover, since in [2] even the first-order model with $\tau = 0.1$ s required $N = 6$ to meet the accuracy requirements, it would be more difficult to meet these requirements for a more precise model under the PN law. The authors of [2] used various values of N for different parts of the kill vehicle's trajectory. It is desirable to have a guidance law with constant parameters for all stages of the kill vehicle's flight or it should be a rigorous rule when and how to change them.

As shown in Chapter 6 (see also Figure 6.3), the use of an additional term (actual missile acceleration) can significantly improve missile performance and decrease its dependence on the missile parameters. The new commanded acceleration a_A (a new guidance law) is formed as a sum of the feedforward signal $G_4(D)a_c$ and the feedback signal $G_4(D)(a_c - a_M)$ [see equation (6.8)].

Figure 11.13 shows the miss distance for the new model for 8-g lunge and 2-g jinking maneuvers when the guidance law with the indicated below parameters was used:

$$a_{Mc}(t) = a_{Mc1}(t) + 4(a_{Mc1}(t) - a_M(t)), \quad a_{Mc1} = 3v_{cl}\dot{\lambda} + N_1\dot{\lambda}^3, \quad N_1 = 6.8 \cdot 10^5$$

$$(11.6)$$

This law satisfies the performance requirements and is more efficient than the law containing only the component a_{Mc1} examined earlier.

The linear approach, based on the assumption that the deviations from a nominal collision course are small, fails when the interception kinematics are highly nonlinear. Guidance system saturation occurs when the system demands (e.g., a commanded lateral acceleration 40-g) exceed the

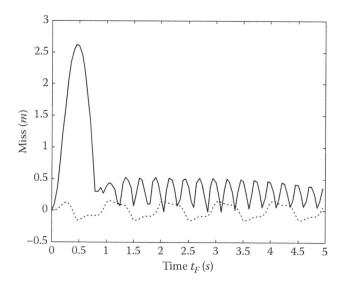

FIGURE 11.13   Miss distance for 8-g lunge and 2-g jinking maneuvers (new model and guidance law (11.6)).

kill vehicle capability (e.g., the KV is only capable of 15-*g*). This situation arises when the KV is far from the nominal collision course and in the case of highly maneuvering targets.

The considered 2-DOF step and sinusoidal maneuvers are a useful starting point for analysis of intercept scenarios that involve ballistic missiles, although the ballistic target dynamics may involve an arbitrary periodic motion in three dimensions. Instead of considering the 3-DOF problem, in many cases we assumed that lateral and longitudinal maneuver planes were decoupled by the means of roll-control, so that the consideration of the 2-DOF problem is justified. It was assumed also that the gravitational component of the total missile lateral acceleration is negligible. Such simplifications are possible only on the initial stage of analysis and design.

11.4.2   3-DOF MODEL. NOMINAL TRAJECTORY

A more precise evaluation of missile performance is based on simulations using 3-DOF and 6-DOF models. Just after booster burnout, axial acceleration, center of gravity, and mass moment of inertia characteristics are changed. These variations should be incorporated in aerodynamic models. The simulation model should analyze the performance of guidance laws in a realistic simulation environment, which accounts for the effects of forces influencing the missile trajectory and flight control system dynamics on the missile's performance. Analysis of the missile's kinematic boundary and other criteria should be used as the measure of effectiveness and basis of comparison. The engagement envelope or kinematic boundary is of paramount importance. The kinematic boundary represents the maximum range at which the missile will achieve a hit, when there is no noise in the system. It can, therefore, be used as a criterion to compare the performance of guidance laws. Among other significant features of guidance systems performance are the miss distance, the time of intercept, maximum rate of turn, and maximum lateral acceleration. The comparative analysis of guidance laws is more restrictive. It includes some of these features (the engagement envelope, miss distance, and the time of intercept), as well as specific features, such as the missile terminal speed and impact angle.

To demonstrate the effectiveness of the considered guidance laws, the initial conditions for the KV in the developed simulation models are chosen so that the intercept can be achieved by using the APN guidance as recommended in [2].

The representative target trajectory was created for a three-stage, solid-propellant model with about 250 *km* altitude at the burn time 188 *s* (see Figure 11.14). The created target module consists of a 3-DOF point-mass presentation of the target motion. The target model is capable of executing

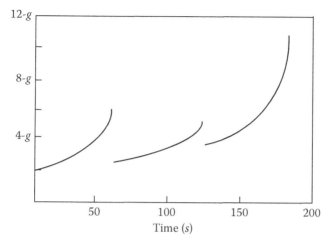

FIGURE 11.14   Acceleration profile of the target.

a maneuver at a given time; time and extent of this maneuver can be adjusted. The most representative maneuvers and the intercept scenarios for the mentioned type of targets will be considered. As indicated in Chapter 9, evasive maneuver design parameters include magnitude, weave period, initiation time, and duration, and the maximum achievable maneuver magnitude-period combination is a function of initiation and duration times. The information about existing ballistic missiles airframe configuration, aerodynamic, and propulsion parameters was used to reflect the target missile dynamics [see equation (11.5)].

Initially, a nominal rather smooth target trajectory was generated that corresponded to the acceleration profile in Figure 11.14. In the model, filters evaluate the target velocity and acceleration based on measurements of the position components [see equations (9.50)–(9.60)]. Tabulated data concerning the target acceleration components were used. By interpolating this data and using integration operations [see equation (9.89)], the target velocity and position components were obtained.

The relative position of the target with respect to the KV is used to compute the actual line-of-sight components $\lambda_s(t)$ and their derivatives $\dot{\lambda}_s(t)$ ($s = 1, 2, 3$) using equations (1.8), (1.11), and (1.12).

Zero-mean Gaussian white noise can be added to the target position or the line-of-sight components, i.e., the LOS can be corrupted by noise directly or indirectly.

The KV's flight control system is represented by the transfer function similar to the considered above planar case [see equation (11.2)]. In the 3-DOF simulation model, the differential equations corresponding to the above-mentioned transfer function should describe the relationship between the coordinates of the commanded and actual accelerations, i.e., the dimension of the system of differential equations is three times higher than in the planar model.

Four types of guidance laws are compared.

i. The PN guidance

$$a_{cs}(t) = 3v_{cl}\dot{\lambda}_s(t) \qquad (s = 1, 2, 3) \tag{11.7}$$

with the effective navigation ratio $N = 3$.

ii. The APN law

$$a_{cs}(t) = 3v_{cl}\dot{\lambda}_s(t) + 2a_{TNs}(t) \qquad (s = 1, 2, 3) \tag{11.8}$$

where a_{TNs} are the orthogonal components of the target acceleration $a_{Ts}(t)$ ($s = 1, 2, 3$).

$$a_{TNs}(t) = a_{Ts}(t) - \lambda_s(t)\sum_{i=1}^{3}a_{Ti}(t)\lambda_i(t) \qquad (s = 1, 2, 3) \tag{11.9}$$

and the guidance laws considered in Chapters 3 and 6.

iii. Three-dimensional variant of equation (11.6)

$$a_{Mcs}(t) = a_{Mc1s}(t) + 4(a_{Mc1s}(t) - a_{Ms}(t)),$$

$$a_{Mc1s}(t) = 3v_{cl}\dot{\lambda}_s(t) + N_1\dot{\lambda}_s^3(t) \qquad N_1 = 6.8 \cdot 10^5 \tag{11.10}$$

$$(s = 1, 2, 3)$$

or

iv. Its modification, where

$$a_{Mc1s}(t) = 3v_{cl}\dot{\lambda}_s(t) + N_1\dot{\lambda}_s^3(t) + a_{Ts}(t),$$

$$N_1 = 2.26 \cdot 10^5 \qquad (s = 1, 2, 3) \tag{11.11}$$

and $a_{Ms}(t)$ ($s = 1, 2, 3$) is a real missile acceleration [see also equation (11.6)].

In addition, the effectiveness of the shaping term [see equation (3.76)]:

$$u_{s2}(t) = -N_{2s}\lambda_s(t)\ddot{r}(t) \tag{11.12}$$

is examined.

For missiles with the controlled part of the commanded acceleration acting orthogonal to the missile's body, it is impossible to reproduce the corresponding components of the missile acceleration precisely without knowledge of the angle of attack. However, for the KV operating at high altitudes it is possible to consider the angle of attack equal to zero. Knowledge of the missile velocity vector enables us to determine the component of the KV acceleration orthogonal to this vector [see equations (9.92)–(9.97)].

The orthogonal part of the commanded acceleration $a_{cNs}(t)$ ($s = 1, 2, 3$; a 15-g acceleration limit was imposed) serves the input of the system of differential equations describing the flight-control system dynamics [see equations (9.100)].

The total KV acceleration $a_M(t) = (a_{M1}, a_{M2}, a_{M3})$ equals:

$$a_{Ms}(t) = a_{cNs}(t) + grav_s \qquad (s = 1, 2, 3) \qquad (11.13)$$

where $grav_s$ ($s = 1, 2, 3$) are the gravitation components [see equation (9.4)].

Since the analytical expressions for the PN, APN, and guidance laws (11.10)–(11.12) were obtained without considering the influence of gravity on the missile trajectory, simulations are made with and without gravity compensation. In the case of gravity compensation, which is widely used in the PN law applications, the commanded acceleration of the developed 3-DOF models contains an additional term to compensate the gravity effect on the actual missile acceleration, i.e., the gravitation components [see equation (9.4)] are added with the opposite sign to the components of the guidance law under consideration. By integrating equation (11.13), as indicated earlier [see equation (9.89)], and using the target position and velocity measurements, all parameters needed to calculate the commanded acceleration [see equations (1.8), (1.11), and (1.12)] are obtained. The sampling period of target information varies depending on the distance between the KV and target. It is selected: 0.2 s if the distance exceeds 250 km, 10^{-2} s if the distance exceeds 50 km and less 250 km, 10^{-3} s if the distance exceeds 100 m and less 50 km, 10^{-4} s if the distance is less than 100 m, and 10^{-5} s if the distance is less than 1 m. The fourth-order Runge-Kutta integration technique is used to solve the system of differential equations. The step of integration coincides with the sampling period for the distances less than 250 km and equals 0.01 s otherwise.

Since the KV's guidance laws are tested, we chose the following initial conditions for the kill vehicle at $t = 100$ s (it is assumed that the interceptor was launched about 80 s later than the target) taking into account that the interceptor's burn time and burnout velocity are about 20 s and 5 km/s, respectively:

$$R_{M1} = = -520 \ km; \quad R_{M2} = 550 \ km; \quad R_{M3} = 90 \ km$$

$$V_{M1} = 2.8 \ km/s; \quad V_{M2} = -2.7 \ km/s; \quad V_{M3} = 2.9 \ km/s$$

At the time $t = 100$ s the target has the following position and velocity generated by a filter, which starts working at $t = 94$ s:

$$R_{T1} = -65.8 \ km; \quad R_{T2} = 68.4 \ km; \quad R_{T3} = 68.3 \ km$$

$$V_{T1} = -1.58 \ km/s; \quad V_{T2} = 1.64 \ km/s; \quad V_{T3} = 1.19 \ km/s$$

This data, as well as about 760 km initial distance of the interceptor from the target's launch-site, is in accordance with the material in [2].

Since we initially considered, similar to [2], the planar first-order model of the flight control system, the three-dimensional case of this system was tested using the developed 3-DOF model. The miss distance and time of intercept (time when the closing velocity $v_{cl} < 0$) for the PN guidance (11.7) with the effective navigation ratio $N = 3$, APN law (11.8) and (11.9) with $N = 3$ and the gain $N_0 = 2$, as it was used in the planar model in [4], and the guidance law (11.10) are presented in Table 11.3. Although the miss distances for the PN, APN, and guidance (11.10) laws satisfy the accuracy requirements, the guidance law (11.10) without gravity compensation shows the best performance: the minimal time of intercept.

Table 11.4 contains simulation results for the more realistic dynamic model (11.2) of the KV's flight control system. Although the PN law accuracy is worse in the case of the model (11.2), the miss distances in all cases satisfy accuracy requirements, so that KV dynamics do not influence significantly KV performance for smooth target trajectories. Using the gravity compensation in the guidance law does not influence missile accuracy meaningfully. The guidance law (11.10) and the APN law give the best performance, the guidance (11.10) gives minimal time of intercept.

The commanded acceleration components of the PN, APN, and (11.10) guidance laws for the model (11.2) (see Table 11.4) are presented in Figures 11.15–11.17 (solid, dashed, and dash-dotted lines correspond to a_{Mc1}, a_{Mc2}, and a_{Mc3}; indices 1, 2, and 3 correspond to axes E, N, and U, respectively). For APN guidance (see Figure 11.16), we considered an "almost ideal filter" that does not distort the form of the target acceleration signal; it only reduces its value by about 10%.

Since the APN formally requires precise information about the target acceleration, significant efforts of researchers are spent developing high accuracy filters reproducing the target acceleration. However, Figure 11.15 shows that such high accuracy can lead to unnecessary expenditure of missile energy resources. Sharp changes of the KV's acceleration at $T = 130$ s are triggered by the target acceleration plunge. This acceleration drop serves as a misleading signal when the APN law is applied, and the better filter—the more misleading the reaction of the missile will be.

TABLE 11.3
Performance of the KV Model With the Time Lag $\tau = 0.1$ s

PN Law With Gravity Compensation	PN Law Without Gravity Compensation	APN Law With Gravity Compensation	APN Law Without Gravity Compensation	Law (11.10) With Gravity Compensation	Law (11.10) Without Gravity Compensation
T_{int} = 179.585 s Miss = 0.037 m	179.381 s 0.05104 m	179.8058 s 0.0455 m	179.5921 s 0.087 m	179.9623 s 0.015 m	178.928 s 0.0043 m

TABLE 11.4

Performance of the KV Model With the More Realistic Dynamics

	PN Law With Gravity Compensation	PN Law Without Gravity Compensation	APN Law With Gravity Compensation	APN Law Without Gravity Compensation	Law (11.10) With Gravity Compensation	Law (11.10) Without Gravity Compensation
$T_{int} = 179.5849s$		$179.3891\,s$	$179.85478\,s$	$179.5999\,s$	$178.9838\,s$	$178.937\,s$
Miss $= 0.0715\,m$		$0.1225\,m$	$0.05621\,m$	$0.08819\,m$	$0.00242\,m$	$0.01972\,m$

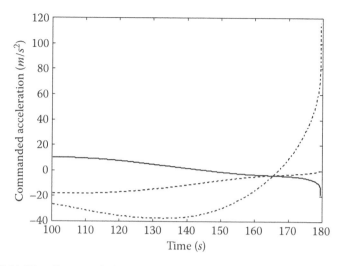

FIGURE 11.15   Commanded acceleration components for PN guidance.

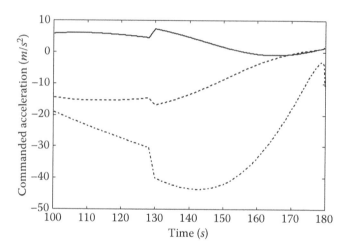

FIGURE 11.16   Commanded acceleration components for APN guidance.

Contrary to the APN law, the commanded accelerations in the case of the PN and guidance (11.10) laws do not distinguish significantly.

REMARK: Although the APN guidance law is widely discussed in the litera-ture, it had no rigorous justification. It was derived from the rephrased miss distance expression for the PN guidance for the planar zero-lag homing loop model (see [4]) assuming a constant target maneuver. In the above-mentioned case, the target position second derivative changes drastically at $T = 130\ s$ (theoretically, if we neglect the target's dynamics, it is the delta-function) so that the APN law should be applied cautiously. From a purely physical con-sideration, information about target acceleration is useful, and by increasing

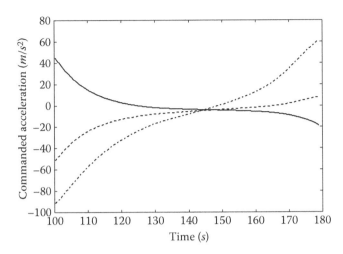

FIGURE 11.17   Commanded acceleration components for the guidance (11.10).

its acceleration so that its velocity would exceed the target velocity, the missile can intercept the target if its acceleration would be kept equal to the target acceleration (their relative motion would correspond to the collision triangle). That is why the gain $N_0 = 1$ can be more accurate than $N/2$ or 2, as recommended in [2,4]. The APN law justification given in Chapter 3 [see equations (3.41)–(3.47)] shows that the target acceleration term depends on the future target strategy, so that a temporarily sharp decrease of the target acceleration should not produce a similar reaction of the kill vehicle.

11.4.3   3-DOF MODEL. STEP AND WEAVING TARGET MANEUVERS

In Section 11.4.1 the miss distance of several planar models was analyzed for 8-g lunge and 2-g jinking maneuvers. Instead of a rather smooth nominal target trajectory considered in the previous section, here we analyze the miss distance for the perturbed trajectory. We consider the maneuver generated at $t = 175\ s$: an 8-g vertical target acceleration and a 2-g sinusoidal fluctuation of the target acceleration in the horizontal plane, i.e.,

$$a_{T1}(t) = a_{T1n}(t) + 2 \cdot 9.81 \cdot \sin(2\pi(t - 175)) \cdot a_{T1n}(t)/\sqrt{a_{T1n}^2(t) + a_{T2n}^2(t)}$$

$$a_{T2}(t) = a_{T2n}(t) + 2 \cdot 9.81 \cdot \sin(2\pi(t - 175)) \cdot a_{T2n}(t)/\sqrt{a_{T1n}^2(t) + a_{T2n}^2(t)}$$

$$a_{T3}(t) = 8 \cdot 9.81; \qquad t \geq 175$$

where the lower index n relates to the nominal trajectory.

The simulation results for the perturbed trajectory are presented in Table 11.5. As seen from Table 11.5, the PN law performance is unsatisfactory;

the miss distance is above 50 m, and without gravity compensation it is above 100 m. The APN law acts with a little bit better accuracy than the guidance (11.10) but worse with respect to the time of intercept.

Figure 11.18 shows the target and missile trajectories for the guidance law (11.10).

The guidance law (11.10) parameters were determined earlier based on the preliminary analysis of the planar model. They can be tuned on the 3-DOF model to satisfy the accuracy requirements even better (miss < 0.5 m). However, we did not do that because miss = 0.41 m is also a satisfactory value taking into account that the total acceleration of the chosen perturbed trajectory exceeds the maximum acceleration level in accordance with Figure 11.14. Moreover, we assumed an almost ideal acceleration filter and ignored the target dynamics. The data of Table 11.5 in parenthesis correspond to the case of the target dynamics (11.5) related to the a_{T3} jump, and the acceleration term of the APN law reflecting the a_{T3} jump at $T = 175$ s is presented as

$$0.9(25.31 + (a_{T3F} - 25.31)(1\text{-}e^{-0.5(t-175)})),$$

where a_{T3F} is a real target acceleration and 25.31 m/s^2 corresponds to a_{T3} at $T = 174.99$ s. It is possible to assume that the APN performance for existing filters is worse.

To examine the effectiveness of the acceleration term of the guidance laws (11.11) simulations were repeated with the gains $N = 3$, $N_{1s} = 2.26.10^5$, $N_{3s} = 1$ ($s = 1, 2, 3$) [see also equation (3.77)]. In contrast to the APN law, here a_{Ts} rather than a_{TNs} ($s = 1, 2, 3$) are used. The miss distance in all cases is below 0.5 m. However, the additional acceleration term with the gain $N_{3s} = 1$ did not produce any meaningful improvement of the KV's performance. More complicated time-varying gains N_{3s} ($s = 1, 2, 3$) [see equation (3.77)] were not considered. As mentioned, using target acceleration information in the guidance law requires additional devices, sophisticated filtering, and so on, so that it should be used only in cases when it can produce substantial positive results.

The tests show that gravity compensation does not influence notably the guidance (11.10) performance. Moreover, in most cases the KV's performance without gravity compensation is better. This can be explained by the effectiveness of the feedback term that also reflects the effect of gravity [see (6.8), (6.20), (11.6), and (11.10)].

In contrast to the considered target maneuvers that act during a short period of time, we also examine the KV's performance in the case of target trajectories perturbed for a significantly longer time. The so-called generalized energy-steering (GEMS) maneuvers started at $t = 150$ s are considered. The components of the acceleration of the third stage are chosen close to [2].

TABLE 11.5

Performance of the KV Model, Target Maneuvers at $T = 175$ s

	PN Law With Gravity Compensation	PN Without Gravity Compensation	APN Law With Gravity Compensation	APN Without Gravity Compensation	Law (11.10) With Gravity Compensation	Law (11.10) Without Gravity Compensation
$T_{int} = $	179.591 s	179.405 s	179.8536 s	179.6038 s	178.995 s	178.9521 s
	(179.5823 s)	(179.393 s)	(179.847 s)	(179.5956 s)	(178.992 s)	(178.9486 s)
Miss =	63.619 m	149.2654 m	0.16735 m	0.081236 m	0.64348 m	0.4115 m
	(52.1016 m)	(121.4533 m)	(0.09775 m)	(0.1748 m)	(0.1446 m)	(0.07315 m)

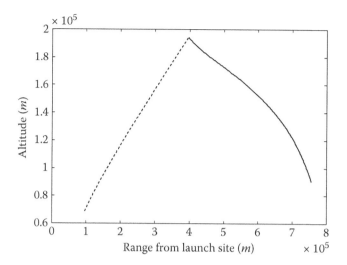

FIGURE 11.18   Target (dashed line) and KV (solid line) trajectories (for the guidance law (11.10) $t_F = 178.95$ s; miss $= 0.41$ m).

The solid and dashed lines of Figure 11.19 show the commanded and real target acceleration components, respectively. Since the PN law produced unsatisfactory results in the case of maneuvering targets, here we compare only the APN law with the guidance laws (11.10). As in many previous tests, here we assume that the kill vehicle has perfect knowledge of a target's acceleration without any time lag. The used assumption of a 10% mistake (less than in reality) in determining the target acceleration value is equivalent to the 10% decrease of N_0. In this way, we reflect only a small distortion of the target acceleration and ignore more influential factors such as time lag and noise, so that the obtained simulation results for the APN law (see Table 11.6) can be considered as optimistic ones. The data in parenthesis corresponds to the real target acceleration (dashed line in Figure 11.19). Target dynamics are presented by the transfer function (11.5). As seen from Table 11.6, the guidance (11.10) enables the KV to intercept the target faster than by using the APN law; the miss distance is also smaller. Moreover, the intercept can be performed without knowledge of the target acceleration.

As seen in Figure 11.18, the curvature of the target trajectory at the initial part of a flight is greater than during the later stage even in a case of the described maneuvers at $t = 175$ s. This enables us to assume that additional information about the second derivative of range, when it is not too small, can be used in the guidance law [see the so-called shaping term (11.12)]. Simulations show that for $N = 3$ and $N_1 = 6.8 \cdot 10^5$ the influence of this term is insignificant because of a 15-g acceleration limit.

The above tests were repeated under the assumption of state-estimate uncertainties for the different sensors related to the target-tracking problem.

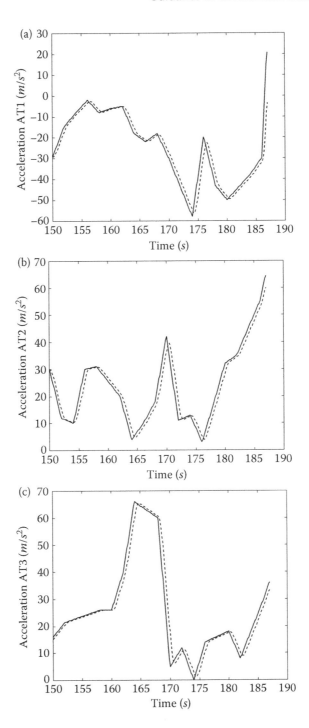

FIGURE 11.19 Components of acceleration during a GEMS maneuver.

TABLE 11.6

Performance of the KV Model, GEMS Target Maneuver

APN Law With Gravity Compensation	APN Law Without Gravity Compensation	Guidance Law (11.10) With Gravity Compensation	Guidance Law (11.10) Without Gravity Compensation
181.136 *s*	180.90 *s*	180.37 *s*	180.358 *s*
(181.24 *s*)	(181.01 *s*)		
0.101 *m*	0.0796 *m*	0.041 *m*	0.01335 *m*
(0.131 *m*)	(0.1 *m*)		

It was assumed the target position uncertainty to be equal to 200 *m* for ranges of more than 50 *km* and 30 *m* for ranges of less or equal to 50 *km*; the last 100 *m*, the measurements are assumed to be ideal. Target acceleration uncertainty was assumed to be 16 *m/s²* and 12 *m/s²* for distances exceeding 50 *km* and less or equal to 50 *km*, respectively; the last 100 *m*, the measurements are assumed to be ideal. In the software program, zero-mean Gaussian distributed numbers and uniform distributed numbers, independent from sample to sample, were added to the target position and acceleration components every 1 *s* for ranges exceeding 50 *km* and 0.01 *s* for smaller ranges; the last 100 *m*, the measurements were not accompanied with noise.

The results of KV performance based on a 100-run Monte Carlo simulation for uniform and Gaussian distribution show that the standard deviation in all simulations has the order $O(10^{-3})$. The simulation model did not take into account the dynamic errors of filters, information delay of IR sensors, and some other factors that should be reflected in the higher stage of design. However, very small miss distances can justify the unsophisticated enough simulation model.

11.5 ADVANCED GUIDANCE LAWS APPLIED TO BOOST STAGE

11.5.1 INTERCEPTOR'S MODEL

Since the authors of [2] conducted an extensive research on the future surface-based boost-phase interceptors, we compared their results with the KV's performance that can be obtained by using the guidance laws discussed in this book.

The KV's design is a part of the boost-phase interceptor's design and it cannot be considered separately. The KV's and boosting motor's weight and velocity are interconnected. The KV's guidance and control system parameters influence the weight of the kill vehicle, which, in turn, has an effect on design requirements to the boosting motor and the guidance and control system of the interceptor. It is desirable to use the same guidance law for the entire flight of boost-phase interceptors. In this case, the whole guidance and control system will be simpler and more reliable.

At the altitude at which surface-based boost-phase interceptors typically burnout, the atmosphere is still too dense for a kill vehicle to start operating; it begins operating autonomously at 80–100 km altitudes [2]. Airborne interceptors have less the burnout time and velocity than surface-based interceptors. But they are launched from UAVs operating at altitudes about 15 km. Their trajectories are predetermined and can be realized as described in Section 8.3.1 [see equation (8.8)]. Here we will build the model of an airborne interceptor with about 3.6 km/s burnout velocity reached in 20 s.

Recommendations concerning the kill vehicle's performance needed to ensure that engagements have a high probability of being successful, based on the analysis of boost-phase engagements using kinetic-energy weapons, are reliable only in the case when the considered dynamic models of the interceptor and the kill vehicle, as its part, are close to reality.

In contrast to the previous sections, here the guidance algorithms are used during all controlled stages of the interceptor's flight. The proposed methodology and algorithm developments constitute a design tool that can be used by the offensive or defensive missile designers to produce, in the initial design stage, an assessment of the threat ballistic missile evasive maneuver capability and to design sophisticated guidance and control systems.

In contrast to the PN and APN laws, the acceleration generated by the considered class of guidance laws is not perpendicular to the line-of-sight. As mentioned earlier, in many missiles its axial component cannot be realized. Since thrust vector control (TVC) motors have such capability, the discussed guidance laws are tested to control the interceptor's motion in its boost phase with and without axial control. The interceptor's performance with the TVC boosting motor is compared with the performance of a more simple boosting motor able to control only the lateral acceleration.

The interceptor is assumed to have two booster stages with a total burnout time 20 s. The acceleration profile (*thrust*) of each stage is presented by a second-order polynomial:

$$thrust = (5 + 0.45t^2)g \tag{11.14}$$

and

$$thrust = (5 + 0.45(t - 10)^2)g \tag{11.15}$$

where t is the time after the interceptor's launch.

The average axial acceleration corresponding to the boost stage is about 20-g, and the vertical burnout velocity is about 3.6 km/s. The E, N, U components $thrust_s$ ($s = 1, 2, 3$) of the axial acceleration equal:

$$trust_s = trust \cdot e_{Ms} \qquad (s = 1, 2, 3) \qquad (11.16)$$

where e_{Ms} are the components of the unit velocity vector.

It is impossible to present the corresponding components of the missile acceleration precisely without knowledge of the angle of attack. If for the KV operating at high altitudes it is possible to consider the angle of attack equal to zero, for the boost stage it does not equal zero. But since we consider a hypothetical acceleration profile (11.14) and (11.15) with a 20 s burnout time, for analysis of the efficiency of the tested guidance laws the assumption of a zero angle of attack is not very restrictive. After the main design parameters of the boost stage are determined, a more precise 6-DOF model should be used.

It is assumed that the boost stage consists of the uncontrolled (up to 3 s) and controlled phases.

First, we consider the boosting motor with an unmovable nozzle— the motor without controlled axial acceleration, and assume that the controlled boost stage lateral acceleration is changed by the guidance laws (11.10)–(11.12) and has a 12-g acceleration limit.

The boosting motor design to realize an additional 12-g lateral acceleration is simpler than in the case of using Lambert guidance, which requires thrust vector control (TVC) and also the ability to cut off the interceptor's engine. A more sophisticated boosting motor with axial control, a gimbaled TVC boosting motor, will be considered later.

The KV's commanded lateral acceleration $a_{cNs}(t)$ (a 12–20-g acceleration limit, depending on the time of flight, was imposed) and the positive axial acceleration $a_{Ls}(t)$ ($s = 1, 2, 3$) (a 3–5-g acceleration limit, depending on the time of flight, was imposed) serve the input of the system of differential equations describing the flight-control system dynamics [see equation (9.100)].

Similarly, the boosting motor commanded lateral acceleration $a_{cNs}(t)$ (it has a 12-g acceleration limit) or the TVC boosting motor acceleration $a_{cMs}(t)$ ($s = 1, 2, 3$) serves as the input of equation (9.100).

For the interceptor during its first two boost stages, we consider in equation (9.100) $\tau = 0.5s$, $\zeta = 0.7$ and $\omega_M = 10 rad/s$. For the KV: $\tau = 0.1s$, $\zeta = 0.7s$ and $\omega_M = 20 rad/s$. The total input interceptor's acceleration $a_M(t) = (a_{M1}, a_{M2}, a_{M3})$ equals:

$$a_{Ms}(t) = a_{cMs}(t) + grav_s + thrust_s \qquad (s = 1, 2, 3) \qquad (11.17)$$

Since the acceleration profile of the boost stage is presented approximately by equations (11.14) and (11.15), in the case of the uncontrolled axial acceleration, for simplicity, we included the axial components directly in equation (11.17).

The acceleration's components related to the boost stage act the first 20 s of the interceptor's flight; after that they equal zero. The components related to the KV's operation act 14 s after the end of the boost stage.

Although the KV's seeker can begin operations after the KV is above a 50 km altitude (at altitudes of less than about 40–50 km and speeds above 2.5–3 km/s there can be seeker window heating issues), assuming that the time to eject the KV is around 3 s and reserving time for related preparatory operations, taking also into account that higher altitudes are more preferable, we chose a 14 s period between the end of the boost stage and the beginning of the KV's operations.

Since the previous analysis shows that the discussed guidance laws are efficient without gravity compensation, which is widely used in the PN law applications, the gravity compensation is not considered.

The interceptor's model, which includes a kill vehicle's model, with the time-varying parameters of the flight control system and time-varying acceleration limits more realistically reflects interceptor's dynamics.

As an additional possibility to improve the KV's performance, the efficiency of a supplementary axial guidance component is discussed, i.e., the KV with axial-thrust capability is considered. The generalized new guidance laws [see equation (3.96)] are applied to this problem. Based on this analysis, the recommendations can be made whether it is worthwhile to use an additional axial thruster.

We use the unchanged guidance law parameters for the control boost stage and the homing stage. The effective navigation ration $N = 3$, which corresponds to the optimal energy efficient mode. The cubic term coefficient $N_{1s} = 6.8 \cdot 10^5$ ($s = 1, 2, 3$), as it was chosen earlier. The shaping term coefficients $N_{21} = N_{22} = 0$ and $N_{23} = 1$. This reflects the fact that during the boost phase the second derivative of range is mostly influenced by the change of a target's altitude. The shaping term acts throughout the controlled boost phase. During the KV's flight, the shaping term is used only for large range distances, when the curvature of the target trajectory may be substantial.

The 3 s uncontrolled part of the interceptor's flight imposes more rigorous requirements to the initial launch parameters of the interceptor. Formally, the situation is similar to the decision a predator makes before starting a pursuit—the initial direction of motion depends on the predator's inner resources and position compared to the target's resources estimate and position. As in the case of SM missiles, the initial elevation and azimuth angles of the interceptors and their related position and velocity during the initial uncontrolled flight should be determined from the developed tables and/or empirical/half-empirical expressions based on approximate calculations accompanied by multiple simulations for various types of targets and tests results of the interceptor's prototype. The mentioned tables and/or expressions to determine these parameters should be developed after the interceptor's acceleration profile has been determined. Necessary corrections should be made for the position and velocity of the interceptor at the beginning of the controlled flight that relate to the speed of the UAV and weather conditions (e.g., wind).

Since the construction of a boosting motor with a fixed axial acceleration is simpler than that of a TVC boosting motor, most of simulations relate to this type of motors. In simulations, the initial elevation El and azimuth Az angles were chosen so that the intercept would take place within a period of time that does not exceed a target's burnout time. For the chosen axial thrust acceleration profile [see equations (11.14) and (11.15)] the interceptor's vertical velocity at the beginning of the controlled boost stage equal $V_M(3) = (5 \cdot 3 + 0.45 \cdot 3^2)g = 186.93$ *m/s*.

The corresponding E, N, and U coordinates of the interceptor's velocity vector (the indices 1, 2, and 3 are used) at the beginning of the controlled flight are determined as:

$$V_{M1} = \cos El \cdot \sin Az \cdot 186.93$$

$$V_{M2} = \cos El \cdot \cos Az \cdot 186.93 \qquad (11.18)$$

$$V_{M3} = \sin El \cdot 186.93$$

The boost motor is controlled by the guidance law:

$$a_{Mc1s}(t) = 3v_{cl}\dot{\lambda}_s(t) + 6.8 \cdot 10^5 \lambda_s^3(t) + N_{2s}\dot{v}_{cl}\lambda_s(t) \qquad (s = 1, 2, 3) \quad (11.19)$$

The KV's flight is controlled by the same law; only in this case the shaping term works at ranges more or equal to 250 *km*. The efficiency of the shaping term and its influence on the time of intercept and the cumulative velocity change will be examined.

11.5.2 SIMULATION RESULTS. NONMANEUVERING TARGET

It is assumed that the interceptor is launched at $t = 75$ *s*, its controllable boost phase starts at $t = 78$ *s* and ends at $t = 95$ *s*, and the kill vehicle starts operating at 109 *s*.

The target position and velocity at $t = 78$ s equal:

$$R_{T1} = -35.845 \ km, \quad R_{T2} = 37.234 \ km, \quad R_{T3} = 43.666 \ km$$

$$V_{T1} = -1.16 \ km/s, \quad V_{T2} = 1.2 \ km/s, \quad V_{T3} = 1,06 \ km/s$$

The interceptor's U-coordinate R_{M3} at $t = 75$ s is 15 km. Its position and velocity at the beginning of the controlled boost stage is determined based on the elevation and azimuth angles at the time of launch [see equation (11.18)].

The simulation results presented in the Tables contain the following parameters: the ground range between target and missile launch sites (km); the interceptor's position (km) at $t = 78$ s; the interceptor's elevation and azimuth reference angles (degree/rad) at $t = 75$ s; the interceptor's velocity components (m/s) at the beginning of a controlled flight $t = 78$ s; the KV's initial position at $t = 109$ s; the time of intercept T_{int} (s), the miss distance (m), and the intercept position (m).

Table 11.7 shows the target intercept for the interceptor's launches situated close to the trajectory plane, so that in this case we have close-to-planar engagements. The standoff distance has its maximum value when the target and interceptor trajectories are in the same plane. About 730 km standoff distance corresponds to about 187 s intercept time, which is close to the last "safe" intercept time. For launch positions closer to the target's launch site, the time of intercept is less; it is about 117 s for the standoff distance of about 220 km. (Assuming that the KV is launched at $t = 109$ s, we do not consider smaller standoff distances here.)

The initial interceptor's direction at the time of launch is the main factor determining a successful intercept. Figure 11.20 shows the target (solid line) and the interceptor (dashed and dotted lines) trajectories that correspond to different elevation and azimuth angles at launch.

Any longitudinal acceleration of the target that is perpendicular to the line-of-sight with the interceptor (or even at angles close to a right angle) appears as a target maneuver to the interceptor. Such positions of the interceptor are possible in the case of nonplanar engagements and, as a result, the last "safe" intercept time corresponds to standoff distances less than for planar engagements. Tables related to nonplanar engagements were built similar to Table 11.7 to determine the operational area from which the intercept is achievable, i.e., the E-N plane projections of the interceptor's launch positions that guarantee intercept.

The data of the tables enable us not only to build the operational area but also to obtain preliminary information concerning the interceptor's main launch operational parameters—elevation and azimuth angles, their values depending on the initial interceptor position.

TABLE 11.7

Simulation Results for Almost Planar Engagements

Ground Range Between Target and Missile Launch Sites (km)	Interceptor Position at $t = 78$ s (km)	Interceptor Elevation and Azimuth Reference Angles at $t = 75$ s (degree/rad)	Interceptor Velocity Components at the Beginning of Controlled Flight $t = 78$ s	Time of Intercept T_{int} (s) (Miss (m))	KV Initial Position $(t = 109$ s) (m)	Intercept Position (m)
$RTM_{gr} = 730.5$	$R_{M1} = -510$ $R_{M2} = 523$ $R_{M3} = 15.2$	46/0.8 135/2.356	$V_{M1} = 92.12$ $V_{M2} = -92.05$ $V_{M3} = 134.1$	186.88 (0.165)	$R_1 = -469,476$ $R_2 = 482,532$ $R_3 = 71,833$	$R_1 = -307,977$ $R_2 = 319,611$ $R_3 = 213,394$
660	$R_{M1} = -460$ $R_{M2} = 473$ $R_{M3} = 15.2$	46/0.8 135/2.356	$V_{M1} = 92.12$ $V_{M2} = -92.05$ $V_{M3} = 134.1$	178.56 (0.037)	$R_1 = -419,768$ $R_2 = 432,763$ $R_3 = 72,104$	$R_1 = -272,898$ $R_2 = 283,237$ $R_3 = 193,944$
449	$R_{M1} = -310$ $R_{M2} = 325$ $R_{M3} = 15.2$	46/0.8 136/2.37	$V_{M1} = 90.73$ $V_{M2} = -93.42$ $V_{M3} = 134.1$	151.61 (0.052)	$R_1 = -273,080$ $R_2 = 286,531$ $R_3 = 74,964$	$R_1 = -181,249$ $R_2 = 188,193$ $R_3 = 142,459$
378.4	$R_{M1} = -260$ $R_{M2} = 275$ $R_{M3} = 15.2$	46/0.8 136/2.37	$V_{M1} = 90.73$ $V_{M2} = -93.42$ $V_{M3} = 134.1$	142.21 (0.025)	$R_1 = -225,865$ $R_2 = 238,907$ $R_3 = 78,593$	$R_1 = -155,020$ $R_2 = 160,982$ $R_3 = 127,083$

(Continued)

TABLE 11.7 (Continued)
Simulation Results for Almost Planar Engagements

Ground Range Between Target and Missile Launch Sites (km)	Interceptor Position at t = 78 s (km)	Interceptor Elevation and Azimuth Reference Angles at t = 75 s (degree/rad)	Interceptor Velocity Components at the Beginning of Controlled Flight t = 78 s	Time of Intercept T_{int} (s) (Miss (m))	KV Initial Position (t = 109 s) (m)	Intercept Position (m)
307.8	$R_{M1} = -210$ $R_{M2} = 225$ $R_{M3} = 15.1$	16/0.28 137/2.39	$V_{M1} = 122.7$ $V_{M2} = -131.2$ $V_{M3} = 51.66$	130.32 (0.063)	$R_1 = -171,493$ $R_2 = 182,291$ $R_3 = 74,644$	$R_1 = -121,117$ $R_2 = 129,948$ $R_3 = 108,830$
279.5	$R_{M1} = -190$ $R_{M2} = 205$ $R_{M3} = 15$	5.7/0.1 137/2.39	$V_{M1} = 112$ $V_{M2} = -136$ $V_{M3} = 18.66$	126.09 (0.042)	$R_1 = -152,236$ $R_2 = 162,361$ $R_3 = 76,091$	$R_1 = -115,330$ $R_2 = 119,789$ $R_3 = 102,631$
222.8	$R_{M1} = -154$ ... $R_{M2} = 161$ $R_{M3} = 15$	-20/-0.35 140/2.45 86/1.5 48.3/0.76	$V_{M1} = 121.1$ $V_{M2} = -135.2$ $V_{M3} = -64.1$ ... $V_{M1} = 9.117$ $V_{M2} = 9.577$ $V_{M3} = 186.5$	116.77 (0.232) ... 186.49 (0.05)	$R_1 = -113,066$ $R_2 = 115,367$ $R_3 = 72,797$ ... $R_1 = -157,598$ $R_2 = 165,118$ $R_3 = 96,805$	$R_1 = -95,645$ $R_2 = 99,351$ $R_3 = 89,733$ ... $R_1 = -306,137$ $R_2 = 317,702$ $R_3 = 212,382$

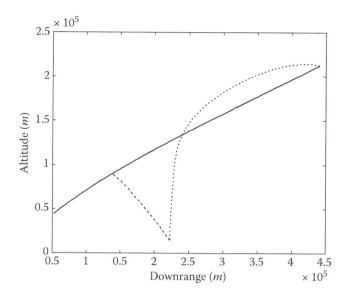

FIGURE 11.20 Two possible interceptor's trajectories for different directions at launch (T_{int} = 186.49 s and 116.77 s).

Here we do not consider the prelaunch function, which determines the elevation and azimuth angles at launch. A firing table is developed from simulation experiments accompanied with kinematical consideration that also determines which shots are kinematically possible. Flyout and firing tables are produced when all components of the interceptor are known and reliable flyout tables require firing tests also.

Based on the data in the mentioned tables, the operational area from which it is possible to intercept the targets is built (see Figure 11.21; for simplicity, only one quadrant is considered). Its boundary is not robust with respect to the input parameters (elevation and azimuth angles), since it corresponds to the best combination of the parameters for which intercept is possible. Inside the bounded area, there exists an admissible domain for these parameters. The deeper inside the bounded area the more freedom in choosing admissible elevation and azimuth angles. Of course, their choice influences the time of intercept. However, it is important that intercept can be achieved.

The determination of initial elevation and azimuth angles is an auxiliary problem that can be solved, for example, in a way similar to the procedure used by the Weapon Control System of SM missiles. Some values and their trend become obvious directly from the study of the mentioned tables. However, the final recommendations require extensive simulations accompanied with firing tests also.

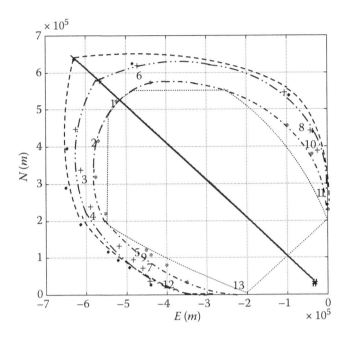

FIGURE 11.21 Operational areas (solid line: a projection of the target trajectory; (#): the target position at $t = 75$ s; (*, +, •): standoff distances corresponding to various guidance algorithms.

In Table 11.8 we indicated the admissible intervals (in square brackets) of the elevation and azimuth angles and the time of intercept with respect to the initially chosen angles of Table 11.7.

It is useful to build the domain (azimuth-elevation) of the admissible angles for the operational area and the most representative angles should be established. Negative elevation angles for short standoff distances are caused by the uncontrolled axial acceleration, which exceeds its "demand." This shows that for short distances less powerful interceptors can be used.

A larger operational area (dash-dotted line) that corresponds to the elevation and azimuth angles chosen better than the earlier (dotted line) is shown in Figure 11.21. The data related to the most representative points of the mentioned area presented in Table 11.9.

11.5.3 SIMULATION RESULTS. INFLUENCE OF SHAPING TERM

Figure 11.21 corresponds to the guidance law (11.19) including the shaping term that acts during the controlled boost stage and during the homing phase for ranges more than 250 km. Since we deal with ascending targets, we use it only in the guidance law shaping component that influences the

TABLE 11.8
Admissible Initial Elevation and Azimuth Angles

Ground Range Between Target and Missile Launch Sites (km)	Interceptor Position at $t = 78\ s$ (km)	Interceptor Elevation and Azimuth Reference Angles at $t = 75\ s$ (degree/rad)	Time of Intercept T_{int} (s) (Miss (m))
$RTM_{gr} = 730.5$	$R_{M1} = -510$	46/0.8	186.88
	$R_{M2} = 523$	135/2.356	(0.165)
	$R_{M3} = 15.2$		
660	$R_{M1} = -460$	46/0.8 [0.55, 1.4]	178.56 [174, 187]
	$R_{M2} = 473$	135/2.356 [1.4, 3.78]	(0.037)
	$R_{M3} = 15.2$		
449	$R_{M1} = -310$	46/0.8 [−0.85, 1.46]	151.61 [150, 166]
	$R_{M2} = 325$	136/2.37 [1.1, 3.7]	(0.052)
	$R_{M3} = 15.2$		
378.4	$R_{M1} = -260$	46/0.8 [−1, 1.25]	142.21 [140, 147]
	$R_{M2} = 275$	136/2.37 [1.45, 3.34]	(0.025)
	$R_{M3} = 15.2$		
307.8	$R_{M1} = -210$	16/0.28 [−1, 0.58]	130.32 [129, 132]
	$R_{M2} = 225$	137/2.39 [4.25, 2.54]	(0.063)
	$R_{M3} = 15.1$		
279.5	$R_{M1} = -190$	5.7/0.1 [−0.9, 0.12]	126.09 [125, 126.1]
	$R_{M2} = 205$	137/2.39 [4.3, 2.48]	(0.042)
	$R_{M3} = 15$		

U-coordinate. To examine the influence of this term on the interceptor's performance we chose the "most difficult" boundary standoff positions of Table 11.7 and a position at the middle of Table 11.7 and excluded the shaping term from the guidance law for the controlled boost stage. (As mentioned earlier, the influence of the chosen shaping term on the KV's performance was not considerable; it decreases the time of intercept T_{int} only slightly.) The simulation results are given in Table 11.10. As seen from Tables 11.7 and 11.10, in the case of the standoff distance 222.8 km, the shaping term is substantial for intercept. Two other cases do not show any specifics.

Similar to [2], we use the term *cumulative velocity change, denoted* $\Delta V(t)$ to refer to the integral of the absolute magnitude of the kill vehicle's acceleration from the time its propulsion system begins to operate until time t. We will use the same term referring to the integral of the absolute magnitude of the boosting motor lateral acceleration during the controlled boost stage and examine the value of $\Delta V(t)$ at the end of the boost stage and at the time of intercept for the most "difficult" standoff distances.

TABLE 11.9
Admissible Initial Elevation and Azimuth Angles Corresponding To a Larger Operational Area

Ground Range Between Target and Missile Launch Sites (km)	Interceptor Position at $t = 78\ s$ (km)	Interceptor Elevation and Azimuth Reference Angles at $t = 75\ s$ (degree/rad)	Time of Intercept $T_{int}(s)$ (Miss (m))
$RTM_{gr} = 730.5$	$R_{M1} = -510$	46/0.8	186.88
	$R_{M2} = 523$	135/2.356	(0.165)
	$R_{M3} = 15.2$		
705.4	$R_{M1} = -574$	51/0.89	186.95
	$R_{M2} = 410$	105.7/1.845	(0.48)
	$R_{M3} = 15.2$		
655.6	$R_{M1} = -575$	46/0.8	186.27
	$R_{M2} = 315$	63.02/1.1	(0.32)
	$R_{M3} = 15.2$		
585.4	$R_{M1} = -540$	48.7/0.85	185.28
	$R_{M2} = 226$	40.1/0.7	(0.19)
	$R_{M3} = 15.1$		
465.7	$R_{M1} = -450$	46/0.8	185.97
	$R_{M2} = 120$	13.18/0.23	(0.1)
	$R_{M3} = 15.2$		
656	$R_{M1} = -440$	43/0.75	186.82
	$R_{M2} = 572$	166/2.9	(0.16)
	$R_{M3} = 15.1$		
407	$R_{M1} = -400$	22.9/0.4	184.63
	$R_{M2} = 75$	5.7/0.1	(0.15)
	$R_{M3} = 15$		
461	$R_{M1} = -100$	45.8/0.8	184.74
	$R_{M2} = 450$	258.5/4.51	(0.18)
	$R_{M3} = 15.2$		
450.7	$R_{M1} = -439$	33.5/0.585	186.78
	$R_{M2} = 102$	9.45/0.165	(0.5)
	$R_{M3} = 15.1$		
376.7	$R_{M1} = -36$	35.5/0.62	186.34
	$R_{M2} = 375$	266/4.6397	(0.2947)
	$R_{M3} = 15.1$		
220	$R_{M1} = 0$	14.1/0.247	144.17
	$R_{M2} = 220$	246.4 /4.3	(0.5)
	$R_{M3} = 15$		

TABLE 11.9   (Continued)
Admissible Initial Elevation and Azimuth Angles Corresponding To a Larger Operational Area

Ground Range Between Target and Missile Launch Sites (km)	Interceptor Position at $t = 78$ s (km)	Interceptor Elevation and Azimuth Reference Angles at $t = 75$ s (degree/rad)	Time of Intercept T_{int} (s) (Miss (m))
352	$R_{M1} = -350$	24.06/0.42	183.71
	$R_{M2} = 38$	3.81/0.06649	(0.199)
	$R_{M3} = 15.1$		
220	$R_{M1} = -220$	4.4/0.077	150.03
	$R_{M2} = 0$	14.3 /0.25	(0.45)
	$R_{M3} = 15$		

Since any additional term of the developed guidance law influences the cumulative velocity change and its maximal value, which is an important performance parameter, we examined the cumulative velocity change with and without the shaping term for the above examples. Figure 11.22 and Figure 11.23 show $\Delta V(t)$ separately for the boost (solid line) and homing (dotted line) stages. Dashed lines correspond to the guidance without the shaping term.

Figure 11.22 corresponds to the standoff position on the boundary of the operational area with the time of intercept close to the target burn time. The shaping term does not influence the KV's maximal value of ΔV, which is about 3 km/s.

As seen from Figure 11.23, the absence of the shaping term can increase the KV's maximal value of $\Delta V(t)$.

The cumulative velocity change depends on the initial azimuth and elevation angles. The properly chosen angles, inside their admissible domains (see Table 11.8), decrease $\Delta V(t)$. Assuming that $\Delta V(t)$ should not exceed, for example, 2.5 km/s, we can determine a real operational area, which will be a subset of the set of standoff distances in Figure 11.21. We can also expect a larger operational area in the case of the uncontrolled boost phase less than 3 s.

The above analysis was made for the deterministic model since its goal was to demonstrate the applicability of the developed guidance laws and the ability to use the same guidance law, easily implemented in practice, both for the boost and homing phases of the interceptor's flight. We can expect that sensors' noise would decrease the effective operational area by not more than 10% since the target acceleration measurements are not used in the guidance law. Later we will examine the interceptor's performance taking into account errors of measurements.

TABLE 11.10

Influence of the Shaping Term

Ground Range Between Target and Missile Launch Sites (km)	Interceptor Position at $t = 78$ s (km)	Interceptor Elevation and Azimuth Reference Angles at $t = 75$ s (degree/rad)	Interceptor Velocity Components at the Beginning of Controlled Flight $t = 78$ s	Time of Intercept T_{int} (s) (Miss (m))	KV Initial Position ($t = 109$ s) (m)	Intercept Position (m)
$RTM_{gr} = 730.5$	$R_{M1} = -510$ $R_{M2} = 523$ $R_{M3} = 15.2$	46/0.8 135/2.356	$V_{M1} = 92.12$ $V_{M2} = -92.05$ $V_{M3} = 134.1$	186.7 (0.097)	$R_1 = -468,828$ $R_2 = 480,231$ $R_3 = 69,818$	$R_1 = -307,210$ $R_2 = 318,815$ $R_3 = 212,969$
378.4	$R_{M1} = -260$ $R_{M2} = 275$ $R_{M3} = 15.2$	46/0.8 136/2.37	$V_{M1} = 90.73$ $V_{M2} = -93.42$ $V_{M3} = 134.1$	142.5 (0.024)	$R_1 = -227,312$ $R_2 = 239,987$ $R_3 = 79,937$	$R_1 = -155,771$ $R_2 = 161,761$ $R_3 = 127,530$
222.8	$R_{M1} = -154$ $R_{M2} = 161$ $R_{M3} = 15$	-20/-0.35 140/2.45		No intercept		

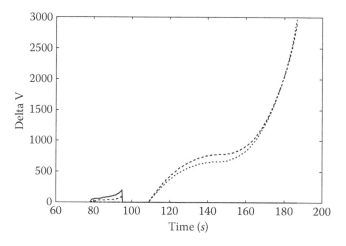

FIGURE 11.22 Cumulative velocity change for the standoff position $R_{M1} = -510\ km$, $R_{M2} = 523\ km$, $R_{M3} = 15.2\ km$.

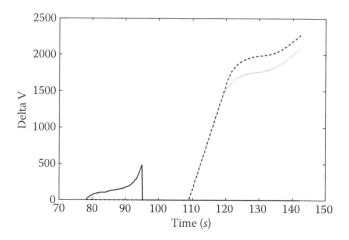

FIGURE 11.23 Cumulative velocity change for the standoff position $R_{M1} = -260\ km$, $R_{M2} = 275\ km$, $R_{M3} = 15.2\ km$.

11.6 INTERCEPTOR'S PERFORMANCE WITH AXIAL CONTROL

11.6.1 AXIAL CONTROL OF KILL VEHICLE

Here we consider the axial control related only to the KV's motion. The chosen boosting motor, which is significantly simpler than in the case of Lambert guidance, does not control the interceptor's axial acceleration.

Initially, it was assumed that the kill vehicle would correct its trajectory by using thrusters generating lateral acceleration. But taking into account the necessity of achieving intercept within the burnout time of boosting targets it is desirable to design kill vehicles with the shortest possible intercept time and the largest possible operational area. We can expect that an additional axial guidance component [see equations (11.19) and (3.96)] can improve the KV's performance decreasing the time of intercept. However, an additional axial thruster would complicate the KV's construction and would increase the KV's weight. This, in turn, would add requirements to other components of the interceptor.

It is of importance to analyze whether the additional axial control can decrease the time of intercept compared to the case of only lateral thrusters (and respectively increase the operational area of the interceptor) at such a degree that a more complicated design would be worthwhile.

The axial thruster of future kill vehicles is assumed to work only in a positive direction (i.e., only as an accelerator) since such a design is simpler. This brings specifics into the guidance problem.

In addition to the law (11.19) we consider the generalized guidance law (3.96):

$$a_{Mc1s}(t) = 3v_{cl}\dot{\lambda}_s(t) + 6.8 \cdot 10^5 \dot{\lambda}_s^3(t) + (1 - N_{2s})r(t)\sum_{s=1}^{3} \dot{\lambda}_s^2(t)\lambda_s(t)$$

$$+ k_1(t)a_{Trs}(t) \qquad (s = 1, 2, 3) \tag{11.20}$$

The simulations related to the generalized guidance law algorithm (3.96) are done with the gain $k_1 = 1$, $N_{21} = N_{22} = 0$, and $N_{23} = 1$. Since the simulation results of Section 11.4 show that the target acceleration term did not decrease substantially the time of intercept, here we analyze only the influence of its axial component, i.e., equation (11.20) does not contain the term $N_{3s}a_{Trs}(t)$ ($s = 1, 2, 3$) of equation (3.96).

The mathematical justification of the efficiency of all terms of equations (11.19) and (11.20) was given without taking into account limits imposed on acceleration. Assuming that the acceleration generated by the KV's axial thruster is limited by 3–5-g [in the simulation model it is presented parametrically $(3 + (t - t_0)/(t_F - t_0) \cdot (5 - 3))g$, where t_0 is the time when the KV starts operating and t_F the burnout time of the target] we will investigate the influence of the axial components of equations (11.19) and (11.20) on the time of intercept. As mentioned, only positive components of the axial acceleration (i.e., acceleration rather than deceleration) are considered.

Since the considered earlier guidance algorithm (11.19) does not generate a "pure" lateral motion, we will consider first the influence of its axial component on the time of intercept and the cumulative velocity change and later compare the efficiency of this and the generalized guidance algorithms.

Table 11.11 contains the simulation results of the algorithm (11.19). Taking into account the relatively small (3–5-g) axial acceleration limit, we do not change the gains of the cubic and shaping terms (such analysis can be useful dealing with more definite information about the kill vehicle at a later stage of design). As mentioned earlier, the shaping term acts for ranges exceeding 250 km and the second derivative of range is determined approximately based on consecutive measurements of the closing velocity.

TABLE 11.11

Simulation Results. Influence of the Axial Component on the Time of Intercept

Ground Range Between Target and Missile Launch Sites (km)	Interceptor Position at $t = 78\ s$ (km)	Interceptor Elevation and Azimuth Reference Angles at $t = 75\ s$ (rad)	Time of Intercept $T_{int}\ (s)$
$RTM_{gr} = 730.5$	$R_{M1} = -510$	(0.8; 2.356)	186.88; [185.56]
	$R_{M2} = 523$	(0.7; 2.356)	No; [184.49]
	$R_{M3} = 15.2$	(0.5; 2.356)	No; [182.11]
		(0.3; 2.356)	No; [180.02]
		(0.; 2.356)	No; [178.72]
660	$R_{M1} = -460$	(0.7; 2.4)	178.09; [176.74]
	$R_{M2} = 473$	(0.6; 1.5)	181.97; [180.82]
	$R_{M3} = 15.2$	(1.2; 3.)	184.77; [184.46]
449	$R_{M1} = -310$	(1.2; 3.)	157.94; [157.89]
	$R_{M2} = 325$	(0.8; 2.37)	151.61; [151.39]
	$R_{M3} = 15.2$	(0.5; 2.)	150.52; [150.57]
680	$R_{M1} = -550$	(0.8; 1.85)	183.04; [181.9]
	$R_{M2} = 400$	(0.5; 1.1)	186.34; [185.26]
	$R_{M3} = 15.1$	(1.; 1.9)	184.71; [184.01]
656	$R_{M1} = -400$	(0.8; 2.95)	179.46; [178.77]
	$R_{M2} = 520$	(0.5; 3.6)	183.3; [182.41]
	$R_{M3} = 15.1$	(1.; 2.5)	180.64; [179.97]
461	$R_{M1} = -100$	(0.8; 4.51)	184.75; [184.6]
	$R_{M2} = 450$	(0.6; 4.1)	No; [170.09]
	$R_{M3} = 15.1$	(1; 4.1)	No; [175.2]

Since the main goal of using the KV's axial control is to extend the operational area, which can be achieved by decreasing the time of intercept obtained without the controlled axial component of the KV's acceleration, it is tested for the standoff positions in Figure 11.21 with ground ranges exceeding 450 *km*. The last column of Table 11.11 compares the time of intercept without and with (data in brackets) the KV's axial acceleration; the miss distances are of order $O(10^{-2})$ *m*.

The analysis of the data of Table 11.11 shows that an additional axial thruster (or a specially designed TVC), enables us to decrease the time of intercept and to extend the operational area. For example, the standoff distances 730.5 *km* and 461 *km* belong (or very close) to the boarder of the operational area, and a notable change of the azimuth or elevation angles makes intercept impossible without the additional axial acceleration, which widens significantly admissible intervals for these angles and extends the operational area as well.

The operational area for the axial control in accordance with the guidance law (11.19) is given in Figure 11.21 (dash-double dotted line). The boundary standoff distances are on average about 10% larger than in the case without axial control (see dash-dotted line in Figure 11.21).

The simulation results for the guidance law (11.20) are given in Table 11.12, which is built similar to Table 11.11 (numbers in brackets show the time of intercept in a case of the generalized law).

The generalized guidance law enables us to decrease the time of intercept by 5–10 *s* and to extend the operational area.

The simulation results (see Figure 11.21) related to axial control reflect the data related to the most representative points of the operational areas. Since the KV's axial control extends the domain of admissible initial elevation and azimuth angles, the operational areas were built assuming the possibility of the interceptor's horizontal launch (i.e., for zero elevation angle). The operational area for the generalized guidance law (11.20) is given in Figure 11.21 (dashed line). The boundary standoff distances are on average about 15% larger than in the case of absence of axial control and about 5% larger than in the case of axial control in accordance with (11.19) (see dash-dotted and dash-double dotted lines in Figure 11.21). The effect of the axial control is more substantial for standoff distances, which are closer to the target trajectory plane, since in this case the axial component of the missile acceleration is larger. For boundary short range standoff distances, the axial component of the missile acceleration is small and, as a result, the effect of the considered more complicated guidance algorithms is smaller.

The indicated above ability of the guidance algorithms (11.19) and (11.20) to decrease the time of intercept and increase the operational area should be evaluated with the analysis of the maximal value of the kill vehicle's

TABLE 11.12
Simulation Results. Generalized Guidance Law

Ground Range Between Target and Missile Launch Sites (km)	Interceptor Position at $t = 78$ s (km)	Interceptor Elevation and Azimuth Reference Angles at $t = 75$ s (rad)	Time of Intercept T_{int} (s)
$RTM_{gr} = 730.5$	$R_{M1} = -510$	(0.8; 2.356)	186.88; [177.26]
	$R_{M2} = 523$	(0.7; 2.356)	No; [175.54]
	$R_{M3} = 15.2$	(0.5; 2.356)	No; [172.3]
		(0.3; 2.356)	No; [170.8]
		(0.; 2.356)	No; [170.46]
660	$R_{M1} = -460$	(0.7; 2.4)	178.09; [168.98]
	$R_{M2} = 473$	(0.6; 1.5)	181.97; [173.44]
	$R_{M3} = 15.2$	(1.2; 3.)	184.77; [177.63]
449	$R_{M1} = -310$	(1.2; 3.)	157.94; [151.95]
	$R_{M2} = 325$	(0.8; 2.37)	151.61; [147.37]
	$R_{M3} = 15.2$	(0.5; 2.)	150.52; [146.33]
680	$R_{M1} = -550$	(0.8; 1.85)	183.04; [174.78]
	$R_{M2} = 400$	(0.5; 1.1)	186.34; [177.91]
	$R_{M3} = 15.1$	(1.; 1.9)	184.71; [177.21]
656	$R_{M1} = -400$	(0.8; 2.95)	179.46; [172.06]
	$R_{M2} = 520$	(0.5; 3.6)	183.3; [174.65]
	$R_{M3} = 15.1$	(1.; 2.5)	180.64; [173.14]
461	$R_{M1} = -100$	(0.8; 4.51)	184.75; [174.53]
	$R_{M2} = 450$	(0.6; 4.1)	No; [166.48]
	$R_{M3} = 15.1$	(1; 4.1)	No; [170.29]

cumulative velocity change. Here the term *operational area* is used for the standoff distances that guarantee intercept, without any restrictions upon ΔV. The real operational area should be determined based on an admissible maximal value of ΔV.

Figure 11.24 corresponds to the standoff position close to the boundary of the three operational areas in Figure 11.21. It compares the cumulative velocity change for three guidance laws:

i. Only the KV's lateral acceleration (dash-dotted line)
ii. The KV's lateral and axial acceleration in accordance with equation (11.19) (solid line)
iii. The KV's lateral and axial acceleration in accordance with the generalized guidance law (11.20) (dashed line)

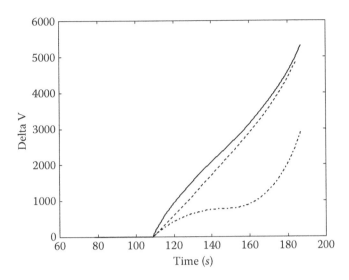

FIGURE 11.24 Cumulative velocity change for the guidance laws (11.19) and (11.20); the standoff positions $R_{M1} = -510/-565/-630\,km$, $R_{M2} = 523/580/643\,km$, $R_{M3} = 15.2/15/15\,km$.

As seen from Figure 11.24, the guidance law controlling only the KV's lateral acceleration has the minimal value of the KV's ΔV. The additional axial control increased the boundary standoff distance about 20% but the maximal value of ΔV increased more than 50%. (The mentioned increase can be less for higher initial elevation angles, but nevertheless it is substantial.)

Since the maximal value of ΔV is an important design parameter, which determines the interceptor's weight, the final recommendation concerning the efficiency of axial control of the KV's motion should be made in a process of solving a more general problem—a multicriterial optimization problem including the optimal interceptor's weight, time of intercept, operational area, and so on. The interceptor's weight (including the KV's weight) and its performance criteria are interconnected. The interceptor's weight, as well as the KV's weight, limited by the UAV's admissible payload, determines dynamic properties of the interceptor and the kill vehicle, respectively. Their dynamics corrected by the designed future autopilots influence the time of intercept and the operational area.

11.6.2 Axial Control of Interceptor

In the previous section we considered the influence of an additional axial thruster of the kill vehicle on the interceptor's performance. As to the boosting motor, we chose its simplest construction—an unmovable nozzle and ability to control the lateral acceleration up to 12-g. Inability to control the interceptor's axial acceleration during the boosting stage puts the kill vehicle in a less favorable situation than in the case of more complicated boosting motors equipped with TVC. In this section, we examine the performance of the interceptor having the gimbaled TVC and analyze the efficiency of the guidance laws applied to this type of boosting motors.

As mentioned earlier, in contrast to the PN and APN guidance, the tested guidance laws have its axial component, which can be easily realized in the case of TVC. Although the chosen boosting motor is more complicated than the one considered earlier, it is still simpler than the boosting motor used in the Lambert guidance, which should have the ability to cut off its engine. To "compensate" a more complex construction of the TVC motor, we will consider a less powerful boosting motor than in the previous sections. Moreover, the simplest possible variant of the TVC boosting motor was chosen. The controlled interceptor's acceleration is realized by positioning a moveable nozzle. Usually, the total TVC nozzle slew angle is limited to about 10°–12°. To simplify the simulation model, we present the boosting motor's acceleration profile by the analytical expressions (11.14) and (11.15). For this profile, a 12-g lateral acceleration limit corresponds to a 14° total TVC nozzle slew angle. However, the chosen acceleration profile has the initial acceleration equal only to a 5-g, which is "compensated " by a 50-g value at the end of each stage. The acceleration profile used earlier and additional angular limits restrict the lateral acceleration more than in the previously considered boosting motor. In the case of more sophisticated TVC motors, we can expect significantly better results.

To apply the guidance algorithm (11.10)–(11.12), the following changes were brought into the developed earlier interceptor's model:

i. The controlled thrust acceleration starts immediately at $t = 75\ s$ (i.e., without a 3 s delay) as it was before.
ii. The boosting motor has the same acceleration profile [see equations (11.14) and (11.15)] and lateral acceleration limit as the earlier considered boosting motor; but because of its movable nozzle the lateral acceleration limit is presented as:

$$LIM = \min(thrust, 12 \cdot 9.81) \qquad (11.21)$$

and if $a_{cN}(t) \geq LIM$ then:

$$a_{cNs}(t) = LIM\, \frac{a_{cNs}(t)}{a_{cN}(t)} \qquad (s = 1, 2, 3) \qquad (11.22)$$

and

$$a_L(t) = \sqrt{thrust^2 - LIM^2}, \quad a_{Ls}(t) = a_L(t)e_{Ms} \qquad (s = 1, 2, 3)\ (11.23)$$

iii. The interceptor's total commanded acceleration a_{cs} consists of the normal a_{cNs} and axial a_{Ls} components:

$$a_{cs}(t) = a_{cNs}(t) + a_{Ls}(t) \qquad (s = 1, 2, 3) \tag{11.24}$$

and instead of equation (11.13) we have:

$$a_{Ms}(t) = a_{cs}(t) + grav_s \qquad (s = 1, 2, 3) \tag{11.25}$$

The simulation results relates to the basic model of the kill vehicle, i.e., to the KV without axial control (see Table 11.13 that contains the data related to the most representative points of the operational area in Figure 11.21). Figure 11.25 presents the operational area (dotted line shows its approximate boundary) for the case of the considered boosting motor and the kill vehicle without an additional axial thruster (i.e., the axial component of the guidance law is not used) both controlled by the guidance law (11.10)–(11.12) [see also equation (11.19)].

The obtained operational area is compared with the operational area obtained earlier for the more powerful boosting motor without axial control and the kill vehicle with (dash-double dotted line) and without (dash-dotted line) axial control and the guidance law (11.19) presented also in Figure 11.21. Since the tested guidance law, even without using its target acceleration term, enabled us to obtain better results than the traditional widely used guidance laws, which require information about the target acceleration and its practical realization is simpler than the guidance laws containing the target acceleration related terms, the larger operational area obtained for the generalized guidance law is not shown in Figure 11.25. The comparison of the presented operational areas enables us to conclude that the use of the guidance law (11.10)–(11.12) in the TVC boosting motors is very effective. The operational area of the considered TVC boosting motor, whose power is less than in the case of the boosting motor considered in the previous sections, is comparable with the operational area obtained for the more sophisticated kill vehicle with axial control.

It is obvious that the combination of these two variants, i.e., both the KV and the boosting motor use the considered guidance law with its axial component, would produce a larger operational area than each separate variant. Moreover, better results can be obtained if the restriction (11.21) would be alleviated. This can be done by choosing better acceleration profile of the boosting motor.

In Figure 11.25, we indicated several standoff distances (see symbols "•") that are far enough from the denoted operational area. For simplicity, they are not included in this area because other positions, not far from the

TABLE 11.13
Simulation Results. Standoff Distances for the TVC Boosting Motor

Ground Range Between Target and Missile Launch Sites (km)	Interceptor Position at $t = 78\ s$ (km)	Interceptor Elevation and Azimuth Reference Angles at $t = 75\ s$ ($degree/rad$)	Time of Intercept T_{int} (s) (**Miss** (m))
$RTM_g = 791.7$	$R_{M1} = -552.8$	13.4/0.2339	186.52
	$R_{M2} = 566.8$	134.6/2.35	(0.18)
	$R_{M3} = 15$		
730	$R_{M1} = -590$	8/0.14	185.75
	$R_{M2} = 430$	106.2/1.85	(0.043)
	$R_{M3} = 15$		
698.4	$R_{M1} = -610$	8/0.14	184.21
	$R_{M2} = 340$	90.4/1.7	(0.053)
	$R_{M3} = 15$		
636.9	$R_{M1} = -590$	8/0.14	186.03
	$R_{M2} = 240$	74.51/1.3	(0.035)
	$R_{M3} = 15$		
574.9	$R_{M1} = -560$	8/0.14	185.87
	$R_{M2} = 130$	63/1.1	(0.2)
	$R_{M3} = 15$		
786	$R_{M1} = -470$	13.75/0.24	186.32
	$R_{M2} = 630$	146.1/2.55	(0.048)
	$R_{M3} = 15$		
508	$R_{M1} = -500$	0/0.	182.21
	$R_{M2} = 90$	55/0.96	(0.28)
	$R_{M3} = 15$		
659.2	$R_{M1} = -110$	0/0.	186.25
	$R_{M2} = 650$	171.9/3.	(0.053)
	$R_{M3} = 15$		
519.7	$R_{M1} = -510$	0/0.	183.32
	$R_{M2} = 100$	54.77/0.956	(0.19)
	$R_{M3} = 15$		
551.7	$R_{M1} = -44$	5.7/0.1	186.34
	$R_{M2} = 550$	213.7/3.75	(0.003)
	$R_{M3} = 15$		
402	$R_{M1} = -40$	5.15/0.09	174.58
	$R_{M2} = 400$	223.4 /3.9	(0.001)
	$R_{M3} = 15$		

(*Continued*)

TABLE 11.13   (Continued)
Simulation Results. Standoff Distances for the TVC Boosting Motor

Ground Range Between Target and Missile Launch Sites (km)	Interceptor Position at t = 78 s (km)	Interceptor Elevation and Azimuth Reference Angles at t = 75 s (degree/rad)	Time of Intercept T_{int} (s) (Miss (m))
601.4	$R_{M1} = -600$	8/0.14	186.35
	$R_{M2} = 40$	72.19/1.26	(0.13)
	$R_{M3} = 15$		
224	$R_{M1} = -220$	5.44/0.095	157.48
	$R_{M2} = 42$	14.78/0.258	(0.2)
	$R_{M3} = 15$		

indicated ones, do not give the intercept. This is explained by the lateral acceleration limit (11.21), which depends on *thrust(t)*. For the indicated positions, the lateral acceleration components influence significantly the interceptor's flight, and more remote positions from the indicated operational area boundary correspond to a more "lucky" time in *thrust(t)*, when the lateral acceleration limit is larger. That is why a more complicated TVC motor with a higher limit (11.21) can increase the operational area significantly, especially the minor radius (assuming its form is similar to ellipse) in Figure 11.25.

The axial component of the considered guidance controlling the acceleration of the TVC boosting motor enables the interceptor to implement parallel navigation more accurately (i.e., to navigate better the kill vehicle), so that its maximal cumulative velocity should be less than in the case when the axial component is not controlled. On the other hand, since for the considered less powerful boosting motor the burnout velocity is less than 3.5 *km/s*, it is natural to expect that the KV's maximal cumulative velocity can be higher than in the considered earlier case of a more powerful boosting motor.

Figure 11.26 shows the cumulative velocity change for the guidance law (11.19) and TVC boosting motor for the standoff positions $R_{M1} = -552.8$ *km*, $R_{M2} = 566.8$ *km*, $R_{M3} = 15$ *km* (solid line; see row 1 in Table 11.13) and $R_{M1} = -110$ *km*, $R_{M2} = 550$ *km*, $R_{M3} = 15$ *km* (dashed line; see row 8 in Table 11.13).

As explained before, some standoff positions were excluded from the operational area presented in Figure 11.25 (for simplicity, to make its shape look similar to the previously obtained areas), so that instead of $R_{M2} = 650$ *km* in row 8 we chose the position with $R_{M2} = 550$ *km* belonging to the operational area; this position coincides with the boundary

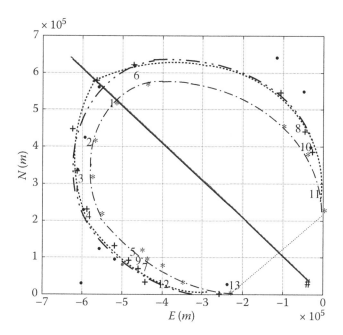

FIGURE 11.25 Comparison of the operational areas for two types of boosting motors (solid line: a projection of the target trajectory; (#): the target position at $t = 75$ s; (*, +, •): standoff distances corresponding to various guidance algorithms.

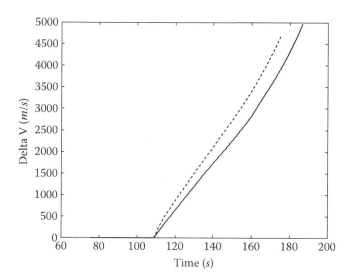

FIGURE 11.26 Cumulative velocity change for the guidance law (11.19) and TVC boosting motor; the standoff positions $R_{M1} = -552.8/-110$ km, $R_{M2} = 566.8/550$ km, $R_{M3} = 15/15$ km.

position of the operational area for the KV with axial control. For the chosen standoff distances the burnout velocity is about 3 *km/s*. Comparing the ΔV change in Figure 11.26 with the corresponding ΔV change in Figure 11.23 (solid lines) for the KV with an additional axial thruster, we can assume that the TVC boosting motor would require less powerful kill vehicles than in the case of the boosting motor without axial control and the KV with axial control. It is obvious that a more powerful and more sophisticated TVC boosting motor would require a less powerful kill vehicle with smaller max ΔV. As indicated, the KV's weight restriction cannot be considered separately from the boosting motor's weight restriction, i.e., there exists the general interceptor's weight optimization problem.

The above simulations for the basic model considered in the previous section were repeated under the assumption of state-estimate uncertainties for the different sensors related to the target-tracking problem. The results of simulation show that the boundary standoff distances for the stochastic case are about 8%–10% less than for the deterministic case.

11.7   COMPARATIVE ANALYSIS WITH LAMBERT GUIDANCE

In recent years significant efforts were directed toward examining the possibility of using Lambert guidance to control the interceptor's motion during its boost phase (see, e.g., [1,3,4]). As indicated in Chapter 9, the mathematically rigorous Lambert problem requires the known initial and final points and the time of flight t_F and its use is reasonable for offensive missiles, in which the boost phase brings them on a gravity field trajectory, which can be calculated in advance.

In the case of boost-phase intercept systems, the interceptor is a defensive missile. In this case, the time of flight and the final intercept point are unknown. As a result, an attempt to reformulate a rigorous mathematical problem to solve the boost-phase intercept problem brings more questions rather than gives a definite positive answer.

Many modern guidance schemes use a combination of guidance laws. For example, SM missiles use the Kappa guidance law early in flight and then switch to proportional navigation for the terminal phase. This approach is often motivated by the dominance of certain forces over others during various phases of the flight. For instance, the Space Shuttle is guided around a precalculated trajectory early in flight, when aerodynamic forces are large, and then switches to a linear tangent law after aerodynamic terms can be safely neglected. Intercontinental ballistic missile guidance is similar, using a simple law based on a precalculated trajectory during early flight and switching to Lambert guidance after departing the atmosphere [1].

The approach recommended in [1] for the ballistic missile interceptor guidance problem and surface-based interceptors based on the solution of the optimal problem used at the initial stage of the interceptor's flight and then Lambert guidance cannot be considered as a real design tool since too many important factors are ignored. The minimal time of intercept is the most important factor for the boost-phase intercept system. But a gravity field free trajectory takes more time than a forced trajectory. Since Lambert guidance deals with such trajectories, the operational area obtained based on Lambert guidance should be smaller than the operational area even for the properly applied APN guidance law.

Despite this obvious fact, in the last years significant efforts were directed toward examining the possibility of using Lambert guidance to control the interceptor's motion during the boost stage. Assuming that air-borne boost-phase interceptors launched by UAVs flying at about 15 *km* altitudes have the capability of acquiring and tracking the target in both angle and range out to approximately 1,200 *km* so that a predicted intercept point (PIP) can be generated with sufficient accuracy, the approach was developed to use Lambert guidance on the initial stage of the interceptor's flight. When the track position, velocity, and acceleration are determined with sufficient accuracy, a prediction (PIP) is made of the target's position at the desired intercept time t_F.

This part of the guidance algorithm is a source of a significant error, which can be even fatal and the reason of inability of the interceptor to hit the target. First of all, it is not clear how to choose the desired intercept time. Various missiles have different burnout times. Usually, in simulations operating with known target trajectories the desired intercept time is chosen about 10 *s* less than the burnout time. But it is not clear how to act in real situations if the type of a missile and its characteristics are unknown. Usually, the predicted intercept point is determined by the Taylor series formula accompanied with the remark that although the Taylor series method is crude, its main virtue is that it does not require a priori information (i.e., knowledge of target type and target intentions) to make a prediction of the intercept point. This is a weak argument.

The problem of using a three-term Taylor series is that functions describing the acceleration profile of multistage ballistic missiles are not continuous and formally the Taylor series method cannot be applied to such functions; only an infinitely differentiable function can be expanded as a Taylor's series. Moreover, some ballistic missiles have similar first-stage profiles or their initial part, based on which the initial PIP is determined, although their range and number of stages can be different, and a wrongly determined t_F would bring a huge initial error. Knowledge, even approximately, of the acceleration profiles of existing ICBM and IRBM missiles and the approximate desirable time of intercept would enable one to evaluate better the PIP based on measurements of the target motion during the boost stage.

Figure 11.20 shows two possible trajectories to intercept the target. But if the intercept time t_F is evaluated based only on the target's known acceleration profile (the burnout time) it is not clear how such an approach can choose the appropriate interceptor's trajectory.

In addition, the target's position, velocity, and especially acceleration are determined with mistakes that also contribute to the PIP error.

It is not a surprise that simulations show the initial PIP error about 100 km even when the time of intercept has an error only about 5%. Since the PIP estimate is constantly changing, the interceptor thrust vector must be steered in order for the interceptor to hit the latest and most refined estimate of the PIP [see equation (9.86)]. When the interceptor burns out, the PIP will still be in considerable error. That is why the KV should be guided to hit the target. The described approach to design the airborne boost-phase interceptors assumes Lambert guidance to be used when the control authority of the interceptor is in the axial direction and augmented proportional navigation (APN) to be used when the control authority of the interceptor is in the lateral direction, i.e., Lambert guidance is used while the interceptor is thrusting and APN is used by the KV during the terminal phase of the interceptor's flight.

Simulations show that maximal potential down range from the target launch site of 5 km/s two-stage potential airborne interceptors with 20 s burn time and around 170 s time of intercept is about 700 km, under the unrealistic condition that the predicted intercept point (PIP) is known perfectly. It is considerably smaller when the PIP is determined approximately based on the three-term Taylor series method.

The simulation results of the previous section show that maximal potential downrange from the target launch site of less powerful 3.5 km/s stage potential airborne interceptors with a 20 s burn time and around 187 s time of intercept is also about 700 km (see dash-dotted line of Figure 11.21). We can conclude that the considered basic model in the previous sections, the KV without axial control and a significantly simpler boosting motor than required by Lambert guidance, can produce much better results (larger operational area) by applying the guidance laws discussed in this book. The use of a more sophisticated kill vehicle with the axial control or/and a more complicated boosting motor (which, nevertheless, is simpler than the TVC motor for Lambert guidance) can bring an additional, up to 20%, increase of the operational area (see Figure 11.21 and Figure 11.25).

What is the most important, the discussed guidance laws that implement parallel navigation and were obtained based on the Lyapunov approach, do not require information about the predicted intercept point. A complicated Kappa guidance algorithm, based on the calculations of the predicted intercept point and used for SM missiles, can be justified since it deals with targets moving slower and having usually smoother trajectories than boosting maneuvering targets. Parallel navigation does not require continuous determination of the predicted intercept point. Predators do not determine the future intercept point. They start their motion based on experience, which includes comparative analysis of their own inner resources and the resources of their victims.

It is also of importance that the recommended realizations of the tested guidance laws do not use information about the target acceleration. The proposed guidance laws, which are algorithmically very simple, can be implemented in both interceptor's components—the kill vehicle and the boosting motor, which will be less complicated than in the case of Lambert guidance. This would significantly simplify the interceptor's design.

This chapter shows how to use in practice the theoretical results presented in the book. The proposed methodology can be used at the initial design stage. The formulation of specific features of the problem under consideration, the choosing of parameters of the offered guidance laws based on simulations using the planar model of engagement, and then a more detailed evaluation of the efficiency of the developed guidance laws by using the three-dimensional simulation model are necessary logical steps to select proper guidance laws and the interceptor's components that can realize these laws. The main parameters of the interceptor and its kill vehicle should be evaluated for each variant of the tested guidance laws (with and without axial control and for different types of boosting motors). The positive effect of the axial acceleration component has been established. However, the simulation results would contain highly reliable information to make a proper design decision only if they are obtained for a properly chosen boosting motor.

As indicated earlier, the interceptor's weight problem should be formulated properly. It cannot be formulated and solved separately for the kill vehicle and boosting motor. It is important to build a proper model of this multicriterial problem and apply proper computational algorithms. Its solution would justify the chosen parameters of the kill vehicle and boosting motor or would indicate the changes needed, whose implementation should be accompanied with additional simulations. Only after that 6-DOF simulation models should be created and tested. The KV's high velocity increases the requirements to the quality of information about the target. Informational time-delay of the kill vehicle's IR sensors can be a factor decreasing the interceptor's performance. It is of importance to examine the influence of the time-delay, and if it decreases the operational area considerably, the modified guidance algorithms should be tested [5].

The desire to create only one class of interceptors able to defeat various types of ICBMs and IRBMs combines the boost-phase intercept problem with the ascent phase intercept problem. The discussed guidance laws should be tested for various types of targets assuming the possibility to hit them (mostly IRBMs) in the ascent phase. The solution for this problem is similar to the above-considered problem and is important for the integrated design of boost (ascent)-phase interceptors.

REFERENCES

1. Dougherty, J., and Speyer, J. A Near-Optimal Approach to the Ballistic Missile Interceptor Guidance Problem, *SIG Technology Review* (1995): 42–57.
2. Kleppner, D., and Lamb, F. K., eds. *Boost-Phase Intercept Systems for National Missile Defense*, Report of the American Physical Society Study Group, July 2003.
3. Wilkening, D. A. Airborne Boost-Phase Ballistic Missile Defense, *Science and Global Security* June (2004): 1–67.
4. Zarchan, P. *Tactical and Strategic Missile Guidance, Progress in Astronautics and Aeronautics.* Vol. 176. Washington, DC: American Institute of Astronautics and Aeronautics, Inc., 1997.
5. Yanushevsky, R. *Missile Guidance*, Lecture Notes, AIAA Guidance, Navigation, and Control Conference, Chicago, IL, 2009.

Glossary

Active homing guidance: A system of homing guidance wherein both the source for illuminating the target and the receiver for detecting the energy reflected from the target as the result of the illumination are carried within the missile.

Actuator: A mechanism that furnishes the force required to displace a control surface or other control element.

Aegis: A computerized combat system used on U.S. Navy ships capable of simultaneous operation against surface, underwater, and air threats.

Aerodynamic missile: A missile that uses aerodynamic forces to maintain its flight path. See also ballistic missile; guided missile.

Aileron: A control surface usually on the trailing edge of the wings used to control roll.

Air-based system: An antimissile system weapon fired from an aircraft.

Aircraft: A vehicle that can travel through the air.

Airfoil: A part or surface, such as a wing, canard, or tail whose shape and orientation control stability, direction, and lift.

Air-launched ballistic missile: A ballistic missile launched from an airborne vehicle.

Altitude: The vertical distance of a level, a point, or an object considered as a point, measured from mean sea level.

Angle of attack: The angle between the missile longitudinal x-axis and the projection of the missile velocity vector on the xz plane.

Aspect angle: The angle between the longitudinal axis of the target (projected rearward) and the line-of-sight to the interceptor measured from the tail of the target.

Attitude: The position of a body as determined by the orientation of its axes with respect to some frame of reference. If not otherwise specified, this frame of reference is fixed to the Earth.

Autopilot: A mechanical, electrical, or hydraulic system used to guide a vehicle without assistance from a human being.

Azimuth: In astronomy, the horizontal angular distance from a reference direction, usually the northern point of the horizon, to the point where a vertical circle, passing through a celestial body, intersects the horizon, usually measured clockwise.

Azimuth angle: An angle measured clockwise in the horizontal plane between a reference direction and any other line.

Ballistic missile: A missile that, after an initial burst of power, coasts toward its target without any significant lift from its surface to alter the course of flight. Part or most of the missile's trajectory is not subject to propulsion or control.

Ballistic trajectory: The trajectory traced after the propulsive force is terminated and the body is acted upon only by gravity and aerodynamic drag.

Booster: An auxiliary or initial propulsion system that travels with a missile and may or may not separate from the parent craft when its impulse has been delivered. A booster system may contain, or consist of, one or more units.

Boost phase: The first phase of a missile's trajectory as the missile flies with its booster still burning. When a part of the boost phase can be controlled, the terms *controlled and uncontrolled boost stages* are used; often the boost phase is identified only with the uncontrolled boost stage, i.e., it is a part of missile flight between initial firing and the time when the missile reached a velocity at which it can be controlled.

Canard: A small surface forward on the body used as an aerodynamic control.

Closing velocity: The negative derivative of the range.

Control surface: Any moveable surface on an aircraft that controls its motion about one of the three principal axes. Ailerons, elevators, and the rudder are examples of control surfaces.

Countermeasures: Measures taken by an attacker to deceive a missile defense system (jammers, decoys, and chaff).

Doppler effect: The phenomenon evidenced by the change in the observed frequency of the reflected wave caused by a time rate of change in the effective length of the path of travel between the source and the point of observation.

Doppler radar: A radar system that differentiates between fixed and moving targets by detecting the apparent change in frequency of the reflected wave due to motion of the target or the observer.

Drag: Force of aerodynamic resistance most influenced by the viscosity of the medium in which the missile is traveling. Drag acts along the velocity vector and impedes the missile's motion.

Early-warning radar: A surveillance radar that provides detection and tracking of approaching missiles or aircraft.

Elevation: The vertical distance of a point or level on or affixed to the surface of the Earth measured from mean sea level. See also altitude.

Elevation angle: An angle measured clockwise in the vertical plane between a reference direction and any other line.

Endoatmospheric: Less than one hundred kilometers above the Earth's surface.

Engagement: In air defense, an attack with guns or air-to-air missiles by an interceptor aircraft, or the launch of an air defense missile by air defense artillery and the missile's subsequent travel to intercept.

Exoatmospheric: One hundred or more kilometers above the Earth's surface.

Fin: A fixed or movable airfoil used to stabilize and control a missile in flight.

Guidance: A dynamic process of directing an object toward a given point that may be stationary or moving.

Guided missile: An unmanned vehicle moving above the surface of the Earth whose trajectory or flight path is capable of being altered by an external or internal mechanism. See also aerodynamic missile; ballistic missile.

Heading: The direction in which the longitudinal axis of the aircraft is pointing, expressed in degrees from North.

Helicopter: A rotorcraft with one or more sets of powered blades.

Hit-to-Kill: A missile defense approach in which an interceptor rams the target, destroying it by force of impact.

Homing guidance: A system by which a missile steers itself toward a target by means of a self-contained mechanism controlled by a certain guidance law.

Homing phase: A part of the missile flight controlled by the missile-contained system.

ICBM: Intercontinental ballistic missile, a land-based missile with a range of more than 5500 kilometers.

Illuminate: Direct radar energy at an object sufficient to obtain radar targeting information.

Inertial guidance: A guidance system designed to project a missile over a predetermined path, wherein the path of the missile is adjusted after launching by devices wholly within the missile and independent of outside information. The system measures and converts accelerations experienced to distance traveled in a certain direction.

Inertial navigation system: A self-contained navigation system using inertial detectors, which automatically provides vehicle position, heading, and velocity. Also called INS.

Inertial reference frame: One that is not accelerating.

Infrared: Wavelengths slightly longer than those forming the color red. Every type of object radiates a unique infrared signature, which can be identified by measuring the received energy.

Infrared imagery: Imagery produced as a result of sensing electromagnetic radiations emitted or reflected from a given target surface in the infrared position of the electromagnetic spectrum (approximately 0.72 to 1000 microns).

Integrated fire control system: A system that performs the functions of target acquisition, tracking, data computation, and engagement control, primarily using electronic means and assisted by electromechanical devices.

Interceptor: A kill vehicle joined with a booster that together are launched against an offensive missile.

Intercept point: The point to which a vehicle is guided to complete an interception.

Kill probability: A measure of the probability of destroying a target.

Kill vehicle: A self-contained package of sensors, thrusters, and navigation gear that, once separated from its booster, can identify a target and maneuver into a collision with it.

Land-based system: An antimissile system that uses locations on land to shoot interceptors at incoming missile.

Lift: A component of the total aerodynamic force acting on a body perpendicular to the undisturbed airflow relative to the body. Lift is directed perpendicularly up with respect to drag and is the main force controlling the flight of an aerodynamic missile.

Line-of-sight: The line that starts at the reference point (e.g., the missile) and passes through the objective of the guidance (the target).

Mach: Speed of sound at sea level (331.46 m/s) that is measured in multiples (Mach 1, Mach 2, etc.).

Mach number: The ratio of the velocity of a body to that of sound in the surrounding medium.

Maneuver: A movement to place a pursuer in a position of advantage over the enemy.

Midcourse guidance: The guidance applied to a missile between termination of the boost phase and the start of the terminal phase of flight.

Midcourse phase: Coming between the boost and terminal phase. A part of missile flight when the missile is guided by an external weapon control system. See also boost phase; terminal phase.

Miss distance: The displacement between the missile and target.

Missile: A weapon that is launched and guided toward a target.

Missile control system: A system that serves to maintain attitude stability and to correct deflections. See also missile guidance system.

Missile guidance system: A system that evaluates flight information, correlates it with target data, determines the desired flight path of a missile, and communicates the necessary commands to the missile flight control system. See also missile control system.

Navigation: A dynamic process of directing an object toward a given stationary point.

Parallel navigation: Guidance when the direction of the line-of-sight is kept constant, i.e., the line-of-sight rate equals zero.

Passive homing guidance: A system of homing guidance wherein the receiver in the missile utilizes radiation from the target.

Pitch: The movement of a missile or an aircraft about its lateral axis.

Propellant: The ejected gas from a rocket.

Proportional navigation: A method of homing guidance in which the missile acceleration commands are proportional to the line-of-sight rate.

Pursuit: An offensive operation designed to catch or cut off a hostile force attempting to escape with the aim of destroying it.

Radar: A radio detection device that provides information on range, azimuth, and/or elevation of objects.

Ramjet: A jet-propulsion engine containing neither compressor nor turbine that depends, for its operation, on the air compression accomplished by the forward motion of the engine.

Range: The distance between any given point and an object or target. In two-point guidance systems, the range means the distance between the missile and target.

Rocket: A vehicle propelled by the recoil force produced when part of its mass is ejected at high velocity; it does not rely on interaction with its environment for propulsion.

Roll: The rotary motion of a missile or an aircraft around its longitudinal axis.

Rotorcraft: An aircraft that derives its lift from rotating lifting surfaces (usually called blades).

Sea-based system: An antimissile system that operates from floating platforms, whether Navy ships or specially outfitted barges.

Seeker: A device used in a moving object (especially a missile) that locates a target by detecting light, heat, or other radiation.

Semiactive homing guidance: A system of homing guidance wherein the receiver in the missile utilizes radiations from the target that has been illuminated by an outside source.

Sideslip angle: The angle between the missile longitudinal x-axis and the projection of the missile velocity vector on the xy plane.

Surface-to-air guided missile: A surface-launched guided missile for use against air targets.

Surface-to-surface guided missile: A surface-launched guided missile for use against surface targets.

Terminal guidance: The guidance applied to a guided missile between midcourse guidance and arrival in the vicinity of the target.

Thrust: The instantaneous recoil force produced by a rocket.

Time-of-flight: Elapsed time from the instant a missile leaves a launcher.

Time-to-go: Calculated time to go until the end of the flight assuming that it will correspond to intercept.

Trajectory: The dynamic path followed by an object under the influence of gravity and/or other forces.

UAV: A powered aerial vehicle that does not carry a human operator, can fly autonomously or be piloted remotely, can be expendable or recoverable, and can carry a lethal or nonlethal payload.

Unmanned aerial vehicle: A space-traversing vehicle that flies without a human crew on board and that can be remotely controlled or fly autonomously.

Warhead: That part of a missile, rocket, or other munitions that contains either the nuclear or thermonuclear system, high explosive system, chemical or biological agents, or inert materials intended to inflict damage.

Yaw: The rotation of a missile or an aircraft about its vertical axis.

Appendix A

A.1 LYAPUNOV METHOD

Control theory, whether it is presented in a classical or modern form, leans on the only and solid foundation—the Lyapunov theory of stability of motion.

Although the Lyapunov theory is the most effective for analysis of stability of processes described by nonlinear differential equations, we will apply the Lyapunov method to analysis of stability of linear differential equations, which are used mostly in this book. Intuitively, the stability of a motion means that under slightly altered initial conditions at t_0 the alteration in the motion will remain slight for all $t > t_0$.

More precisely, the solution $x_0(t)$ of the differential equation:

$$\dot{x} = Ax, \quad x(t_0) = x(0) \tag{A1}$$

is said to be stable (or the system described by the differential equation (A1) is stable about the equilibrium point $x_0 = 0$), if for every $\varepsilon > 0$ there exists such $\delta(\varepsilon, t_0) > 0$ that for every solution $x(t)$ and for all $t \geq 0$ we have $\|x(t) - x_0(t)\|^2 < \varepsilon$ provided $\|x(0) - x_0(0)\|^2 < \delta$, where $\|x\|^2 = \sum x_i^2$. (In the case of linear differential equations with constant coefficients δ does not depend on t_0.)

The system (A1) is said to be asymptotically stable, if it is stable and $\lim_{t \to \infty} x(t) \to 0$.

Stability and asymptotic stability are determined based on the Lyapunov method, which assumes the utilizations of the so-called positive definite and positive semidefinite functions $V(x) \geq 0$. The positive definite $V(x)$ is positive for all $x \neq 0$. The negative definite function has the opposite sign.

Theorem: The system (A1) is asymptotically stable, if there exists such positive definite function $V(x)$ ($V(0) = 0$) that its derivative along (A1) is negative definite.

The derivative of $V(x)$ along (A1) equals:

$$\frac{dV}{dt} = \frac{\partial V^T}{\partial x} Ax \tag{A2}$$

By choosing $V(x) = x^T W x$, where W is a symmetric positive definite matrix, instead of (A2) we have $x^T(WA + A^T W)x$, so that the asymptotic stability condition is:

$$WA + A^T W = -R < 0 \qquad (A3)$$

i.e., the matrix (A3) must be negative definite [2].

The physical interpretation of the above theorem is the following: $V(x)$ is bowl shaped. The condition (A3) implies that $V(x(t))$ decreases monotonically with time along any trajectory of (A1). Hence $V(x(t))$ will eventually approach zero as $t \to \infty$ Since $V(x)$ is positive definite, we have $V(0) = 0$ only at $x = 0$. Hence, if we can find positive definite matrices W and R that are related by (A3), then every trajectory of (A1) will approach zero as $t \to \infty$ The function $V(x)$ is called a Lyapunov function of the system (A1).

There exist various modifications of the Lyapunov method, various definitions of stability for special types of dynamic systems [3,4]. Here we discuss the possible application of the Lyapunov method to the stability analysis of systems operation on a finite interval $[0, t_F]$. By introducing:

$$\tau = \frac{1}{t_F - t} \qquad (A4)$$

the interval $[0, t_F]$ with respect to t is transformed into the interval $[1/t_F, \infty]$ with respect to τ. Taking into account $d/dt = \tau^2 d/dt$, the (A1) can be presented as:

$$\frac{dx}{d\tau} = \frac{1}{\tau^2} A x \qquad (A5)$$

If $V(x)$ is the Lyapunov function with respect to equation (A5), the solution of equation (A5) is stable on τ-interval. Since the transformation (A4) does not change the sign of

$$\frac{dV}{d\tau} = \frac{1}{\tau^2} \frac{dV}{dt}$$

the solution of equation (A5) is stable also on t-interval, i.e., for each stable trajectory on τ-interval there exist a stable trajectory on the finite interval, in a sense that $V(x)$ will decrease, when $t \to t_F$. However, because for $\tau \to \infty$ $dV/d\tau$ is always zero, the decrease of x cannot be asymptotic.

A.2   BELLMAN-LYAPUNOV APPROACH

Let us consider a dynamic system described by the following equation:

$$\dot{x} = Ax + Bu, \quad x(t_0) = x(0) \tag{A6}$$

where x is an m-dimensional state vector, u is an n-dimensional control vector, A and B are matrices of appropriate dimensions.

We will determine the control law u that minimizes the cost functional:

$$I = \frac{1}{2}\left(x^T(t_F)C_0x(t_F) + \int_{t_0}^{t_F}(x^T(t)Rx(t) + \|u(t)\|^2)dt \right) \tag{A7}$$

where C_0 and R are symmetric positive semidefinite matrices.

To find the optimal control we will use the dynamic programming approach [1]. The derivation of the Bellman functional equation is given according to the optimality principle: every tail of the optimal trajectory is the optimal trajectory.

Let the optimal functional value be:

$$\varphi(x(t_0),t_0) = \min_{u(t)} I \tag{A8}$$

Then in accordance with the optimality principle, it can be written:

$$\varphi(x(t_0),t_0) = \min_{u(t)} \frac{1}{2}\left\{ x^T(t_F)C_0x(t_F) + \int_{t_0}^{t_0+\delta}\left(x^T(t)Rx(t) + \|u(t)\|^2\right)dt \right.$$

$$+ \int_{t_0+\delta}^{t_F}\left(x^T(t)Rx(t) + \|u(t)\|^2\right)dt \bigg\}$$

$$= \min_{u(t)}\left\{ \frac{1}{2}\left[x^T(t_F)C_0x(t_F) + \int_{t_0}^{t_0+\delta}\left(x^T(t)Rx(t) + \|u(t)\|^2\right)dt \right] \right.$$

$$+ \varphi(x(t_0+\delta),t_0+\delta) \bigg\} \tag{A9}$$

Suppose that δ is small enough and that there exist partial derivatives of $\varphi(x)$ for $x \in [x(t_0), x(t_0+\delta)]$. Then expanding $\varphi(x(t_0+\delta), (t_0+\delta))$ into the Taylor series in the vicinity of $x(t_0)$, after appropriate transformations we obtain:

$$\varphi(x(t_0), t_0) = \min_{u(t)} \left\{ \frac{1}{2} (x^T(t_0) R x(t_0) + \|u(t_0)\|^2) \delta + \varphi(x(t_0), t_0) \right.$$

$$\left. + \frac{\partial \varphi}{\partial t} \delta + \frac{\partial \varphi^T}{\partial x} (Ax(t) + Bu(t)) \Big|_{\substack{x=x_0 \\ u=u_0}} \delta + O(\delta) \right\} \qquad \text{(A10)}$$

where

$$\frac{\partial \varphi^T}{\partial x} = \left(\frac{\partial \varphi}{\partial x_1}, \dots, \frac{\partial \varphi}{\partial x_m} \right)$$

is a row vector, and it is assumed that $\lim_{\delta \to 0} O(\delta)/\delta = 0$.

Tending δ to zero and taking into account that, in accordance with the optimality principle, the strategy must be optimal regardless of the state in which the system is at the actual instant, i.e., $x(t_0)$ and $u(t_0)$ can be treated as the current values of the vectors $x(t)$ and $u(t)$, we obtain the required functional equation as follows:

$$\min_{u(t)} \left\{ \frac{1}{2} (x^T(t) R x(t) + \|u(t)\|^2) + \frac{\partial \varphi}{\partial t} + \frac{\partial \varphi^T}{\partial x} (Ax(t) + Bu(t)) \right\} = 0 \quad \text{(A11)}$$

For the existence of the minimum of the expression in brace brackets, its derivative with respect to $u(t)$ ($d/du\{\ \}$) must be equal zero, i.e.,

$$u(t) = -B^T \frac{\partial \varphi}{\partial x} \qquad \text{(A12)}$$

Substituting equation (A12) in equation (A11), we obtain:

$$\frac{1}{2} x^T(t) R x(t) + \frac{\partial \varphi}{\partial t} + \frac{\partial \varphi^T}{\partial x} Ax(t) - \frac{1}{2} \frac{\partial \varphi^T}{\partial x} BB^T \frac{\partial \varphi}{\partial x} u(t) = 0 \quad \text{(A13)}$$

The solution of the considered problem reduces to finding the function $\varphi(x)$ satisfying the Bellman functional equation (A13) [or the equivalent equation (A11)].

The solution will be sought in the form:

$$\varphi(x) = \frac{1}{2} x^T(t) W(t) x(t) \qquad \text{(A14)}$$

Its substitution in equations (A12) and (A13) gives:

$$u(t) = -B^T W(t) x(t) \qquad (A15)$$

and

$$\dot{W} + A^T W + WA - WBB^T W + R = 0 \qquad (A16)$$

This is the so-called Riccati differential equation. Comparing equations (A7) and (A14) for $t = t_F$, we conclude that $W(t_F) = C_0$. For the quadratic integral criterion with the infinite upper limit [see equation (A7)], W is a constant matrix and instead of equation (A16) we have the so-called algebraic Riccati equation, which corresponds to the stationary solution of equation (A16) [5]:

$$A^T W + WA - WBB^T W + R = 0 \qquad (A17)$$

Comparing equation (A17) with equation (A3), we can see that equation (A17) is the Lyapunov equation (A3) for the closed-loop system with control (A15), and W is the Lyapunov function for this system.

The above-detailed analysis was focused on establishing the linkage between the Lyapunov method, which is used in this book to design new guidance laws, and the optimal approach, more precisely, a class of optimal systems based on minimization of the integral quadratic cost functional. The discrete analog of Riccati equations, applied to the optimal filtering problem, is given in Chapter 9. The Lyapunov-Bellman approach, in accordance with the above-described optimality principle, was considered in Chapter 3 [see equations (3.17)–(3.49)].

In conclusion, we will obtain the expression for the optimal PN (proportional navigation) guidance law (2.56) given in Chapter 2. For equations (2.54) and (2.55) the matrices in (A6) and (A7) are:

$$A = \begin{bmatrix} 0 & 1 \\ 0 & 0 \end{bmatrix} \quad B = \begin{bmatrix} 0 \\ 1 \end{bmatrix}, \quad C_0 = \begin{bmatrix} C & 0 \\ 0 & 0 \end{bmatrix}, \quad R = 0 \qquad (A18)$$

so that equation (A16) can be presented as:

$$\begin{bmatrix} \dot{w}_{11} & \dot{w}_{12} \\ \dot{w}_{12} & \dot{w}_{22} \end{bmatrix} + \begin{bmatrix} 0 & w_{11} \\ 0 & w_{12} \end{bmatrix} + \begin{bmatrix} 0 & 0 \\ w_{11} & w_{12} \end{bmatrix} - \begin{bmatrix} w_{12}^2 & w_{12} w_{22} \\ w_{12} w_{22} & w_{22}^2 \end{bmatrix} = 0$$

$$w_{12}(t_F) = w_{22}(t_F) = 0, \quad w_{11}(t_F) = C$$

The solution of the nonlinear matrix Riccati equation presents significant difficulties, even for this relatively simple problem. It is easy to check that

$$w_{11}(t) = \frac{3}{3/C + (t_F - t)^3}, \quad w_{12}(t) = \frac{3(t_F - t)}{3/C + (t_F - t)^3}, \quad w_{22}(t) = \frac{3(t_F - t)^2}{3/C + (t_F - t)^3}$$

satisfy the obtained Riccati equation, so that the expression of the optimal control $u(t) = -a_M(t) = -w_{12}x_1 - w_{22}x_2$ coincides with equation (2.56).

REFERENCES

1. Bellman, R. *Dynamic Programming.* Princeton, NJ: Princeton University Press, 1957.
2. Bellman, R. *Introduction to Matrix Analysis.* New York, NY: McGraw-Hill, 1960.
3. Martynyuk, A., ed. *Advances in Stability at the End of the 20th Century (Stability, Control, Theory, Methods and Applications).* Boca Raton, FL: CRC Press, 2002.
4. Rumyantsev, V. V. On Asymptotic Stability and Instability of Motion with Respect to a Part of the Variables, *Journal of Applied Mathematics and Mechanics* 35, no. 1 (1971): 19–30.
5. Yanushevsky, R. *Theory of Optimal Linear Multivariable Control Systems.* Moscow, Russia: Nauka, 1973.

Appendix B

B.1 LAPLACE TRANSFORM

For a function $f(t)$ defined on $0 \le t \le \infty$, its Laplace transform, denoted as $F(s)$, is obtained by the following integral:

$$L\{f(t)\} \equiv F(s) = \int_0^\infty f(t)e^{-st}dt$$

where s is real and L is called the Laplace transform operator.

The Laplace transform exists and is defined for $s > \sigma$ if $f(t)$ is a function piecewise continuous on $[0, K]$ (for every $K > 0$) and does not grow asymptotically faster than $Me^{\sigma t}$, i.e., $|f(t)| \le Me^{\sigma t}$.

In actual physical systems the Laplace transform is often interpreted as a transformation from the time-domain point of view, in which inputs and outputs are understood as functions of time, to the frequency-domain point of view, where the same inputs and outputs are seen as functions of complex variable. There is a unique "mapping" between functions in the "t-domain" and the corresponding functions in the "s-domain."

The inverse Laplace transform is determined as:

$$L^{-1}\{F(s)\} = f(t) = \frac{1}{2\pi i}\int_{\sigma+i\infty}^{\sigma+i\infty} F(s)e^{st}ds$$

Conditions for the existence of the inverse Laplace transform are:

i. $\lim F(s) = 0$, $s \to \infty$;
ii. $\lim sF(s)$, $s \to \infty$, is finite.

Usually, the Laplace transform is used for the solution of linear differential equations with constant coefficients, and we deal with rational functions of the complex variable s with real-valued coefficients (i.e., with single-valued functions).

B.2   PROOF OF THEOREM

The integral (5.13) and the related expression (5.53) for $P(t_F, s)$ present multiple-valued functions. We will show that $P(t_F, s)$ is the Laplace transform of $P(t_F, t)$, which is bounded and is tending to 0, if the condition (5.64) is satisfied.

The function $P(t_F, s)$ is a multiple-valued function that has infinitely many branches, which are obtained, if we fix a branch of each factor in (5.34). By denoting the last complex exponent factor as $x^p = e^{p\ln|x| + pi(\arg x + 2\pi k)}$, it can be presented as $x_0^p e^{p2k\pi i}, k = 0,1,2,...$, where x_0^p corresponds to $k = 0$ (see (5.38)–(5.41) given for $k = 0$ and $s = i\omega$). It follows from (5.34) that for real s $\ln|x| = 0$, so that, since p is pure imaginary, $x^p = e^{p_1(\arg x + 2k\pi)}$, $k = 0, 1, 2,...$, where

$$p_1 = \frac{N\omega_j(D_j - \zeta_j\omega_jC_j)}{2\sqrt{1 - \zeta_j^2}}$$

i.e., values of x^p are real for real s.

By fixing branches of all other factors of (5.34) in such a way that they are real for real s and taking an arbitrary k branch of the last factor, we obtain infinitely many branches $P_k(t_F, s), k = 0,1,2...$, real for $s > 0$. Evidently, $P_k(t_F, s) = P(t_F, s)e^{2\pi p_1 k}$, so it suffices to consider only $P(t_F, s)$.

The function $P(t_F, s)$ defined by (5.34) is analytic in the region $C_v = \{s : \mathrm{Re}s > -\sigma\}$, where $\sigma = \min(1/\tau_k, \zeta_j\omega_j), k = 1,..., l; j = 1,..., m$. It is analytic in the right-half plane ($\mathrm{Re}s \geq 0$), real-valued for real s and

$$P(t_F, s) = O\left(\frac{1}{|s|^\alpha}\right), \quad P'(t_F, s) = O\left(\frac{1}{|s|^{\alpha+1}}\right), \quad s \to \infty, \ s \in C_v \quad (B1)$$

for some $\alpha > 0$.

Define for $t > 0$ and $0 < \gamma < \sigma$:

$$y_0(t) = \frac{1}{2\pi i}\int_{\gamma - i\infty}^{\gamma + i\infty} P(t_F, s)e^{t_Fs}ds = \lim_{\alpha \to \infty}\frac{1}{2\pi i}\left(\frac{1}{t}P(t_F, s)e^{ts}\Big|_{\gamma - ia}^{\gamma + ia}\right.$$
$$\left. -\frac{1}{t}\int_{\gamma - ia}^{\gamma + ia} P'(t_F, s)e^{ts}ds\right) \quad (B2)$$

$y_0(t)$ can be rewritten in the form:

$$y_0(t) = \frac{1}{2\pi} \int_{-\infty}^{\infty} P(t, \gamma + iz) e^{(1+iz)t} dz \qquad (B3)$$

Since $P(t_F, s)e^{ts}$ is real for real s, then, by the symmetry principle, the numbers $P(t_F, \gamma + iz)^{(1 \pm iz)t}$ are complex conjugates. Therefore, the imaginary part of the integrand of (B3) is an odd function of z. Hence, the integral does not change, if one replaces the integrand with its real part, so that $y_0(t)$ is real-valued.

It follows from (B1) and (B2) that $y_0(t)$ tends to 0 as $t \to \infty$. Based on established relationships between the considered branches $P(t_F, s)$, we can conclude that each of these branches is the Laplace transform of a real-valued function $y_k(t)$ tending to 0 as $t \to \infty$ and $y_k(t) = y_0(t)e^{2\pi p_1 k}$, $k = 0, 1, 2\ldots$, and $y_0(t)$ is absolutely integrable on $[0, \infty)$.

We used the principal branch, $k = 0$, because it satisfies the zero condition for the lower limit of integration of equation (5.13) [see equations (5.35)–(5.37), (5.39), and (5.40)].

The above consideration corresponds to the case when $N > 2$ is an integer. If N is not an integer, the factor s^{N-2} of equation (5.34) is a multiple-valued function. In this case, the exponent s^{N-2} is not well defined in the neighborhood of zero, so that instead of $C_v = \{s : \text{Res} > -\sigma\}$ we have $C_v = \{s : \text{Res} > -\sigma/\{s : -\sigma < s \leq 0\}\}$ and the contour of integration in (B2) should be replaced with the contour consisting of four intervals, i.e.,

$$y_0(t) = \frac{1}{2\pi i} \left(\int_{-\gamma - i\infty}^{-\gamma} + \int_{-\gamma}^{-0} + \int_{+0}^{-\gamma} + \int_{-\gamma}^{-\gamma + i\infty} \right) P(s, t_F) e^{ts} ds \qquad (B4)$$

It follows from equations (B1) and (B2) that $y_0(t)$ tends to 0 as $t \to \infty$ (see also [1]).

REFERENCE

1. Yanushevsky, R. Frequency Domain Approach to Guidance System Design, *IEEE Transactions on Aerospace and Electronic Systems*, 43 (2007).

Appendix C

C.1 AERODYNAMIC REGRESSION MODELS

For a chosen range of altitudes and Mach numbers, Missile Datcom provides tabulated data about the lift, drag, axial, and normal force coefficients as a function of trim angles of attack, i.e., for each pair ij of the Mach number and altitude values $(Mach(i), Alt(j))$ we have a table with lines containing coefficients C_{Lk}, C_{Dk}, C_{Nk}, and C_{ak} corresponding to a certain angle of attack α_{Tk} ($k = 1, 2,...,m_{ij}$), where m_{ij} indicates that the trim angles of attack are bounded and depend upon $Mach(i)$ and $Alt(j)$.

To present, for example, the relationship between α_T and C_N by the second-order polynomial $\alpha_T = k_{10} + k_{11}C_N + k_{12}C_N^2$, we should substitute α_{Tk} and C_{Nk} in this equation. As a result, we obtain $m_{ij} > 3$ linear equations that should be solved with respect to unknown coefficients k_{10}, k_{11}, and k_{12} (i.e., we should solve the system of linear equations):

$$Ck = \alpha \qquad (C1)$$

where $k = (k_{10}, k_{11}, k_{12})$ is an unknown vector; $\alpha = (\alpha_{T1},...,\alpha_{Tm_{ij}})$ is the vector of trim angles (in obvious cases, here and earlier in the book we do not specify whether it is a row or column vector); C is the $m_{ij} \times 3$ matrix of the following form [2]:

$$C = \begin{vmatrix} 1 & C_{N1} & C_{N1}^2 \\ 1 & C_{N2} & C_{N2}^2 \\ ... & .... & ... \\ 1 & C_{Nm_{ij}} & C_{Nm_{ij}}^2 \end{vmatrix} \qquad (C2)$$

In various problems with finding the functional relationship based on experimental data, we have overdetermined systems of linear equation, similar to equations (C1) and (C2).

The unknown coefficients k_{10}, k_{11}, and k_{12} are determined by minimizing the sum of squares of the deviations of the data from the regression model (i.e., $\min_{k}\|\alpha - Ck\|^2$). The optimal solution is presented as [1]:

$$k = C^+\alpha, \quad C^+ = (C^TC)^{-1}C^T \qquad (C3)$$

where the so-called pseudoinverse matrix C^+ is written, assuming that the columns of C are linearly independent. The more general expression of C^+ is given in [1].

Using Matlab, the least square solution can be found with the backslash operator (i.e., $k = \alpha \backslash C$).

REFERENCES

1. Albert, A. *Regression and the Moor-Penrose Pseudoinverse*. New York, NY: Academic Press, 1972.
2. Phillips, G. M. *Interpolation and Approximation by Polynomials*. New York, NY: Springer Verlag, 2003.

Appendix D

D.1 RUNGE-KUTTA METHOD

Most of the differential equations of this book have no closed-form analytical solutions, so that numerical integration techniques should be used to solve or simulate these equations. We will describe the Runge-Kutta method, which is simple, accurate, and widely used in practice.

For a differential equation of the form:

$$\dot{y} = f(y,t) \tag{D1}$$

we will describe the Runge-Kutta numerical integration procedure.

The fourth-order Runge-Kutta method is one of the standard algorithms to solve differential equations. Before we give the algorithm of the fourth-order Runge-Kutta method, we will derive the second-order Runge-Kutta method, which is also used in many applications.

We start with the original differential equation and integrate it formally:

$$y_{n+1} = \int_0^{t_n} f(t,y)dt + \int_{t_n}^{t_{n+1}} f(t,y)dt = y_n + \int_{t_n}^{t_{n+1}} f(t,y)dt \tag{D2}$$

where $y_n = y(t_n)$.

Various computational procedures depend on how the integral at the right side of equation (D2) is calculated. By changing this integral to $h\dot{y}(t)$, we obtain the Euler formula, which has accuracy $O(h^2)$:

$$y_{n+1} = y_n + h\dot{y}_n + O(h^2) = y_n + hf(t_n,y_n) + O(h^2) \tag{D3}$$

where $h = t_{n+1} - t_n$ is the integration interval.

Integrating equation (D2) using the trapezoid formula we obtain:

$$y_{n+1} = y_n + 0.5h(f(t_n,y_n) + f(t_{n+1},y_{n+1})) + O(h^3) \tag{D4}$$

so that, based on equation (D3), we have:

$$y_{n+1} = y_n + 0.5h(f(t_n,y_n) + f(t_{n+1},y_{n+1})) + O(h^3), \quad y_{n+1} = y_n + hf(t_n,y_n) \tag{D5}$$

Integrating equation (D2) using the rectangle formula we obtain:

$$y_{n+1} = y_n + hf(t_n + 0.5h, y(t_n + 0.5h)) + O(h^3) \qquad \text{(D6)}$$

where

$$y(t_n + 0.5h) = y_n + 0.5hf(t_n, y_n)$$

Instead of two components presented in the second-order Runge-Kutta method, the fourth-order formula requires knowledge of four terms:

$$
\begin{aligned}
k_1 &= f(t_n, y_n) \\
k_2 &= f(t_n + 0.5h, y_n + 0.5k_1) \\
k_3 &= f(t_n + 0.5h, y_n + 0.5k_2) \\
k_4 &= f(t_n + h, y_n + hk_3)
\end{aligned}
\qquad \text{(D7)}
$$

and

$$y_{n+1} = y_n + \frac{h}{6}(k_1 + 2k_2 + 2k_3 + k_4) + O(h^5) \qquad \text{(D8)}$$

We demonstrate the fourth-order Runge-Kutta method by considering the system of differential equation (9.100). According to equations (D7) and (D8) we have:

$$k_{1i}^1 = x_{2in}, \quad k_{2i}^1 = x_{3in}$$

$$k_{3i}^1 = -\frac{\omega_M^2}{\tau} x_{1in} - \frac{\omega_M^2 + 2\zeta\omega_M}{\tau} x_{2in} - \frac{2\zeta\omega_M\tau + 1}{\tau} x_{3in} + \frac{\omega_M^2}{\tau} a_{MTin}$$

$$\boldsymbol{k}_1 = (k_{1i}^1, k_{2i}^1, k_{3i}^1)^T, \quad \boldsymbol{y}_n = (x_{1in}, x_{2in}, x_{3in})^T$$

$$x_{jin}^1 = x_{jin} + 0.5hk_{ji}^1$$

$$k_{1i}^2 = x_{2in}^1, \quad k_{2i}^2 = x_{3in}^1$$

$$k_{3i}^2 = -\frac{\omega_M^2}{\tau} x_{1in}^1 - \frac{\omega_M^2 + 2\zeta\omega_M}{\tau} x_{2in}^1 - \frac{2\zeta\omega_M\tau + 1}{\tau} x_{3in}^1 + \frac{\omega_M^2}{\tau} a_{MTin}$$

$$\boldsymbol{k}_2 = (k_{1i}^2, k_{2i}^2, k_{3i}^2)^T$$

$$x_{jin}^2 = x_{jin} + 0.5hk_{ji}^2$$

$$k_{1i}^3 = x_{2in}^2, \quad k_{2i}^3 = x_{3in}^2$$

$$k_{3i}^3 = -\frac{\omega_M^2}{\tau}x_{1in}^2 - \frac{\omega_M^2 + 2\zeta\omega_M}{\tau}x_{2in}^2 - \frac{2\zeta\omega_M\tau + 1}{\tau}x_{3in}^2 + \frac{\omega_M^2}{\tau}a_{MTin}$$

$$\boldsymbol{k}_3 = (k_{1i}^3, k_{2i}^3, k_{3i}^3)^T$$

$$x_{jin}^3 = x_{jin} + hk_{ji}^2, \quad \boldsymbol{y}_n^3 = (x_{1in}^3, x_{2in}^3, x_{3ni}^3)^T$$

$$k_{1i}^4 = x_{2in}^3, \quad k_{2i}^4 = x_{3in}^3$$

$$k_{3i}^4 = -\frac{\omega_M^2}{\tau}x_{1in}^3 - \frac{\omega_M^2 + 2\zeta\omega_M}{\tau}x_{2in}^3 - \frac{2\zeta\omega_M\tau + 1}{\tau}x_{3in}^3 + \frac{\omega_M^2}{\tau}a_{MTin}$$

$$\boldsymbol{k}_4 = (k_{1i}^4, k_{2i}^4, k_{3i}^4)^T \qquad (i = 1, 2, 3)$$

$$\boldsymbol{y}_{n+1} = \boldsymbol{y}_n + \frac{h}{6}(\boldsymbol{k}_1 + 2\boldsymbol{k}_2 + 2\boldsymbol{k}_3 + \boldsymbol{k}_4) \qquad \text{(D9)}$$

Index

For Product Safety Concerns and Information please contact our EU
representative GPSR@taylorandfrancis.com Taylor & Francis Verlag GmbH,
Kaufingerstraße 24, 80331 München, Germany

Printed and bound by CPI Group (UK) Ltd, Croydon, CR0 4YY
01/05/2025
01858482-0003